Michael Has

Bio-Based Resources: Rethinking Sustainable Supply

Also of Interest

Sustainable Products.
Life Cycle Assessment, Risk Management, Supply Chains, Ecodesign
Michael Has, 2024
ISBN 978-3-11-131482-2, e-ISBN 978-3-11-131546-1

Science, Engineering, and Sustainable Development.
Cases in Planning, Health, Agriculture, and the Environment
Robert Krueger, Yunus Telliel, Wole Soboyejo (Eds.), 2024
ISBN 978-3-11-075749-1, e-ISBN 978-3-11-075760-6

Sustainable Process Engineering
Gyorgy Szekely, 2024
ISBN 978-3-11-102815-6, e-ISBN 978-3-11-102816-3

Energy and Sustainable Development
Quinta Nwanosike Warren, 2021
ISBN 978-1-5015-1973-4, e-ISBN 978-1-5015-1977-2

Machine Learning for Sustainable Development
Kamal Kant Hiran, Deepak Khazanchi, Ajay Kumar Vyas, Sanjeevikumar
Padmanaban (Eds.), 2021
ISBN 978-3-11-070248-4, e-ISBN 978-3-11-070251-4

Michael Has

Bio-Based Resources: Rethinking Sustainable Supply

Social Metabolism, Resource Scarcity, Raw Material
Substitution, Circular Economy

DE GRUYTER

Author
Prof. Dr. habil. Michael Has
Langgasse 11
84559 Kraiburg
Bavaria, Germany
e-mail: has@monopteros.net

ISBN 978-3-11-914334-9
e-ISBN (PDF) 978-3-11-221874-7
e-ISBN (EPUB) 978-3-11-221905-8

Library of Congress Control Number: 2025942103

Bibliographic information published by the Deutsche Nationalbibliothek
The Deutsche Nationalbibliothek lists this publication in the Deutsche Nationalbibliografie;
detailed bibliographic data are available on the internet at http://dnb.dnb.de.

© 2025 Walter de Gruyter GmbH, Berlin/Boston, Genthiner Straße 13, 10785 Berlin
Cover image: koya79/iStock/Getty Images Plus
Typesetting: Integra Software Services Pvt. Ltd.

www.degruyterbrill.com
Questions about General Product Safety Regulation:
productsafety@degruyterbrill.com

Preface

The European Green Deal drew significant international attention to efforts made to reduce the consumption of fossil resources. It aimed at facing the challenges going along with the replacement of resources and the reduction of consumption of energy and raw materials by the support and implementation of technological innovation, policy frameworks, and economic incentives (European Commission, 2020). In this context, substitution of fossil and mineral raw materials with biogenic alternatives offers promising potential. Successful approaches already exist as technological developments, economic growth, and regulatory targets are set to enable a broader implementation of biogenic alternatives. However, saving resources is, aside of these changes, of utmost importance. The matter has been brought to the attention of a wide audience to Mairea Sameda, the later deputy secretary of the Food and Agriculture Organization of the United Nations (FAO). During a conference, she has been quoted saying "60 more harvests, then it's over" – a quote that refers to a serious issue: the dramatic loss of soil fertility worldwide. It is an alarming statement intended to draw attention to the urgency of sustainable agriculture and soil protection. The quote was uttered a few years ago (2015) at a UN conference on soil health in Berlin. At the time Sameda was the UN representative, who had worked on agriculture, climate change, and soil erosion.

If the above statement holds true and from today (2025) onwards we continue with our current way of farming, we will only have around 50 harvests left from our arable land – then these soils will be so degraded that they will no longer be able to produce to the degree they do today. This would imply that the world's soils could be unable to support the population with sufficient nutrition within around 50 years if there is no fundamental change in farming practices. And, of course, this perspective puts an end to the option to regard biogenic resources as a widely used option to replace fossil (and mineral) by biogenic resources.

The reasons for this gloomy forecast are complex:

Excessive use of the soil through monocultures and intensive agriculture

Erosion is caused by wind and water, but is enabled by the way the soil is treated

Chemical pollution from pesticides and synthetic fertilizers

Lack of organic matter in the soil (loss of humus)

Compaction of the soil by heavy machinery

Deforestation and associated loss of protection for the soil

Climate change, which favours soil dehydration, extreme weather, and thus increased erosion

The above individual statement has also been taken up by institutions such as the FAO. It is not a precise figure, but rather a metaphorical warning based on a period estimated on studies on soil quality worldwide. According to the FAO, we are losing around 24 billion tonnes of fertile soil worldwide yearly, and around a third of the

https://doi.org/10.1515/9783112218747-202

soils are already considered degraded. The situation outlined above is exacerbated by another aspect: fossil fuels and mineral resources are becoming critical – in other words, their supply range is limited. In around 50–70 years, natural gas and crude oil are likely to be in very short supply, and some of the raw materials that are vital to our economy will be used up. This will result in massive price increases – as is always the case when needed materials become scarce and this will happen ahead of the resource being used up. It can safely be predicted that the search for resources will become more hectic and ruthless, and the use of materials will increase because these raw materials will be more difficult to access. In principle, this leads to the following main alternatives to prologue the supply:

- Recycling
- Saving
- The attempt to replace mineral and fossil raw materials by other minerals or biogenic materials

Here the focus is on the latter option, which is by no means a dream of the future in some environments (see Table 1). Biogenic raw materials (also called *biomaterials*) are widely used today, as listed in Table 1 (the figures are coarse guidelines and vary significantly from region to region).

Table 1: Contribution of biogenic raw materials to the supply of raw materials in various industries.

Industrial sector	Share of biogenic raw materials worldwide	Typical biogenic raw materials
Paper and pulp industry	~95–100%	Cod, cellulose, and waste paper
Textile industry	~30–40%	Cotton, linen, hemp, and viscose
Construction industry	~10–15%	Wood, cork, straw, and natural insulation materials
Chemical industry	~10–15%	Starch, sugar, vegetable oils, cellulose, and glycerine
Cosmetics and personal care	~20–30%	Vegetable oils, fats, extracts, and essential oils
Plastics industry	~2–5%	Biogenic plastics from starch, PLA, PHA, and cellulose
Energy (biomass and biogas)	~15–20% *(regionally higher)*	Wood, energy crops, liquid manure, and bio-waste
Pharmaceutical industry	~5–10%	Plant-based active ingredients and fermentation products

Of course, this contribution of biogenic raw materials to the supply of raw materials does not come without land consumption (at the expense of food production). Land use for biofuels is already significant today.

> According to the estimates by the FAO, OECD, and other institutions, the proportion of agricultural land used for biofuels worldwide is around 5–6% of global arable land (as of approx. 2022–2023). This corresponds to around 70–100 million hectares worldwide, depending on the source and definition. To put this into perspective, the global arable land area is around 1.5–1.6 billion hectares. Around 8–10% of the EU's agricultural land is used (directly or indirectly) for biofuels, although the EU now wants to limit the proportion of food in biofuels. Palm oil, for example, will be gradually banned in the EU starting 2030.
>
> The most common energy crops are:
> **Corn bioethanol** (mainly in the USA)
> **Sugarcane bioethanol** (mainly in Brazil)
> **Rapeseed biodiesel** (e.g. in the EU)
> **Palm oil and soy biodiesel** (mainly in Southeast Asia and South America)

> The carbon footprint of biofuels is not automatically positive. Although at first glance these fuels are seen as a climate-friendly alternative to fossil fuels, their actual carbon footprint depends heavily on the type of production, the use of fertilizers, and land use. In many cases, indirect emissions – for example through land use changes – can even cancel out the supposed climate benefit.

An essential point of criticism of the use of agricultural land for the production of biofuels is the so-called *competition for land.* Land, on which energy crops are grown, is often no longer available for food production. This is particularly problematic in regions where food is already scarce. In addition, the cultivation of energy crops such as palm oil or soy often goes hand in hand with deforestation (e.g. in countries such as Indonesia or Brazil). Rainforests are cleared for new plantations – with devastating consequences for the global climate or local biodiversity and indigenous communities. Against this backdrop, the so-called tank versus plate debate is already taking place, as it is ethically questionable to use arable land for energy production while millions of people around the world are affected by hunger. The prioritization of biofuels over food production raises both environmental and social or ethical questions.

This dichotomy applies to all biogenic raw materials, and Ms. Sameda's warning about the possible massive limitations to agricultural production is relevant not only to food but also to all biogenic raw materials. This foreseeable intensification of competition and its consequences are the subjects addressed below. The development of the traditional supply of raw materials will be described along with emerging alternatives in the area of renewable raw materials. This positive trend is contrasted by the ecological and social consequences of overexploitation of nature, which are also described below.

Acknowledgements

As always, help is essential when working on a larger project. Special thanks deserve Dr. Uwe Has, Tom Lee, Dr. Bernard Pineaux (INP Grenoble), Andy, and Julie Plata, and Ute Skambraks (from de Gruyter-Brill) for support and constant encouragement, as well as many companions and friends who supported along the way.

Digital writing assistance tools were employed in the preparation of the manuscript, particularly for linguistic and stylistic refinement when transferring from a German to the English version.

Literature

I hope the references provided are useful and that the quotations are accurate. If I have inadvertently quoted the ideas of others or have unintentionally reproduced their prior work, please bear with me. I welcome any comments so that I can make appropriate corrections in future editions.

Contents

Chapter 1
Introduction

The use of alternative raw materials may be crucial for a sustainable production supply. However, both biogenic and synthetic materials present specific advantages and disadvantages. Biogenic raw materials are characterized by their availability and, often, but not necessarily, potential biodegradability. Still, they face challenges such as land consumption and limited technological performance (Rosenboom et al., 2022). Synthetic materials offer potential for optimization, but they are often energy-intensive to produce and may lead to disposal problems. The scientific discussions of recent years have led to the insight that neither of the two alternatives represents a universal solution for the existing industry. Therefore, in cases of limited resources, replacements and altered processes must be identified; otherwise, respective products will become unaffordable over time. An open approach to technology that combines biogenic and synthetic materials, optimized recycling processes, and integrated realistic circular economy concepts is necessary (UBA, 2020). Future research must focus more on the development of high-performance and sustainable materials in order to ensure an environmentally friendly and resource-efficient supply of raw materials in the long term.

Recycling and the circular economy are vital ways to reduce resource consumption and minimizing environmental impact (Ellen MacArthur Foundation, 2019). Recycling allows for the fast return of materials to the economic cycle, whereas the circular economy may take a longer-term approach to conserving resources. Both concepts must be integrated with appropriate product design and consumer awareness. However, both concepts face significant challenges that require technological innovation, policy frameworks, and economic incentives (European Commission, 2020). Increasing recycling rates and the effective implementation of the circular economy will largely depend on advancements in material separation processes, new business models, and innovative approaches to substituting critical raw materials. The transition to a circular economy requires close cooperation between science, industry, and politics in order to develop sustainable solutions.

The substitution of fossil and mineral raw materials with biogenic alternatives presents promising potential. Successful approaches already exist in areas such as plastics, building materials, energy sources, and chemical products (Carus et al., 2021). Nevertheless, further technological developments, along with economic and regulatory incentives, are needed to facilitate a broader implementation of biogenic alternatives. A sustainable bioeconomy must ensure that land-use conflicts and ecological risks are minimized to promote a resource-conserving and climate-friendly economy (BMEL, 2020). The substitution of mineral and fossil raw materials is already well advanced in many areas, such as bioplastics production and vegetable oil-based fuels. Despite the potential to reduce environmental impact and dependence on fossil feedstock, challenges remain regarding costs, technical adaptations, and environmental

https://doi.org/10.1515/9783112218747-001

side effects (Rosenboom et al., 2022). Scaling these alternatives necessitates further technological innovation and a holistic view of the environmental and social impacts.

Although many sectors have already made strides in successfully implementing substitution strategies in industry – such as the construction industry utilizing biogenic and recycled materials or in the chemical industry with sustainable plastics – challenges remain (IEA, 2021). The automotive industry is promoting greater use of lightweight materials and alternative energy sources. Nevertheless, the full implementation of these substitution strategies will only be feasible through continuous research, investment, and appropriate political framework conditions in order to achieve both ecological and economic goals (European Commission, 2020).

The scalability and economic efficiency of alternative raw materials present significant challenges for sustainable industrial production. Biogenic materials, recycling processes, and innovative technologies offer promising solutions, but notable economic and technological hurdles persist (UBA, 2020). In the long run, a blend of technological advances, political measures, and evolving market conditions will be essential for the large-scale utilization of alternative raw materials. A sustainable transformation of the industry will require considerable investment and a thorough rethinking across the entire value chain.

Securing a sustainable supply of raw materials is one of the key challenges faced by modern industrial societies. Given the increasing scarcity of conventional raw materials, rising environmental regulations, and geopolitical uncertainties, the use of alternative raw materials is increasingly becoming the focus of scientific and political discussions (BMUV, 2021). A distinction is made between biogenic raw materials and synthetic alternatives. While *biogenic raw materials are obtained from renewable natural sources, synthetic alternatives originate from fossil sources* – both are processed through chemical or technical processes that enable the targeted modification or creation of new materials. Both offer potential for conserving resources but also face specific challenges in ecological, economic, and technological aspects (BMEL, 2020).

Alternative raw materials encompass all substances that can either replace or augment traditional fossil, mineral, or metallic raw materials. The objective is to mitigate resource scarcity, diminish environmental pollution, and foster a sustainable raw materials economy in the long run.

1.1 Social Metabolism and Resource Availability

1.1.1 The Concept of the Socio-economic System (SES)

The concept of the socio-economic system (SES) has been introduced to emphasize the profound interactions between intertwined social and ecological processes that co-evolve and adapt. Originally initiated by Karl Marx in his writings on human nature and the use of natural resources, this model was further developed in resilience research by Elinor Ostrom and Carl Folke in the 1990s and 2000s. Their focus has been on studying complex systems where social and ecological elements are inextricably linked and constantly interact through mutual feedback mechanisms and threshold values (Folke, 2006; Ostrom, 2009).

The SES thus refers to a network of processes in which both social structures and natural environmental conditions continuously interact with each other and influence society and economic systems (Berkes et al., 2003). This interdisciplinary approach does not view the term "nature" in isolation, but as an integral part of the socio-economic system in which economic, ecological and social processes are inextricably linked.

The exploitation of natural resources was already addressed in the works of Marx and Engels, particularly concerning the relationship between cities and agriculture. They described how cities consume food and raw materials from nature, but these are returned to the environment as waste, without a sustainable return of the raw materials contained in the food to the system (Marx, 1867/2005; Engels, 1872/1973). This concept of resource use, which is based on the interactions between human society and nature, was later expanded in the context of resilience research and used as a basis for understanding sustainable development processes within SESs (Folke, 2006) as sketched in Figure 1.1.

In this context, resilience refers to the ability of a SES to adapt to change, cope with crises, and return to a stable state after experiencing a disruption or collapse (Walker et al., 2004). The term is not only significant in psychology, where it describes an individual's capacity to recover psychologically after traumatic experiences, but it also plays a crucial role in the fields of ecology, sociology, and economics. Here, resilience suggests that a system – be it an ecosystem, a society, or an economy – can respond flexibly to disruptions, adapt to new conditions, and ideally emerge stronger from a crisis (Folke et al., 2010).

As displayed in Figure 1.2, three important strategies can be used to promote resilience:

Maintain robustness: This strategy emphasizes the system's ability to endure stresses and external disturbances without compromising its fundamental structural integrity.

Develop adaptability: This involves the ability to flexibly adapt to changing environmental conditions so that the system can still function effectively, even if the initial conditions change.

Maintain flexibility: When adaptation is insufficient to meet demands, the system must develop the capacity to fundamentally reorganize itself and discover innovative solutions to respond to long-term changes (Walker et al., 2004; Folke et al., 2010).

These three dimensions of resilience promotion are essential for describing how SESs can cope with both short-term disruptions and long-term, profound changes without jeopardizing their livelihoods. Balanced and flexible actions and the use of the three

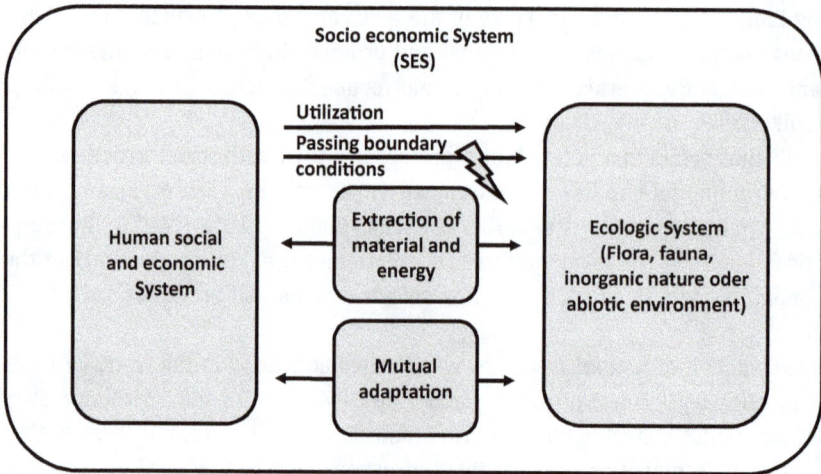

Figure 1.1: Dynamics within the socio-economic system (SES).
Conceptual model illustrating the dynamic interactions between human socio-economic systems and ecological systems within the overarching framework of a socio-economic system (SES). The diagram depicts two primary subsystems: the human social and economic system and the ecologic system, which includes flora, fauna, inorganic nature, and abiotic environmental components. These subsystems are interconnected through two main processes: the extraction of material and energy and mutual adaptation. The human social and economic system extracts resources from the ecologic system to sustain economic activities and social needs, as depicted by the arrow labelled "Extraction of material and energy." Simultaneously, the ecologic system and human system engage in mutual adaptation, reflecting bidirectional feedbacks and adjustments that influence both social behaviours and ecological conditions. Additionally, the concept of "Passing boundary conditions" highlights thresholds or limits in resource utilization, beyond which systemic stress or degradation may occur, indicated by a lightning bolt symbolizing disruption or critical transitions. This model emphasizes the complex, interdependent relationship between human development and environmental sustainability, underscoring the importance of balancing resource use within ecological limits to maintain system resilience and long-term viability.

parameters enable SES not only to respond to current challenges but also to proactively prepare for future risks and adapt to changes in environmental conditions over the long term.

1.1.2 Social Metabolism

In order to ensure the survival of SESs, a continuous supply of raw materials and energy is required. The intake of these resources, it's digestion, and the exhaust of residuals are often referred to as "societal metabolism" or "social metabolism" (Marx, 1867/ 2005; Weizsäcker, 2002). In this context, the availability of raw materials and energy can be directly linked to the intensity of this metabolism and the associated tipping points (Hubbert, 2002) as sketched in Figure 1.3.

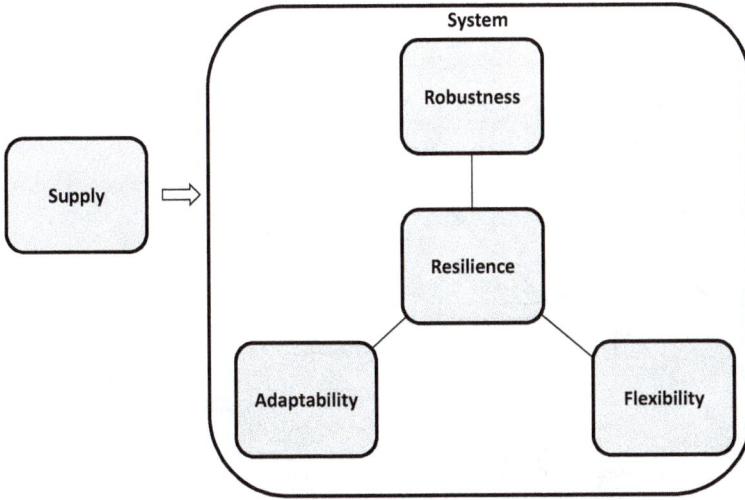

Figure 1.2: Strategies for promoting the resilience of a system.
This figure illustrates the conceptual framework for enhancing the resilience of a system in response to supply inputs. Central to the system's resilience is the ability to maintain and recover critical functions under varying conditions. The diagram identifies three primary strategies contributing to resilience: robustness, adaptability, and flexibility. Robustness refers to the system's capacity to withstand disturbances without significant degradation. Adaptability highlights the system's ability to adjust its structure or operations in response to changing external or internal conditions. Flexibility emphasizes the system's capability to shift between different states or modes of operation to effectively cope with disruptions. Together, these interconnected strategies form a comprehensive approach to strengthening system resilience, ensuring sustained performance despite challenges or uncertainties. The flow from supply into the system underscores the importance of external inputs as factors influencing system stability and resilience measures.

The Hubbert curve represents the typical production profile of an individual mine for finite resources, particularly for fossil fuels such as crude oil. It is based on the assumption that the production of raw materials initially increases, reaches a maximum, and then decreases again. The curve is divided into three main phases:

- **Rise:** In this phase, production grows gradually as more and more resources are extracted, accompanied by an increase in efficiency and investment.
- **Peak** (maximum production): Once about half of the total resource has been extracted, production reaches its peak and then declines.
- **Decline:** After the peak, the extraction rate begins to fall as the remaining resources become more difficult to access, and energy expenditure increases (Hubbert, 2002).

This curve illustrates that the maximum production rate is not reached at the end of resource availability, but rather in the middle of extraction. This model was developed by Marion King Hubbert in the 1950s and has been successfully applied to the US oil production.

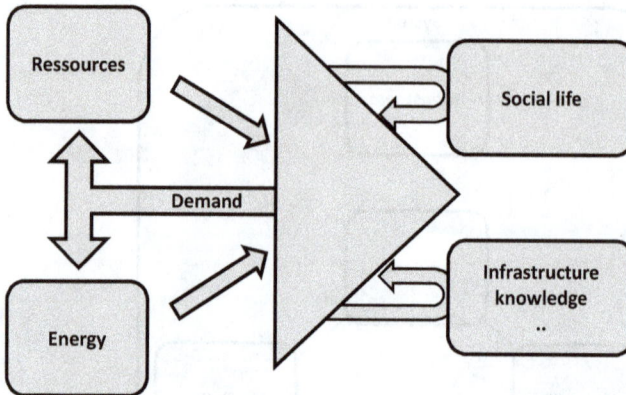

Figure 1.3: Social metabolism.
This diagram illustrates the concept of social metabolism, highlighting the dynamic flow of energy and resources required to sustain and influence social life and infrastructure. In order to maintain social life, raw materials and energy are needed, which are converted within the economic system that society has established in such a way that society can develop according to its own ideas. The figure emphasizes how societies extract natural resources and energy inputs from the environment, which are then processed and transformed through various economic, technological, and social systems. These inputs support the development and maintenance of social infrastructure such as transportation networks, housing, communication systems, and public services. The diagram further depicts the reciprocal relationship between social metabolism and social life, showing how resource consumption patterns shape cultural practices, economic activities, and social organization. By visualizing these interdependencies, the figure underscores the integral role of energy and material flows in shaping societal functions, the built environment, and overall human well-being, while also hinting at the environmental pressures and sustainability challenges inherent in maintaining these metabolic processes.

The concept of the harvest factor (energy returned on energy invested – EROI) describes the relationship between the energy extracted and the energy used to extract it. Without taking this factor into account, the Hubbert curve only refers to the physical availability of the resource. However, if the harvesting factor is included, it accounts for the fact that net energy gains decrease over time. In the early phase of resource extraction, the EROI is high, as easily accessible deposits are used first. However, as extraction progresses, energy costs increase, for example, due to deeper drilling or the development of deposits that are difficult to access. This causes the EROI to fall, and although gross production may remain high, the actual usable energy may decline earlier (Murphy & Hall, 2010). A sharp decline in EROI can make production economically unviable, leading to an accelerated decline in production after peak production (see Figure 1.4). At extremely low EROI, a resource becomes economically unusable despite its physical presence, as is the case with oil sands or deep-sea oil, where the energy required for extraction is disproportionately high (Heinberg, 2009).

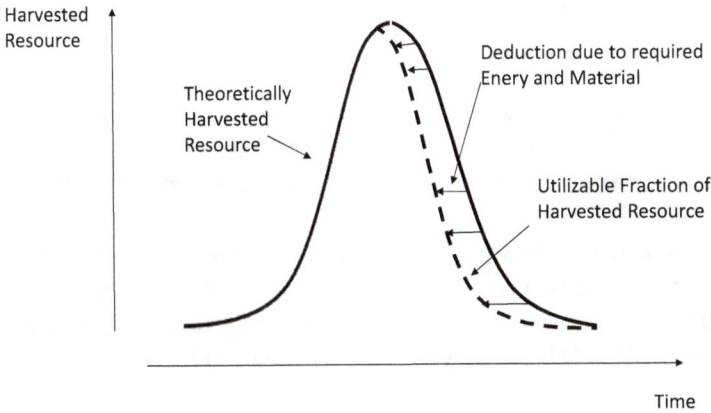

Figure 1.4: Hubbert curve – harvested material out of one source over time with correction due to the material and energy required to extract the raw material (EROI).
This figure depicts the classic Hubbert curve, illustrating the rate of harvested material extracted from a single resource source over time. The curve shows an initial phase of increasing extraction as the resource becomes more accessible and demand grows, reaching a peak when the maximum extraction rate is achieved. Following this peak, extraction rates decline due to resource depletion and increasing difficulty in accessing remaining reserves. Importantly, this version of the Hubbert curve incorporates a correction factor accounting for the energy and material inputs required to extract the raw material, reflecting the net available yield rather than gross extraction. This adjustment highlights how increasing extraction costs – both energetic and material – impact the effective availability of the resource over time, shifting the peak and altering the decline trajectory. By including these extraction costs, the figure provides a more realistic representation of resource depletion dynamics, emphasizing the growing energy and material investment necessary as resources become scarcer, which can significantly influence economic viability, environmental impact, and the long-term sustainability of resource-dependent systems.

Therefore, the supply of raw materials follows a dynamic development that reflects an EROI-corrected Hubbert curve. As soon as the supply of resources collapses for any reason, the SES loses the ability to interconnect its links, resulting in a loss of ability to sustain economic and social life. The point at which this supply can no longer be guaranteed, and a collapse becomes possible, can be described as a *tipping point* (Tainter, 1988). The time required for collapse may be relatively short on a historical scale.

By applying this theory to the SES and its supply of natural resources, aspects of the dynamics of the "metabolism" of the SES can be well explained:

> If the necessary resource is missing – be it in the form of fossil or mineral raw materials or other products of nature – the SES will no longer be able to maintain its cohesive operation. The lack of basic resources leads to the collapse of the entire system, as its structure and functionality are closely linked to the continuous availability of these resources.

The availability of resources will, therefore, be examined in more detail below.

1.2 Economic Growth Versus Ecological Stability

The current economic model, from which most of the world's wealth derives, is geared towards continuous growth, whether through higher production volumes, increasing consumer spending, or the expansion of markets. However, this growth pattern is at odds with the Earth's biophysical limits of what it can sustainably deliver (Meadows et al., 1972; Rockström et al., 2009).

Most modern economic systems are based on a linear approach to resource use – the so-called take, make, dispose principle, in which raw materials are extracted, processed, and ultimately discarded instead of being kept in sustainable cycles (Ellen MacArthur Foundation, 2013). This linear approach is illustrated in Figure 1.5.

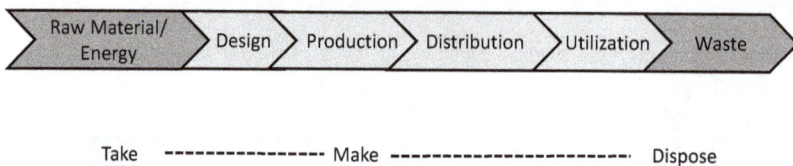

Take ---------------- Make ------------------------- Dispose

Figure 1.5: Linear value flow without recycling according to the guiding principle: "take-make-dispose". Diagram illustrating the linear value flow model based on the guiding principle of "take-make-dispose". In this system, raw materials are extracted from natural resources ("take"), processed and manufactured into products ("make"), and subsequently consumed by end users. After use, these products are discarded as waste without any efforts toward recycling or recovery, leading to a one-way flow of materials that contributes to resource depletion and environmental pollution. This linear approach contrasts with circular economy models that emphasize reuse, recycling, and sustainable resource management. The figure highlights the unsustainable nature of the traditional production and consumption system, underscoring the need for more regenerative alternatives.

This value flow leads to an accelerated consumption of resources, which cannot be regenerated to the same extent as they are consumed (UNEP, 2011). The proportion of raw materials recycled varies greatly depending on the material, region, and economic sector. Nevertheless, the global economy remains predominantly linear, and the transition to a circular economy is slow (Circle Economy, 2023).

The global recycling rate for plastics is around 9% (Geyer et al., 2017); in Germany, it is some 15% (Umweltbundesamt, 2023), while in Switzerland only around 6.9% of the raw materials used consist of recycled material (Circular Economy Switzerland & Deloitte Schweiz, 2023). This makes it clear that the majority of materials used, particularly plastics, still comes from primary raw materials.

In comparison, materials such as metals, especially aluminium and steel, achieve significantly higher recycling rates. Around 30–40% of the material used in steel production worldwide already comes from recycling processes, the share is increasing

(World Steel Association, 2022). Aluminium can achieve recycling rates of up to 75% due to its good recyclability (International Aluminium Institute, 2021).

However, recycling rates for rare earths and other critical raw materials that are essential for technologies such as electromobility and renewable energies are often below 1%, as the recovery of these materials is technologically complex and not economically viable (Binnemans et al., 2013).

If annual resource consumption continues to grow unaltered, it will become increasingly impossible to stay within the planet's ecological limits (see Section 1.8) (Rockström et al., 2009). A sustainable economic system would therefore need to rely on concepts such as *post-growth* (Paech, 2012), the circular economy, and ecological efficiency (see Figure 1.6). However, these approaches are still in their early stages and face political resistance in many countries, as they challenge established growth models and would require comprehensive economic adjustments. The impacts of the described resource conflicts often become visible even before they lead to acute shortages – similar to a bow wave that a ship generates before reaching its destination (Has, 2025). These long-term developments are frequently exploited in financial markets, where options are traded based on speculations about changes over a 5-year period (Heinberg, 2011).

The optimized material flow is certainly more complicated than in this illustration, because the design must ensure the longest possible and lowest-consumption product life (with reparability, serviceability, reuse of parts, etc.) across all phases of the product life. In addition, only raw materials that can be recovered as easily as possible, with high recycling rates and for which a recycling infrastructure has been established, should be used. In addition, one or more further product lives should be aimed for.

1.3 Effects of Stagnating Economic Performance

Government debt plays an important role in a country's economic capacity to act, as it has both short-term and long-term effects on fiscal policy and economic stability. In principle, government debt is a tool used by governments today to finance investments, cushion economic fluctuations, or support social and infrastructural measures (Blanchard & Johnson, 2017). As shown in Figure 1.7, moderate levels of debt can indeed have positive effects, especially when the borrowed funds are used productively – for example, for education, research, or the expansion of infrastructure (IMF, 2020).

If a country's economic performance declines due to external factors, such as stagnating supplies of raw materials and energy, this has profound consequences for the state's ability to repay its debts. In such a situation, societal metabolism also declines along with the weakening economy, and the increasing risk that government revenues will fall due to decreasing tax income (Tainter, 1988; Heinberg, 2011). If ex-

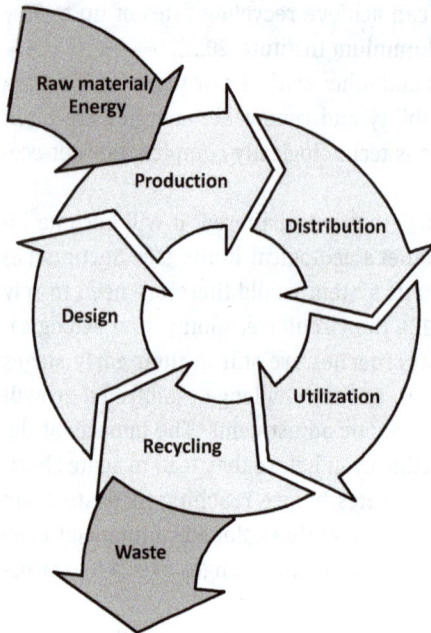

Figure 1.6: Simple circular economy with the aim to minimize waste by minimizing disposal and energy consumption.

Schematic of a simple circular economy model aimed at reducing environmental impact by minimizing waste disposal and energy consumption. This system encourages the continual use and regeneration of resources through strategies such as reuse, remanufacturing, and recycling, which help close the material loop. By limiting the extraction of virgin materials and decreasing the amount of waste sent to landfills or incineration, the circular approach fosters resource efficiency and sustainability compared to traditional linear models. The figure highlights the core objective of the circular economy: to create a regenerative system that conserves energy and reduces waste generation across the entire value chain.

penditures remain unchanged while existing debt burdens persist, the debt-to-GDP ratio rises accordingly. This can lead to a debt trap, in which the state must allocate an increasingly large share of its revenues to interest payments, rather than investing in productive or social sectors (Reinhart & Rogoff, 2010).

An economic downturn can lead to the destabilization of the entire state apparatus. If the government is forced to implement drastic austerity measures, this can trigger social unrest, especially if social benefits or public services are cut (Stiglitz, 2012). At the same time, a loss of confidence in the country's economic and political leadership can result in capital flight, further worsening the economic situation (OECD, 2022). Another critical issue is the potential erosion of state sovereignty: countries that are no longer able to service their debts are often compelled to turn to external institutions such as the International Monetary Fund (IMF) or other creditors. In many cases, such assistance comes with strict conditions requiring far-reaching economic policy reforms. This can limit national autonomy and lead to political tensions,

German Governmental
Debt in % of GNP

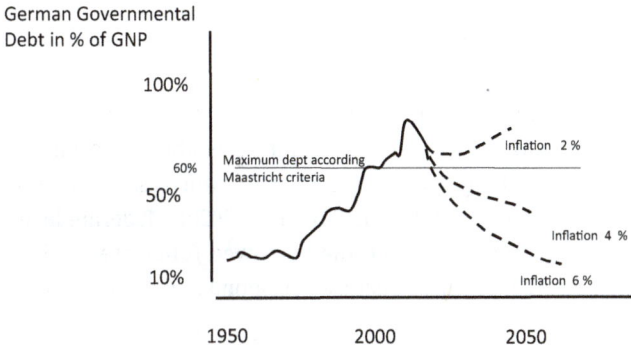

Quelle: IMF Historical Dept Database, IMF World Economic Outlook

Figure 1.7: Development of the government debt ratio in Germany since 1950 as a percentage of GDP
(IMF, 2024).
The figure highlights the evolution of Germany's government debt ratio as a percentage of GDP from
1950 to the present, depicting the country's fiscal position across seven decades. It shows that around
the year 2000, Germany decided to accumulate higher debt than suggested in the Maastricht criteria. For
the future development, it indicates that with an inflation rate between 2% and 3%, it will be possible to
pay back the debt accumulated. This long-term perspective offers insights into the sustainability of public
finances and the challenges faced by Germany in managing its sovereign debt burden over time.

particularly if the population perceives the imposed measures as socially unjust or
economically harmful – and in the past, this has been received as taking away national sovereignty by an anonymous and not locally elected institution.

1.4 Public Debt and Populism

High public debt can significantly limit a government's economic capacity to act. On
one hand, rising debt levels lead to higher interest payments, which consume an increasing share of the national budget. As a result, fewer financial resources may be
available for essential expenditures such as social welfare, education, or healthcare
(European Commission, 2021). On the other hand, a high debt-to-GDP ratio can impair
a country's credit rating, prompting investors to demand higher interest rates for new
government bonds (Standard & Poor's, 2020).

Declining confidence in a state's ability to repay its debts may also trigger capital
flight and currency depreciation, putting additional pressure on economic stability.
Moreover, a heavy debt burden restricts a state's fiscal policy flexibility. During economic crises – precisely when expansionary fiscal policies are needed to stabilize demand – highly indebted governments may be forced to adopt austerity measures instead of making productive investments (Alesina & Ardagna, 2010). This can result in

procyclical policies that exacerbate economic downturns rather than alleviate them. In extreme cases, an unsustainable debt load may lead to a full-blown debt crisis, potentially culminating in sovereign default (Krugman, 2009).

Positively speaking, public debt is an important instrument of fiscal policy and influences a country's short and long-term economic stability. It enables governments to make investments, smooth out economic fluctuations, and finance essential social and infrastructure initiatives (Blanchard & Johnson, 2017; IMF, 2020). Moderate levels of debt can have positive effects, especially when the borrowed funds are used for purposes such as education, research, or infrastructure development (OECD, 2022).

1.4.1 Impacts of Public Debt on the Capacity of States to Act

When a country's economic performance stagnates due to external factors, such as a declining supply of raw materials and energy, it has profound consequences for the state's ability to service its debt (Tainter, 1988; Heinberg, 2011). This economic downturn leads to reduced societal output and a decrease in government revenues due to falling tax income.

High public debt can significantly restrict a country's economic capacity to act. At the same time, the relative debt burden may increase as expenditures remain constant and economic performance declines. This can lead to what is known as a "debt trap," in which the state must allocate an increasing portion of its revenues to interest payments rather than investing in productive areas (Reinhart & Rogoff, 2010; Alesina & Ardagna, 2010). Rising debt leads to higher interest payments, which again tie up a larger share of the national budget, leaving fewer resources available for essential expenditures such as social welfare, education, or healthcare (Standard & Poor's, 2020). Such declining confidence can also result in capital flight and currency devaluation, further jeopardizing economic stability (Krugman, 2009). Procyclical policies can exacerbate the downturn rather than mitigate it and, in extreme cases, lead to a sovereign default.

As displayed in Figure 1.8, an economic downturn also risks undermining political and social stability. If the government is forced to implement drastic austerity measures, social unrest may arise, especially if public services or social benefits are cut (Stiglitz, 2012). A loss of confidence in political leaders can also lead to capital flight, further exacerbating the economic situation (European Commission, 2021).

Another issue is the potential erosion of state sovereignty.

Countries that can no longer service their debt are often forced to seek assistance from international institutions like the IMF. Such aid is frequently tied to strict conditions that require extensive economic policy reforms (Blyth, 2013). In these cases, economic autonomy may be lost, creating additional tensions, especially if the population perceives the IMF-imposed measures as unjust or economically harmful. Moreover, high levels of debt can reduce confidence in a country's creditworthiness, leading investors to demand higher interest rates for future bonds.

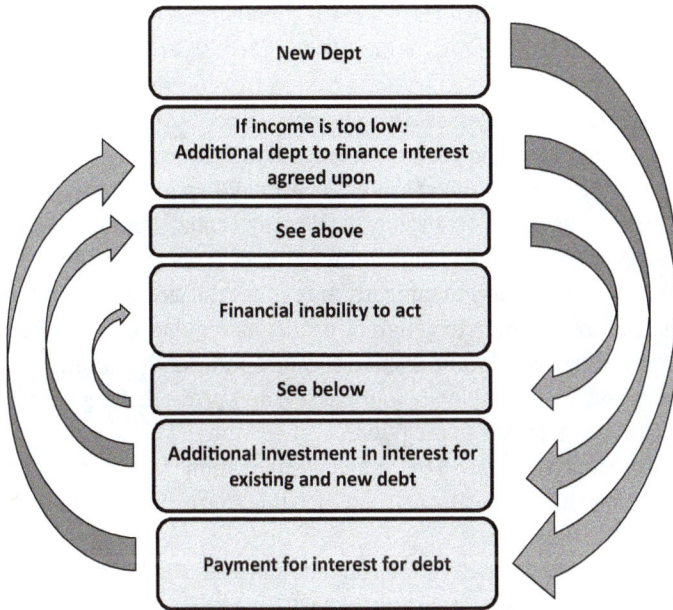

Figure 1.8: Visualization of the debt spiral phenomenon, highlighting the feedback loop between mounting debt and interest costs that can lead to unsustainable fiscal trajectories.
Schematic of the debt spiral process, illustrating how an initial increase in debt raises interest expenses, which necessitates further borrowing to cover costs. This cycle perpetuates itself, causing debt levels and interest burdens to escalate over time. The figure emphasizes the risk of unsustainable public or private debt growth when borrowing outpaces repayment capacity, potentially leading to fiscal crises or financial instability.

Innovation, Administration, versus Societal and Ecological Limits – A Conflict of Economic Objectives

Innovation is one of humanity's greatest strengths and has often been underestimated in the past – for example, by Thomas Malthus, whose theory predicted a long-term imbalance between population growth and food production. However, technological progress, particularly in agriculture and through societal change, has disproven many of his dire predictions. Nevertheless, his core idea – the finiteness of natural resources – remains relevant in today's considerations about planetary boundaries.

Innovations have significantly contributed to global economic growth. A considerable share of global growth since 1975 can be attributed to innovation – an estimated 40–55% of long-term economic growth in industrialized nations (e.g. Jorgenson et al., 2008). This is especially evident in the IT and healthcare sectors. On a global scale, between 1975 and 2025, innovation-driven gains in prosperity are estimated at US$150–300 trillion – partly through new markets, products, or business models.

In contrast, substantial administrative costs have accumulated. Globally, companies spend around US$1.5–2 trillion annually on bureaucracy. Over 50 years, this amounts to a cumulative cost of US$80–120 trillion– caused by reporting obligations, compliance efforts, or inefficient processes. In Germany alone, bureaucracy costs have amounted to €1,500–2,000 billion in real terms since 1975. Public administration itself has also become more expensive: its share of government spending has risen from around 5–8% in the 1970s to 10–15% today (McKinsey Global Institute, 2021; OECD, 2020). It is reasonable to say that a significant fraction of these expenditures is due to regulations which come into place to mitigate and reduce the side effects of the misuse of societies and nature in order to generate profits as regulations are often reactive to market externalities, regulatory frameworks are inherently administrative, bureaucracy often grows with complexity and misuse, and profit-seeking behaviour, unless regulated, can shift burdens to the public.

In a realistic scenario, approximately 44% of global innovation gains have been offset by administrative costs. In pessimistic calculations, this share rises to as much as 80%.

Additionally, ecological crises intensify the economic burden. The Stern Review (2006) and studies such as those by the Swiss Re Institute (2021) forecast annual losses of 10–30% of global gross domestic product (GDP) due to climate change, biodiversity loss, and other environmental crises – caused by crop failures, extreme weather events, or the loss of ecosystem services. These effects impact not only the Global South but also spread globally through markets, supply chains, and geopolitical instability.

On top of that, the interest payments that governments must make on their public debt are increasing: In 2023, these payments amounted to approximately 0.87% of Germany's GDP. By contrast, the average for OECD countries stood at 3.0% of GDP, rising to 3.3% in 2024 (International Monetary Fund, 2023; Organisation for Economic Co-operation and Development, 2024). Paying back these debts – provided it happens at all – is often re-financed ranges at some 1% of the gross national product (GNP).

To summarize: While the accumulated growth of the BIP contributes to an estimated 40–55% of long-term economic growth in industrialized nations, the accumulated effects of additional administration, ecologic and social burdens due to overuse of the environment and societies, as well as the interest rates of debt, accumulate to 22–49%, with a strong trend toward increase. If, as argued below (Section 1.10), the availability of resources and energy causes a reduction of the GNP, innovation will not provide for an as large increase of the GNP any more while the remaining costs are still in place.

1.4.2 Populism and the Need for Power Projection

As mentioned, public debt can prevent governments from providing social benefits in the usual manner. When a sense of entitlement has been promoted, and a large portion of the population depends on social benefits, high debt and unmet social benefits will lead to disappointment and frustration (Mounk, 2018). In this situation, populist rhetoric may gain influence, as populist parties often promise to maintain unaffordable benefits without addressing the financial and economic consequences.

The connection between resource and energy shortages, rising public debt, and the emergence of populist movements is not coincidental, as illustrated in Figure 1.9. Populist movements often arise from a sense of national or societal powerlessness (Mudde & Rovira Kaltwasser, 2017). A common feature of these movements is the desire for external power projection, which not only pursues geopolitical goals but also helps stabilize the internal political situation by strengthening the collective sense of identity.

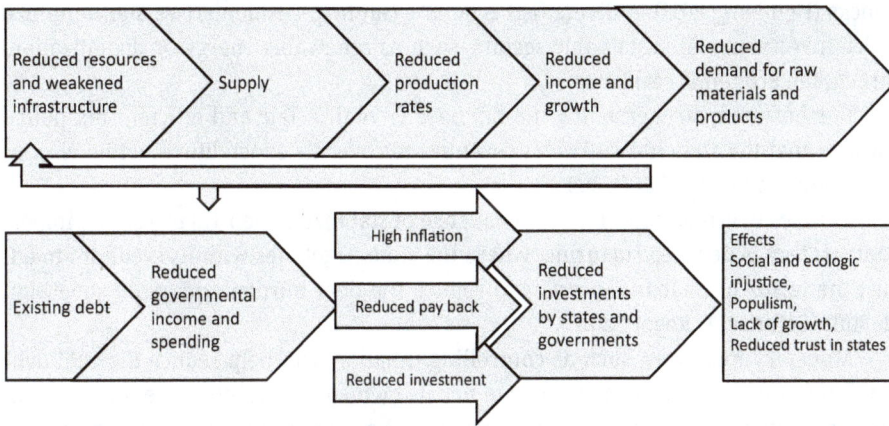

Figure 1.9: Consequences of public debt and declining resources.
The figure is intended to offer an overview of the multifaceted consequences arising from escalating public debt combined with the challenge of declining natural and economic resources. The figure illustrates how increasing government indebtedness can lead to higher interest payments, reduced fiscal space, and constraints on public investment, which, in turn, may slow economic growth. Simultaneously, the depletion of critical resources – such as fossil fuels, minerals, and arable land – exerts additional pressure on economies by raising production costs, limiting raw material availability, and increasing environmental degradation. Together, these factors create a complex interplay that can exacerbate financial instability, reduce the capacity for sustainable development, and necessitate structural adjustments in economic policy. The diagram highlights the urgent need for integrated strategies that address both fiscal sustainability and resource management to ensure long-term economic resilience.

However, this form of power projection can have a destabilizing effect. Politicians often use populist rhetoric to satisfy the demand for national strength and thus gain political support. In the long term, this often leads to decisions that are not based on rational, long-term interests but rather aim to fulfil short-term, emotional needs for recognition and power (Norris & Inglehart, 2019).

1.4.3 A Few Preventive and Reactive Measures by States

A sustainable fiscal policy is essential for ensuring long-term financial stability. This involves disciplined budgeting, ensuring that debt remains within a stable framework, and that the state does not live beyond its means (IMF, 2020). A balanced fiscal policy means that government spending remains within long-term revenues. In times of economic prosperity, states should aim for surpluses or at least limit new borrowing to have sufficient financial room to manoeuvre during times of crisis (European Commission, 2021).

Another important aspect is economic diversification. States that depend on a limited number of raw materials or sectors are particularly vulnerable to economic shocks (Heinberg, 2011). A diversified economy can help cushion crises and minimize risks. Investments in sustainable sectors, such as renewable energy or digitalization, strengthen economic resilience.

Furthermore, strengthening the tax base is vital. A fair and efficient tax policy ensures that the state has sufficient revenues to cover its expenditures while also investing in the future (OECD, 2022).

In times of crisis, states face the challenge of stabilizing their finances. One important strategy is debt restructuring, where the state negotiates with its creditors to adjust the terms of its loans in order to reduce the debt burden and avoid sovereign default (Reinhart & Rogoff, 2010).

Monetary measures, such as controlling inflation, can help reduce the real debt burden in the short term. When a state has its own currency, moderate inflation can relieve finances, as the nominal debt remains the same but its real value decreases (Krugman, 2009).

States may also seek international support in particularly severe crises. The IMF or the World Bank offers financial assistance, but such help is often tied to strict conditions that require far-reaching economic reforms (Blyth, 2013).

Ultimately, it becomes clear that a state's capacity to act in a debt crisis depends on its ability to take appropriate preventive and reactive measures to ensure long-term stability and minimize the social and economic consequences of debt crises.

1.5 The Necessity of Structural Changes

Regardless of the chosen theoretical approach, there is broad consensus that long-term ecological sustainability cannot be achieved without profound structural economic changes (Steffen et al., 2015). This understanding is based on the fact that existing economic systems and modes of production, especially in industrialized countries, significantly contribute to current ecological crises such as climate change, biodiversity loss, and resource depletion. To avoid exceeding the ecological boundaries of the planet (see Section 1.8.2), comprehensive adjustments to the economic system are required in terms of economic actions, as well as political and social structures.

The existing economic system, largely based on a model of unchecked growth, is increasingly coming into conflict with the planet's natural resources. The assumption that economic growth and ecological integrity are compatible is facing growing criticism, as resource consumption, CO_2 emissions, and pollution rise exponentially, while the Earth's capacity to regenerate is limited (see Section 1.8.2). The traditional growth paradigm often relies on the assumption of inexhaustible resources and overlooks planetary boundaries, which are outlined in concepts such as the "planetary boundaries" framework. These boundaries, including climate stability, biodiversity, and biogeochemical cycles, define the limits within which human activities should occur to ensure the long-term viability of our planet.

The challenge for science and politics lies in overcoming the current economic model, which is based on linear production processes, resource waste, and the pursuit of maximum profit. Instead, there is an increasing call for a transition to a circular economy and a "post-growth" perspective that places ecological and social values at the centre and questions the pursuit of unlimited growth. Such a transformation requires not only technological innovations and the promotion of sustainable consumption patterns but also a profound redesign of the social and political frameworks that shape economic actions (Steffen et al., 2015).

Recognizing of the social dimension of ecological sustainability is an important aspect of these considerations. As stated in the Brundtland report, the current economic model leads to enormous social imbalances, both within states and between industrialized nations and developing regions. Sustainability cannot be considered in isolation from social justice issues. The introduction of fair and inclusive economic models that distribute wealth more equitably while respecting natural resources is, therefore, inseparable from ecological sustainability.

This interdisciplinary perspective demonstrates that the path to a sustainable future cannot be achieved solely through technical or market-based solutions. Instead, a fundamental transformation of the economy is necessary, one that not only changes production methods but also involves profound societal and political changes. Only through such a holistic approach can an ecological and social balance be achieved that has the potential to ensure the well-being of future generations and enhance the quality of life for all people worldwide.

Regardless of the chosen approach, there is broad consensus that long-term economic sustainability cannot be achieved without profound structural changes. These include:

Circular economy: Reducing waste and improving product lifespan through recycling contributes to reducing resource consumption (see Section 1.7).

Socially just transformation: A sustainable economic model must ensure social justice, distribute wealth more widely, and foster societal acceptance (Steffen et al., 2015).

New indicators of prosperity: Shifting away from GDP as the only measure could align political decisions more closely with long-term sustainability (Stiglitz, 2012).

The question of whether growth is compatible with sustainability is complex and evaluated differently. While the established model considers growth a key driver of wealth, ecological and social challenges highlight the limits of a purely quantitative growth model. Concepts like "green growth" attempt to combine economic growth with ecological responsibility, while alternative approaches such as "degrowth" or "post-growth" call for a fundamental transformation of the economic culture (Has, 2024). In the long run, the question of whether sustainable growth is achievable or if a new economic culture is required will significantly shape the ecological and societal future.

1.6 Energy, Emissions, and Economic Growth

The historical development of global gross national product (now often referred to as gross national income) is closely linked to the availability and use of fossil energy sources, as well as mineral and biological resources (Krausmann et al., 2009). This connection is explained by the fundamental role of energy and raw materials as production factors and drivers of economic growth. During the Industrial Revolution, the exponential growth of the global economy began, closely tied to the increasing consumption of fossil fuels such as coal, oil, and natural gas (Smil, 2017). The transition from manual labour to machine production significantly boosted industrial productivity, resulting in a rapid increase in global GDP and all associated phenomena from public wealth in some economies, extended lifetime to the over-exploitation of natural resources and the globalization of economies and cultures:

- **Eighteenth Century–Nineteenth Century:** The Industrial Revolution (ca. 1750–1850) was primarily based on the use of coal for steam engines, which revolutionized transportation (railroads, steamships) as well as industrial manufacturing (Malm, 2016).
- **Twentieth Century:** The widespread use of oil and natural gas enabled the rise of automobile culture, mass production, and the emergence of global trade networks (Smil, 2017).
- **After 1950 ("Great Acceleration"):** The economic expansion following World War II, particularly in industrialized nations, was facilitated by cheap access to fossil fuels, which led to a massive increase in both GDP and global CO_2 emissions (Steffen et al., 2015).

As can already be assumed from the analysis of social metabolism (see Section 1.1), there is a close relationship between economic growth and energy consumption: Generally, an increase in economic activity is accompanied by a rise in energy demand, which further intensifies fossil fuel use (Krausmann et al., 2009). This relationship between energy availability and economic performance is crucial, as energy is a key factor of production across almost all sectors of the economy. A reliable and affordable energy supply is essential for the competitiveness and growth of economies (IEA,

2021). In many countries, high per capita energy consumption correlates with a higher GDP per capita, indicating that energy plays a limiting role in economic development, especially in emerging countries with inadequate energy infrastructure (Stern, 2011).

Subsidies for fossil fuels, granted by many governments, promote inefficient structures and delay the transition to renewable energy sources (IEA, 2021). These subsidies continue to foster growth based on emission-intensive processes, thereby maintaining the trajectory of climate change. The IMF estimated that global fossil fuel subsidies in 2020 amounted to around US$5.9 trillion – a clear indication of the continued support for fossil fuels (IMF, 2021).

The close coupling of economic growth and the use of fossil fuels inevitably leads to increasing greenhouse gas emissions as the economy expands. Therefore, without decoupling economic growth from fossil fuel consumption, climate change will remain one of the greatest challenges. Potential strategies to reduce this dependence include cutting fossil fuel subsidies, promoting renewable energy, and encouraging technological innovations for the decarbonization of industry and transportation (IEA, 2021; IMF, 2021).

1.7 Utilization of Raw Materials and Economic Growth

Access to and utilization of natural resources play a critical role in nations' economic development strategies. Depending on the availability and type of resources, countries adopt different approaches to shaping their economic structures. However, the finite nature of many raw materials imposes an inherent limit to growth. Once these resources are depleted, the boundaries of economic expansion are reached, beyond which further growth becomes unfeasible (Meadows et al., 2004). Consequently, the choice of a specific resource strategy has profound implications for a country's long-term economic stability, as well as its ecological and social sustainability (Bleischwitz et al., 2012).

1.7.1 Extractivist Strategy – Maximizing Short-Term Resource Exploitation

A common approach in resource-rich but economically less diversified countries is the strategy that involves maximizing resource extraction. The primary objective is to generate short-term economic growth and wealth through intensive exploitation and rapid utilization of natural resources. However, this often leads to a strong dependence on commodity exports and the neglect of long-term resource security.

Characteristics of this strategy include:
- A strong emphasis on exploiting new resource deposits, such as through offshore drilling or lifting protections on conserved areas.
- Promotion of the intensive use of fossil fuels, often backed by government subsidies and tax incentives (IEA, 2021).

- Neglect long-term environmental consequences in favour of short-term economic gains.
- High economic dependence on the export of raw materials.

Example: During Donald Trump's second presidential term, the United States pursued a policy of intensified fossil fuel extraction, famously encapsulated in the slogan "Drill, Baby, Drill!" cited in his inaugural address. This strategy was advanced earlier through deregulation and the opening of new extraction zones (U.S. EIA, 2020). While aimed at boosting energy independence and economic growth, it also accelerated resource depletion and contributed to severe environmental challenges that are likely to burden future generations.

1.7.2 Sustainable Resource Management – Conservation and Circular Economy

Some countries adopt a more sustainable approach to resource use by optimizing extraction and processing over the long term. The primary goal is to preserve resource availability, promote recycling, and establish a circular economy to minimize environmental impacts (European Commission, 2020).

Key characteristics of this strategy include:
- Limiting resource extraction through environmental regulations and sustainable production quotas.
- Promoting renewable energy sources as substitutes for fossil resources.
- Developing and advancing recycling technologies to reduce the consumption of primary raw materials.
- Investing in research and innovation to boost resource efficiency.

Example: Germany's *Raw Materials Strategy 2020* aims to ensure a long-term sustainable supply of raw materials. As part of this strategy, the country is expanding its circular economy, increasing resource efficiency, and intensifying international cooperation to secure access to critical raw materials (BMWK, 2020).

1.7.3 Resource Diversification and Import Strategies – Reducing Dependency

Countries with limited domestic raw material reserves often depend on diversification and strategic imports to strengthen their economic resilience against fluctuations in global commodity markets (Kooroshy et al., 2015). This strategy seeks to lessen dependency on a narrow range of resource suppliers and ensure a stable supply of critical materials.

Key characteristics of this strategy:
- Establishment of strategic reserves for critical raw materials.
- Securing a stable resource supply through international trade agreements and partnerships.
- Promotion of research into alternatives for scarce materials, such as through materials science.
- Investments in technologies that reduce raw material consumption.

Example: Japan, a country with limited availability of domestic resources, pursues a diversification strategy focused on technological advancement. Through strategic trade agreements, the development of resource stockpiles, and an emphasis on resource-efficient high-tech industries, Japan has strengthened its economic resilience against global resource shortages (METI, 2021).

1.7.4 National Control and Protectionism – Securing Resources for the Domestic Market

Some countries implement protectionist measures to reserve their raw material resources primarily for domestic use. This strategy aims to reduce dependence on international markets while strengthening the national economy (Scholz & Wellmer, 2013).
Characteristics of this strategy:
- State control over strategically important raw materials (e.g. rare earth elements or lithium).
- Export bans or high tariffs to reserve raw materials for the domestic market.
- Mandatory domestic value creation, requiring companies to increase processing and production within the country.
- Promotion of the national mining and processing sector through subsidies and industrial policies.

Example: China implements a raw material security strategy, particularly regarding rare earth elements that are essential for high-tech industries. Through export quotas and strict state control over mining operations, China ensures its domestic industry is preferentially supplied with these critical materials (Wübbeke, 2013).

1.7.5 Long-Term Implications of Different Resource Strategies

The choice of resource-related strategies carries far-reaching economic, social, and environmental implications. Each approach entails both opportunities and risks:

- **Maximizing Resource Extraction (Extractivism):** While this strategy can stimulate rapid economic growth in the short term, it accelerates the depletion of natural resources and may cause significant environmental damage, placing a burden on future generations (Auty, 2001).
- **Sustainable Resource Management:** This strategy promotes long-term economic and ecological stability, but it requires substantial investments in research and technology, as well as strong political commitment to implement sustainable practices (European Commission, 2020).
- **Diversification and Import Strategies:** These approaches help mitigate the risks of resource scarcity but can increase depending on international supply chains, which are vulnerable to global market fluctuations (Kooroshy et al., 2015).
- **National Control and Protectionism:** While such strategies can strengthen domestic industries and the economy in the short term, they may lead to trade conflicts and result in the country's isolation from global markets (Scholz & Wellmer, 2013).

Countries adopt different resource-related strategies depending on their economic structures and natural resource endowments. Some prioritize maximum resource utilization to achieve rapid growth, while others pursue sustainable and long-term solutions. The key challenge lies in striking a balance between economic development, resource security, and environmental and social sustainability to avoid passing the consequences of today's decisions onto future generations.

It appears to be advisable to restrict growth to what becomes available from sun and geothermal sources plus by immaterial products (such as services), and to minimize the use of mineral and fossil resources, as indicated in Figure 1.10. It is noteworthy that, as the figure indicates that, as the fossil and mineral resources decline with respect to their availability, growth becomes limited.

Sufficiency and Limits

As indicated in Figure 1.10, growth is dependent on the local availability of resources. In the word "local", the insight is hidden that, for local or regional discussions, a global average is meaningless. Furthermore, a statement about local availability alone does not provide a clear indication of the upper or lower limits of consumption. The sufficiency discourse offers such an indication for the lower limit and thereby delivers impulses for resource conservation by emphasizing not technical but rather cultural, social, and political solutions. Instead of asking how societies can consume, this discussion challenges whether it is required to consume as much as today in the first place (Paech, 2012). This fundamental reorientation toward basic needs has the potential to gain widespread public support, particularly as a promise that everyone's essential needs can be adequately met. However, in the social realm, this requires a stronger role for the state. This raises questions about the need for fiscal measures

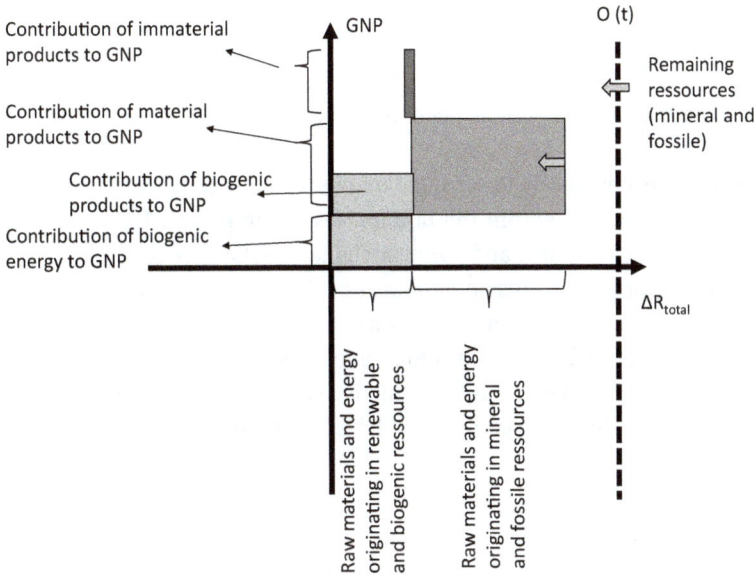

Figure 1.10: Contributions to gross national product.
Gross national product (GNP) consists of various contributing sectors. Intangible products – such as services, banking, insurance, and intellectual property rights – account for approximately 20–30% of the total and require virtually no physical resources. In contrast, tangible products are manufactured from either biogenic or non-renewable raw materials and require energy for their production, regardless of the energy source. While some raw materials are irreplaceable, others can be substituted with biogenic alternatives. In the strictest sense, one could envision a complete renunciation of mineral and fossil resources – not as a practical objective, but rather as a guiding principle or normative ideal.

such as wealth-based taxation, especially since excessive public debt increases dependency on financial markets (Schneidewind & Zahrnt, 2013). At its core, sufficiency calls for a limitation of resource use through modest, conscious consumption and lifestyle changes.

Key elements of the sufficiency discourse include several interrelated strategies.

First, **behavioural change** plays a central role: for example, reducing car usage in favour of cycling or public transportation, consuming less meat, avoiding frequent air travel for holidays, and limiting the purchase of clothing and electronics (Sachs, 1993).

Second, **supportive infrastructural and political frameworks** are essential. These include urban planning that promotes short distances (the so-called "city of short distances"), spatial policies that favour decentralization and local provisioning, as well as regulatory measures such as speed limits, advertising restrictions, and progressive taxes on high consumption (Lorek & Fuchs, 2013).

Third, the **development of social innovations and new social norms** is a significant component. This involves promoting shared-use systems (the "sharing economy"), strengthening local economic cycles, and fostering a de-commercialized and decelerated lifestyle. Fourth, time prosperity and reduced working hours can lower consumption pressure and create more time for care work, education, and volunteerism – resulting in broader societal benefits (Coote, Franklin, & Simms, 2010).

The goal of sufficiency is not merely to consume more efficiently or more "green," but to reduce consumption overall within the boundaries of the planet. This objective goes along with a range of political and cultural challenges. Sufficiency is often unpopular as a political approach because it implies restraint or behavioural changes. In societies dominated by the cultural logic of "more is better," significant resistance from prevailing growth paradigms and consumption norms – and their stakeholders – should be expected (Brand & Wissen, 2017). Nonetheless, the sufficiency debate still lacks a concrete and societally accepted proposal for an upper limit to consumption.

1.8 Global Boundary Conditions

Planetary boundaries (also referred to as "global boundary conditions") constitute a crucial framework derived from studies originating in various scientific disciplines. A simple explanation may be obtained by first looking into a different concept: In the late 1790s, Thomas Malthus concluded that, as mankind grows exponentially while the agraric supply must, due to the limits of the surface of the earth, be limited. Based on that consideration, he estimated the time available until what is harvested will not suffice humans needs (Malthus, 1798). Although striking, the number suggested turned out to be wrong. The deeper reason for that was that Malthus underestimated the human ability to invent – in this case artificial fertilizer in the 1860s. More or less the same concept has been used in Meadows' studies "the limits to growth" of 1972/1973. Also in that study, human needs were in the centre of attention and the focus has been the mismatch between the possible supply with raw materials, overpopulation and pollution, to name a few parameters. The models used in the respective simulations indicated an overuse of resources available between the years 2050 and 2100. However, although the logic used is striking also this approach suffers the weakness that human genius is not considered.

In a fundamental way, the considerations discussed here are tightly linked to those thoughts: If raw materials and fossil resources become scarce as predicted, could biogenic resources serve as their replacement? As shown below, this will not easily be possible.

The models introduced by the Club of Rome and by Malthus set humans and their needs in the centre of attention. One may suggest a different viewpoint: With or without humans – if one would regard the Earth as a system capable of self-regulation . . . which parameters are required to describe the self-regulating abilities, and what are the operating windows for those parameters. In that sense, the concept of planetary boundary conditions refers to the physical, biological, and geochemical limits within which the Earth system remains stable and avoids transitioning into chaotic or unpredictable states.

1.8.1 Planetary Boundaries as a Model

Rockström et al. (2009, 2023) developed the concept of the planetary boundaries framework, which defines nine critical ecological thresholds.

These boundaries primarily represent the aspect to be considered and tipping points. Tipping points are the transgressions or violations that could lead to irreversible damage to the Earth system and potentially result in the collapse of the SES. Scientific discourse focuses on the identification of these boundaries, their complex interdependencies, and the societal consequences that could arise from exceeding them.

The nine planetary boundaries are as follows (Rockström et al., 2009; Steffen et al., 2015):

- **Climate Change** (greenhouse gas emissions)
- **Biosphere Integrity** (biodiversity loss)
- **Land-System Change** (land-use alterations)
- **Freshwater Use** (consumption and distribution of freshwater resources)
- **Biogeochemical Flows** (disruption of nitrogen and phosphorus cycles)
- **Ocean Acidification** (increased CO_2 absorption by oceans)
- **Atmospheric Aerosol Loading** (concentration of particulate matter in the atmosphere)
- **Introduction of Novel Entities** (e.g. microplastics, synthetic chemicals, and other anthropogenic substances)
- **Stratospheric Ozone Depletion** (the reduction of the ozone layer in the upper atmosphere)

These parameters (see Figure 1.11) can be understood as boundary conditions of the Earth system. Thresholds are chosen by the lowest value possible mankind has experience with. Hence, the thresholds are set below the tipping points for the individual conditions that define the system's resilience and capacity limits. A key aspect of this concept is that the Earth system possesses a certain degree of resilience – that is, the ability to absorb disturbances, adapt to change, and return to a stable state after disruption (Folke et al., 2016). However, if a planetary boundary "crosses" the tipping point, the system may shift into a new, potentially less hospitable but stable state – or even enter a chaotic regime in which the characteristics of the SES become unpredictable (Lenton et al., 2008).

Examples of such transitions include a global temperature rise exceeding 2 °C above pre-industrial levels, widespread deforestation, or the depletion of freshwater sources (Rockström et al., 2021).

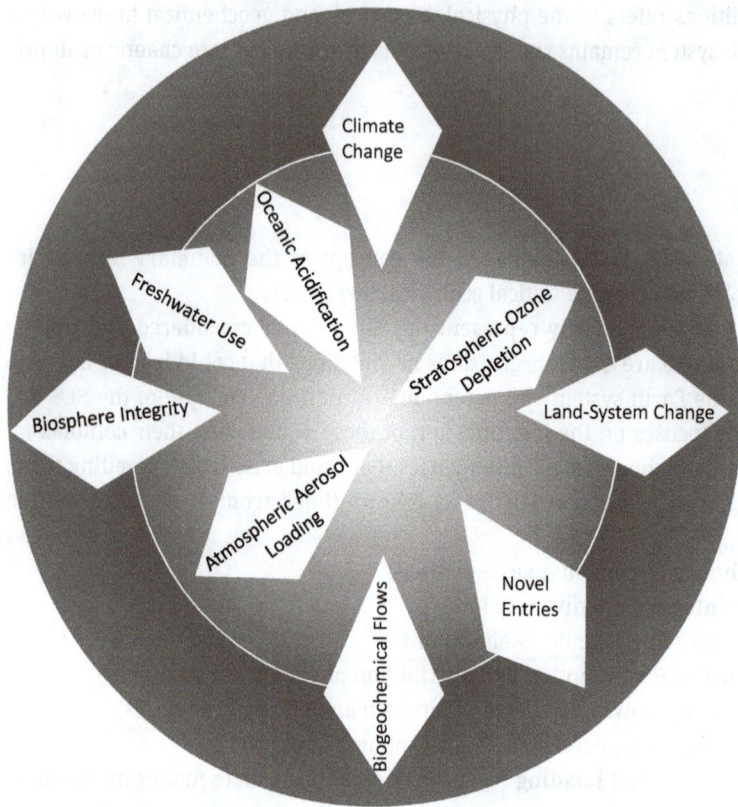

Figure 1.11: Planetary boundary conditions.
Overview of the nine planetary boundaries that define the safe operating space for humanity within Earth's environmental limits. Of the currently established parameters, six have already been exceeded, while three are still considered to be within the "safe operating space" according to current assessments. These boundaries represent critical thresholds in key Earth system processes, including climate change, biodiversity loss, biogeochemical flows, and land-system change, among others. According to current scientific assessments, six of these boundaries have already been exceeded, signalling that human activities are pushing the planet beyond its sustainable limits and increasing the risk of irreversible environmental change. The remaining three boundaries remain within the safe operating space, offering a window of opportunity for corrective action. The figure underscores the urgent need for global efforts to reduce ecological impact and maintain planetary stability to safeguard Earth's resilience and ensure sustainable development for future generations.

1.8.2 Uncertainties in Defining Planetary Boundaries

The definition and quantification of planetary boundaries involve considerable uncertainties, stemming from limited empirical data, complex system interactions, and varying scientific methodologies. These uncertainties challenge the precision of threshold values and necessitate continuous refinement as new knowledge emerges.

A key aspect of the scientific discourse on planetary boundaries concerns the uncertainty surrounding their precise definition. While the framework proposed by Rockström et al. (2009, 2021) establishes explicit thresholds for nine critical Earth system processes, some researchers argue that these boundaries should not be viewed as fixed thresholds but rather as dynamic corridors (Lenton et al., 2008). This perspective considers the fact that the Earth's resilience is not solely determined by individual environmental variables but also by their complex interactions and the adaptive capacity of the Earth system itself (Rockström et al., 2021):

> One example is the resilience of ecosystems to biodiversity loss, which may depend on local environmental conditions or the presence of additional stressors (Folke et al., 2016). Another argument is that historical thresholds can only be reliably defined if all other environmental parameters remain at their historical levels as well. Rigid adherence to fixed thresholds may therefore lead into misinterpretations or oversimplifications of the system's true state and dynamics. (Steffen et al., 2015)

Furthermore, there are regional differences in the stress placed on the Earth system. While some planetary boundaries, such as atmospheric CO_2 concentrations, have a global dimension, other stresses are distributed unevenly across different regions (Gerten et al., 2013):

> Water scarcity, for instance, affects regions such as the Mediterranean or large parts of India, while it is less pronounced in other areas (Steffen et al., 2015). Likewise, biogeochemical cycles reveal regional disparities: Industrial agriculture in Europe and North America leads to excess nitrogen and phosphorus, whereas other regions face challenges such as malnutrition and soil degradation. (Gerten et al., 2013)

The regional differences raise the question of whether the planetary boundaries, in their current form, should be considered in a more differentiated and locally adapted manner (Gerten et al., 2013).

Another issue is the incomplete data available, particularly regarding novel pollutants such as microplastics or per- and polyfluoroalkyl substances (PFAS). While well-documented data exist for traditional environmental stressors like CO_2 emissions or nitrogen cycles, information on the occurrence, concentrations, and long-term impacts of these emerging pollutants remains scarce (Schwarzenbach et al., 2019). Since many of these substances accumulate in the environment, determining the concentration threshold at which they cause ecological or health damage is challenging. This complicated the definition of clear planetary boundaries and underscores the need for ongoing research to reduce uncertainties (Schwarzenbach et al., 2019) – however, the framework is solid, and the parameters identified are solid.

1.8.3 Interactions Between Planetary Boundaries

The planetary boundaries are not isolated from one another but rather interdependent and interconnected. The interactions between these boundaries can amplify or mitigate environmental stressors, creating complex feedback loops that can either exacerbate or buffer the impacts of exceeding certain thresholds. Understanding these interactions is crucial for accurately assessing the risks associated with crossing multiple boundaries simultaneously. A particularly challenging aspect of considering planetary boundaries is understanding their complex interactions. While exceeding a single planetary boundary can already pose a threat to the Earth system, interactions between multiple factors can amplify the effects and lead to unexpected cascading consequences. From what is known so far, these interactions are often nonlinear, meaning that even small changes in one area can have far-reaching impacts in other areas (Steffen et al., 2015).

One example of such interactions is the link between climate change and biodiversity loss:

> The rise in global temperatures is altering climatic conditions worldwide, with severe consequences for many ecosystems. Rising temperatures and changing precipitation patterns can render habitats uninhabitable and lead to the extinction of numerous species. At the same time, the ecosystem becomes less resilient to climatic changes due to the loss of biodiversity. Rainforests, such as the Amazon, play a critical role in carbon storage and help stabilize the regional climate. However, as biodiversity is diminished through deforestation and species loss, the forests' capacity to absorb CO_2 declines, further exacerbating climate change – a classic example of a tipping point, where the Earth system may shift into a new, potentially irreversible state. (Rockström et al., 2009)

Also, biogeochemical cycles are closely interconnected with other planetary boundaries: Excessive nitrogen and phosphorus inputs from industrial agriculture lead to long-term disruptions in water cycles and ecosystems. Excess nitrogen enters water bodies, resulting in eutrophication and oxygen depletion, which harms aquatic habitats. Additionally, nitrogen can escape into the atmosphere in the form of nitrous oxide (N_2O), a potent greenhouse gas, further exacerbating climate change (Gerten et al., 2013):

> Another example of the coupling of planetary boundaries is the interaction between deforestation, water cycles, and agricultural productivity. Forests are crucial for regulating regional climate and precipitation patterns. However, when forests are cleared for agriculture or settlement, this function is diminished. Particularly in regions such as the Amazon Basin, deforestation leads to a reduction in evapotranspiration and consequently rainfall, which can result in long-term water scarcity and agricultural degradation. (Nepstad et al., 2008)

These examples illustrate that individual planetary boundaries cannot be viewed in isolation. An integrative approach that considers the complex interactions between various environmental processes is essential for developing sustainable solutions. If, for example, climate change is addressed without simultaneously tackling biodiversity loss or land-use changes, unforeseen side effects may arise, potentially exacerbating the problem (Steffen et al., 2015). Therefore, the sustainable management of planetary boundaries requires a holistic approach that incorporates all relevant interactions and feedback effects (Rockström et al., 2009).

1.8.4 Exceeding the Planetary Boundaries

Exceeding planetary boundaries not only has immediate environmental impacts but can also bring long-term risks that destabilize the Earth system in irreversible ways. Particularly concerning are the crossing of critical tipping points in the climate system, the loss of biodiversity, and pollution by novel substances. These developments can not only trigger abrupt changes in natural processes but also have profound consequences for human society (Lenton et al., 2008).

The Earth's climate system is an extremely complex structure, whose stability is influenced by various feedback mechanisms. As indicated, tipping points represent thresholds beyond which the climate system may undergo fundamental and often irreversible changes. A particularly significant tipping point is the melting of the Greenland ice sheet, which is amplified by the ice-albedo feedback: As ice melts, less sunlight is reflected and more heat is absorbed, further accelerating the melting process (Lenton et al., 2008). Researchers suggest that a local warming of 1.5–2 °C above pre-industrial levels could trigger an irreversible process, leading to a significant rise in sea levels over the long term (Rockström et al., 2009).

Other tipping points closely associated with climate change include:

- The dieback of the Amazon rainforest due to increasing droughts, deforestation, and fires, which could cause the rainforest to shift from a CO_2 sink to a CO_2 source (Nobre et al., 2016).
- The melting of the West Antarctic Ice Sheet, which could also lead to a dramatic rise in sea levels (Mercer, 1978).
- Changes in the Gulf Stream and the Atlantic Meridional Overturning Circulation (AMOC), which significantly influence the climate in Europe and, if weakened, could lead to extreme weather events (Broecker, 1997).
- A particular risk lies in the potential coupling of these tipping points. Once one tipping point is crossed, it could trigger other tipping points, potentially shifting the entire Earth system into a new, more unstable state (Lenton et al., 2008).

1.8.5 Loss of Biodiversity – Weakening of Ecosystem Resilience

Biodiversity is essential for the proper functioning of ecosystems and the provision of crucial ecological services such as pollination, water cycle regulation, carbon sequestration, and climate regulation (Cardinale et al., 2012). The ongoing loss of biodiversity is occurring at a rate unprecedented in history (Pimm et al., 2014).

The primary causes of this loss include the destruction of natural habitats through deforestation, agricultural expansion, urbanization, and infrastructure projects. Tropical rainforests, which are considered biodiversity hotspots, are particularly affected (Barlow et al., 2018). Other significant drivers of species extinction include climate change, over-exploitation of species through activities such as fishing and poaching, pollution, and the introduction of invasive species that displace native ecosystems (Sala et al., 2000).

Another critical issue is the introduction of novel chemical substances into the environment, such as plastics, industrial chemicals, pesticides, pharmaceuticals, and PFAS. These substances are often non-biodegradable and accumulate in organisms, soils, water bodies, and the atmosphere (Schwarzenbach et al., 2019).

A particularly concerning example is microplastics, which have now been detected worldwide – from the deepest ocean trenches to the human placenta (Rochman et al., 2019). Microplastics have the ability to bind toxic substances, which are then absorbed through the food chain and are, when released, potentially harmful to human or animal health (Browne et al., 2008). Equally worrying are persistent chemicals such as PFAS, which are used in industry, for example, in waterproof clothing, Teflon pans, or firefighting foams (Gauthier et al., 2015). These substances are virtually non-biodegradable and have been detected in drinking water sources as well as in human blood (Hu et al., 2016). The risks associated with these chemicals in the food chain are not yet fully understood, but there is evidence suggesting that they can:

–   cause hormonal and neurological disorders (Andersen et al., 2017),
–   sustainably damage ecosystems by affecting the behaviour and reproduction of animals, and
–   have toxic effects on aquatic life and land animals at high concentrations (Patel et al., 2013).

Since many of these substances are still insufficiently regulated or studied, their long-term accumulation in the environment could trigger an uncontrollable environmental and health crisis (Sunderland et al., 2019).

Political and Economic Challenges in Adhering to Planetary Boundaries

Adhering to planetary boundaries is not only a scientific or technological challenge but also a profound political and economic issue. While research increasingly agrees that exceeding these boundaries can jeopardize the long-term stability of the Earth's system, political and economic interests often conflict with the necessary changes. Three key areas shape this debate: global justice, the balance between economic growth and ecological sustainability, and issues of governance in international environmental management.

Global Justice: Who Bears Responsibility?

A significant issue of justice arises from the fact that industrialized countries have historically been the main contributors to planetary stresses, while developing countries often bear the brunt of the consequences (Piguet et al., 2017). Wealthy regions in North America, Europe, and parts of Asia have burned large amounts of fossil fuels, deforested large areas, and released pollutants over decades (Boden et al., 2017). At the same time, it is often developing countries that suffer the most from the impacts, such as rising sea levels, droughts, or biodiversity loss, despite having made a relatively small historical contribution to these issues.

> This apparent inequality raises difficult moral and political questions, such as: Who should bear the costs of the countermeasures necessary? How can a fair distribution of burdens be achieved?

> While many developing countries argue that industrialized nations bear a historical responsibility for addressing the climate crisis, the financial commitments made by the Global North so far are still far from what is required (Schalatek & Schmalz, 2018). For example, the climate fund of $100 billion annually for developing and emerging countries was pledged in 2009, yet it has been only partially implemented, with a significant portion of these funds being provided as loans rather than direct support.

Governance Issues: How Can Environmental Agreements Be Enforced?

Another significant obstacle to adhering to planetary boundaries lies in the lack of global governance. Although there are numerous environmental agreements in place, such as the 2015 Paris Agreement on climate change or the 1987 Montreal Protocol to protect the ozone layer, their enforcement is often insufficient (Falkner, 2016). National governments frequently pursue short-term economic or political interests that conflict with global environmental objectives. For instance, many countries continue to rely on fossil fuels to secure economic growth and social stability, despite these contradicting emission reduction goals.

In addition, effective sanction mechanisms are lacking to consistently penalize violations of environmental agreements. While international trade rules are often strictly enforced (e.g. in the case of violations of WTO rules), environmental agreements are typically equipped with only voluntary commitments that do not carry direct legal consequences (Biermann, 2009). This results in some countries formulating ambitious climate goals but failing to implement them, as there are no immediate and tangible consequences for non-compliance.

1.8.6 Boundaries Passed

The previously cited source by Rockström et al. (2009) shows that six of the nine identified planetary boundaries have already been exceeded (see also Figure 1.11).

Climate Change: Exceeding CO_2 Limits and Their Impacts

The development of surface temperatures and the underlying causes are well known and well documented. Detailed records of global temperature trends have existed since the second half of the nineteenth century, enabling precise comparisons between model-based predictions and observed climate dynamics (Hansen et al., 2006). Particularly since the late 1980s and early 1990s, there is indisputable evidence that the predicted effect of global warming, caused by specific gases in the atmosphere, is indeed occurring (IPCC, 2014). This development is not only supported by atmospheric measurements and experiments but is also corroborated by astronomical studies. Based on the temperature changes of planets with atmospheres, such as Venus, the impacts of the greenhouse effect were precisely modelled and later observed (Houghton, 2009). The measured temperatures confirm the theoretical models.

In 2017, Xu and Ramanathan (2017) predicted in a study that a CO_2 concentration of approximately 430–440 ppm in the atmosphere would be reached within this decade, leading to an increase of about 1.6 °C. By 2024, the global average temperature has already exceeded the 1.5 °C threshold set by the Paris Agreement (IPCC, 2023). These values highlight the urgency of achieving global climate goals. However, warming varies significantly by region. For instance, by 2018, the temperature increase in Germany was already 1.92 °C on average, representing a noticeable rise compared to long-term average temperatures (Bocksch, 2021). In some regions, the increase is even higher: In Brandenburg, the temperature rise has already exceeded 2.2 °C, while in Kassel (Hesse), the average increase was about 1.5 °C.

For the Arctic, a dramatic average temperature rise of more than 15 °C is expected by the end of this century, which, given the current average temperatures of around −35 °C, represents an extraordinary change (AMAP, 2019). This temperature rise could lead to Arctic summers with no ice cover by around 2050, which would have significant implications for the global climate system and the ecosystems in this region (Overland et al., 2019).

The predictions from Xu and Ramanathan (2017) suggest a global warming of about 2.4 °C by 2050. Another concerning effect is the thawing of permafrost, which could release greenhouse gases stored in the soil. This process further enhances the greenhouse effect and could accelerate the warming by an additional 0.6 °C by 2050, resulting in a total increase in global average temperature of approximately 3 °C (Schuur et al., 2015). In pessimistic scenarios, which predict a warming of more than 3.5–4 °C by 2050, this trend is further amplified. A scenario with a median warming of about 3 °C is now considered the most likely (IPCC, 2021). This development indicates a pressing need for action to mitigate the catastrophic consequences of the climate crisis and to meet global warming targets.

Biosphere Integrity: Loss of Biodiversity and Ecological Instability

The loss of biodiversity is driven by several interconnected human activities, primarily the destruction of natural habitats. The main causes include:

1. **Habitat Destruction and Fragmentation**: Deforestation, urbanization, and agriculture lead to the destruction of natural habitats. Tropical rainforests, considered biodiversity hotspots, are particularly affected. Through this destruction, many species lose their habitats, increasing their risk of extinction.
2. **Climate Change**: Climate change alters the habitats of many species and shifts climatic zones. Some species cannot adapt or migrate quickly enough, putting their survival at risk.
3. **Overuse of Resources**: Excessive hunting, fishing, and agriculture lead to the overexploitation of species. These activities place considerable stress on many species, reducing their long-term survival prospects.
4. **Pollution**: Environmental pollution, including chemicals, plastics, and other waste, threatens both biodiversity and the functioning of ecosystems. Toxic chemicals, such as pesticides, industrial waste, and heavy metals, have detrimental effects on plants and animals, disrupting food webs.
5. **Invasive Species**: The introduction of non-native species into ecosystems through human activities can displace native species. These invasive species, lacking natural predators in their new environment, can spread unchecked, endangering local flora and fauna.

Hence, biosphere integrity, encompassing the diversity and functionality of global ecosystems, is under serious threat. The current extinction rate is at least 100–1,000 times higher than the natural background rate, which has dramatic consequences for the stability of many ecosystems (Barnosky et al., 2011). Pollinators, such as bees and butterflies, which are crucial for food production, are particularly affected (Gallai et al., 2009). Additionally, the loss of forests and seagrass meadows reduces carbon sequestration and exacerbates climate change (Díaz et al., 2019). Many of these losses are irreversible (Wilson, 2016).

Biodiversity not only refers to species diversity but also to genetic diversity and ecosystem diversity. High biodiversity strengthens the adaptability of ecosystems to environmental changes. This adaptability can be measured through parameters such as genetic variability, species richness, and habitat diversity (Pimm et al., 2014). Protecting biodiversity is critical for maintaining the ecological functions of habitats and ensuring resilience to future environmental changes (Mace et al., 2014).

Biodiversity is monitored through various methods, such as species inventories, diversity indices, genetic analyses, and remote sensing (Vellend et al., 2013). These measurements are important for detecting ecological changes and developing appropriate conservation measures. In the European Union, biodiversity is an integral part of sustainability reporting (EU, 2020).

Impacts of Biodiversity Loss

Biodiversity forms the foundation of stable and resilient ecosystems, which are essential for human well-being, food production, and the global economy. The loss of biodiversity has far-reaching consequences, affecting ecological, economic, and social dimensions (Sala et al., 2000):

– **Reduced Ecosystem Stability:** As biodiversity declines, key species that perform important ecological functions may be lost. This reduces the resilience of ecosystems to environmental changes and extreme events, such as natural disasters or climate fluctuations (Oliver et al., 2015).

– **Declining Agricultural Productivity:** Pollinators, such as bees, are responsible for pollinating many crops and play a vital role in agriculture. Their loss would lead to reduced yields (Klein et al., 2007).

– **Restricted Medical Development:** Many medications are derived from natural substances, and the loss of plant species could hinder the discovery of new therapeutic agents (Newman & Cragg, 2016).

Thus, the loss of biodiversity is not only an ecological issue but also a significant economic and social challenge. The role of biodiversity in climate resilience and agriculture is particularly important. Diverse ecosystems are better equipped to respond to climatic changes, as they provide a complex network of interactions that counteracts temperature fluctuations, extreme weather events, and pest infestations (Barton et al., 2015). In agriculture, high biodiversity results in more stable yields and reduces dependence on chemical fertilizers and pesticides, as natural pollinators and beneficial organisms help protect crops (Bengtsson et al., 2005).

The destruction of ecosystems also has substantial economic impacts, such as reduced agricultural yields and polluted drinking water (Costanza et al., 2014). Regenerative approaches, such as regenerative agriculture and habitat restoration, help mitigate these negative effects (Gomiero, 2018). Sustainable land-use strategies, like agro ecology and agroforestry, can help preserve biodiversity while ensuring food security (Altieri et al., 2012).

The conservation of biodiversity requires a combination of protected areas, sustainable land-use practices, and nature-based solutions. Political measures are essential for ensuring the long-term security of biodiversity. At the EU level, the goal is to protect 30% of the total land area, with at least 10% of agricultural land remaining fallow to preserve biodiversity (EU, 2020).

Land Use Changes: Destruction of Natural Ecosystems through Deforestation and Agriculture

Land use has significantly altered the Earth system, particularly through extensive deforestation and the conversion of natural ecosystems into agricultural land. Approximately 50% of the original forests have already been destroyed or severely degraded, meaning that the remaining forest areas no longer meet the threshold of at least 75% intact forests (FAO, 2018). These changes have serious ecological consequences: forests not only act as carbon sinks but also play a critical role in regulating both regional and global water cycles (Schlesinger & Bernhardt, 2013). Forest destruction leads to the release of CO_2 and triggers negative feedback mechanisms that can alter local climates and exacerbate droughts (Lewis et al., 2015). Particularly dramatic is the loss of tropical rainforests, such as the Amazon rainforest, often referred to as the "lungs of the Earth," which may undergo an irreversible transition from a humid forest ecosystem to a savanna-like landscape (Nobre et al., 2016).

Introduction of Novel Substances: Chemical Pollution and the Issue of Plastics

The increasing environmental burden caused by novel substances such as plastics, industrial chemicals, heavy metals, and pesticides poses a significant threat to ecosystems and human health (Scheringer et al., 2018). These substances are produced and dispersed at a scale that exceeds the natural regenerative capacities of the environment (Hahladakis et al., 2018). Of particular concern is plastic pollution, as millions of tonnes of plastic waste enter the oceans every year, where it can accumulate over decades to centuries (Jambeck et al., 2015). Microplastic particles have been detected in nearly all environmental compartments, ranging from the deepest ocean trenches to human blood (Gouin et al., 2011). The long-term health effects of these substances are not yet fully understood, but preliminary studies suggest potential toxic effects and hormonal disruptions (Rochman et al., 2013). Additionally, persistent chemicals such as PFAS present a growing problem (Buck et al., 2011). These substances, found in many everyday products, can accumulate in the environment over decades and have now been detected even in remote regions (Liu et al., 2017).

Biogeochemical Fluxes: Disruption of the Nitrogen Cycle

Sustainable freshwater use is of crucial importance for life on Earth. Due to overuse, inefficient irrigation practices, and climate change, water resources in many regions worldwide are increasingly at risk (Postel, 2000). Mountainous countries and agriculturally dominated regions are particularly affected, where intensive irrigation leads to a drastic decline in groundwater levels (Richey et al., 2015). In many areas, the planetary boundary for green water availability has already been exceeded, particularly in semi-arid and arid zones such as the Mediterranean, parts of India, and the southwestern United States (Foley et al., 2011). The long-term consequences of this development include desertification, crop yield declines, and rising social tensions related to the distribution of water resources (Vörösmarty et al., 2000)

Freshwater Use: Overuse and Water Scarcity

Sustainably produced freshwater use is crucial for life on Earth. Due to overuse, inefficient irrigation practices, and climate change, water resources in many regions worldwide are increasingly at risk (Postel, 2000). Mountainous countries and agriculturally dominated regions are particularly affected, where intensive irrigation leads to a drastic decline in groundwater levels (Richey et al., 2015). In many areas, the planetary boundary for green water availability has already been exceeded, particularly in semi-arid and arid zones such as the Mediterranean, parts of India, and the southwestern United States (Foley et al., 2011). The long-term consequences of this development include desertification, crop yield declines, and rising social tensions related to the distribution of water resources (Vörösmarty et al., 2000).

1.8.7 Boundary Conditions Not Passed Yet

Despite the exceeding of several planetary boundaries, there are still some areas where the Earth system remains within safe limits – at least on a global scale. These include ocean acidification, atmospheric aerosol pollution, and stratospheric ozone depletion. However, these areas are regionally affected in varying degrees, and it is possible that critical thresholds could be exceeded in the future as well (Rockström et al., 2009; Steffen et al., 2015).

Ocean Acidification: Still Within the Boundaries, but with Increasing Risk

The oceans absorb a significant portion of the CO_2 emitted by humans, leading to a decrease in the pH of seawater – a process known as ocean acidification (Rhein et al., 2013). Although the global pH of the oceans is still within safe limits, a concerning trend of acidification is emerging. Since the Industrial Revolution, the pH has decreased from around 8.2 to approximately 8.1 (Feely et al., 2009). While this may seem minimal at first glance, it has far-reaching effects on the chemistry of the oceans, particularly on calcifying marine organisms such as corals and shellfish (Doney et al., 2009). Some regions are already more affected, and acidification could accelerate in the coming decades if CO_2 emissions continue to rise (Orr et al., 2005).

Atmospheric Aerosol Pollution: A Regional Issue, but Affecting Global Stability

Aerosols, which originate from both natural and human sources, play a complex role in the climate system (Andreae & Rosenfeld, 2008). Globally, aerosol pollution remains within safe limits; however, there are regional hotspots, particularly in heavily industrialized areas (Hansen et al., 2005). High concentrations of aerosols can impact the climate, for example, by altering precipitation patterns and solar radiation (Schwartz, 2004). Furthermore, fine particulate matter poses a significant health risk, as it is associated with respiratory diseases and cardiovascular problems (Pope & Dockery, 2006).

Stratospheric Ozone Depletion: A Success Through International Action

Stratospheric ozone depletion was one of the largest global environmental issues in the 1980s (Farman et al., 1985). The international response, particularly through the 1987 Montreal Protocol, led to the successful banning of ozone-depleting substances (Molina & Rowland, 1974). This has resulted in a slow recovery of the ozone layer, and global ozone levels are currently within safe limits (WMO, 2018). While challenges persist, particularly due to extreme weather events or unregulated chemicals in some regions (Shao et al., 2017), the recovery of the ozone layer demonstrates that international action can be crucial in stabilizing planetary boundaries (Newman et al., 2016).

1.9 Adaptive Cycles

Published by B.S. Holling in 1973, the adaptive cycle model provides a theoretical framework for deriving and explaining tipping points in complex systems (Holling, 1973). The theory was developed in the early 1970s and was based on studies of animal populations in northern Canada (Holling, 1973). Since its introduction, this model has proven to be extremely useful not only in the analysis of ecosystems but also in the study of SESs, particularly regarding their development in the context of resource availability and use (Walker et al., 2004). It describes the dynamic, cyclical progression of systems, which go through four recurring phases: exploitation, conservation, release, and reorganization (Holling, 1973) (see Figure 1.12).

The first phase, exploitation, describes the initial phase of a system in which resources are abundant, leading to rapid growth and increasing connectivity (Holling, 1973). In the following phase, conservation, the system eventually reaches a state of stability, as resources are efficiently utilized and connectivity is optimized (Walker et al., 2004). However, when resources become scarce or external disturbances occur, the system enters the release phase. Here, there is a sudden and often dramatically chaotic collapse of structures, which may also be long-lasting (Walker et al., 2004). While the system in the conservation phase returns to equilibrium after a disturbance, this is not necessarily the case in the release phase, as the system reacts chaotically (Walker et al., 2004).

Finally, the reorganization phase follows, in which the system may develop new structures and mechanisms to adapt to the altered conditions (Holling, 1973). This cyclical process illustrates how complex systems, whether in nature or in social systems, continuously change and reshape (Walker et al., 2004). A characteristic feature of this model is that it does not assume the system returns to a state similar to its original

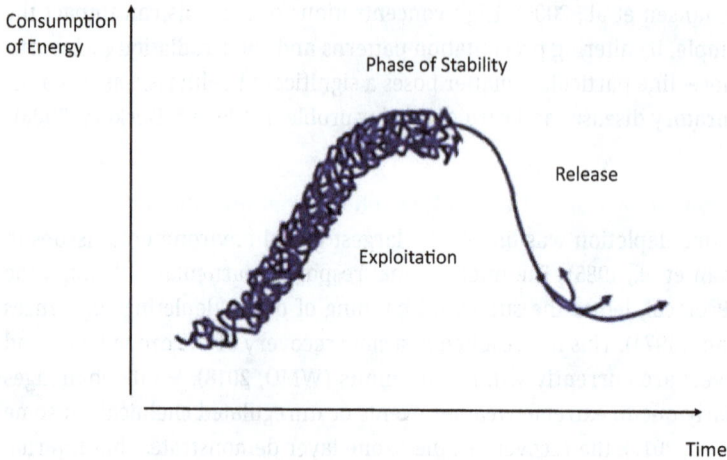

Figure 1.12: Model of the adaptive cycles.
In a phase of reorganization, the system begins to build structures, but it does not begin a teleological growth in the exploitation phase, which involves the consumption of raw materials and energy. The exploitation phase is not linear but rather characterized by leaving equilibrium and going back into it. This growth culminates in a phase of stability, where the system develops rigid structures and can no longer respond flexibly to external disturbances or return to its original state. The system collapses during a release phase.

one after completing the cycle. Rather, it describes the transformation of a system and its ongoing adaptation to changing environmental conditions (Walker et al., 2004). The connectivity in the model is an expression of resource consumption and energy turnover, which are necessary for the system's functioning (Holling, 1973).

The model demonstrates how the energy and resource turnover of a system is linked to its resilience – that is, its ability to adapt to changes (Folke et al., 2004). A system in the conservation phase is characterized by high stability and resistance. However, in the release phase, the system becomes more vulnerable to disturbances that can lead to the crossing of a tipping point (Folke et al., 2004). Tipping points are critical thresholds beyond which the system suddenly and unpredictably shifts into a chaotic state (Lenton et al., 2008). This transition is often nonlinear and typically occurs in response to gradual but cumulative changes (Lenton et al., 2008). Crossing a tipping point means that the system no longer operates under the same laws and structures that previously defined it (Lenton et al., 2008). Once a tipping point is crossed, it is difficult to reverse, and it can have long-term, often irreversible, consequences (Lenton et al., 2008). Prominent examples of tipping points include the collapse of ecosystems such as the Great Barrier Reef or the potential tipping of the Gulf Stream, which influences climate and weather conditions in Europe and North America (Lenton et al., 2008).

The concept of boundaries and tipping points is of crucial importance in the discussion of climate change and the stability of SES (Folke et al., 2004). Boundaries are conditions that the system must not exceed in order to ensure stable development (Rockström et al., 2009). These boundaries are often linked to the ecological and climatic conditions that humanity has experienced throughout history. Typically, these boundaries refer to the conditions that prevailed during the last Ice Age, which serves as a reference period for understanding the Earth's carrying capacity (Rockström et al., 2009). The challenge in applying this knowledge to the modern socio-ecological system, however, lies in the difficulty of precisely determining how close we currently are to these boundaries or tipping points (Rockström et al., 2009).

Another issue is the slow response of nature (compared to the lifetime of humans): Often, the impacts of a tipping point being passed may only become clearly visible years or decades later (Lenton et al., 2008). This makes predicting and responding in time to potential tipping points particularly difficult. For example, there is ongoing debate about whether the Great Barrier Reef has already crossed a tipping point, which could lead to irreversible coral loss (Berkes et al., 2006). Similarly, the Gulf Stream, a critical component of the global climate system, may have already entered a phase of tipping, which would have catastrophic consequences for the climate in Europe and North America (Lenton et al., 2008).

Given these uncertainties, it is crucial to take preventive measures to avoid crossing critical tipping points. This means that we must not only prevent these boundaries from being reached but also acknowledge that we do not know how far we can go before irreversible and potentially catastrophic changes occur. It is the responsibility of humanity to ensure that the limits of ecosystem and societal resilience are not exceeded in order to minimize the risk of unpredictable and dramatic changes (Rockström et al., 2009).

1.10 Resource Availability

The availability of resources can be graphically represented in a model developed by McKelvey (1996). In this model, resources are divided into three main categories:
–   Known resources that are already being extracted
–   Known resources that have not yet been extracted
–   Unknown resources, whose existence has not yet been confirmed (McKelvey, 1996)

Unknown resources are inherently difficult to assess, as their actual availability and potential for exploitation are still unclear (McKelvey, 1996). Even for known resources that have not yet been extracted, there is uncertainty, particularly regarding their actual economic feasibility. The question arises as to how much of these resources can actually be extracted at reasonable costs. Several factors come into play that further restrict access to resources:

- **Political Restrictions**: Some countries have restricted the export of certain raw materials or only allow the export of processed products (Dube & Matusz, 2000).
- **Infrastructure and Logistics**: When raw materials are located in hard-to-reach regions – such as remote parts of Africa without adequate transportation links – access to these resources is significantly hindered (Amin & Kline, 2004).
- **Energy Costs and Transport Effort**: Even if a raw material is physically accessible, the costs associated with its extraction or transport may be so high that it becomes economically unfeasible to exploit it (Dube & Matusz, 2000).

For these reasons, only a portion of the existing raw materials is actually accessible and usable (see Figure 1.13).

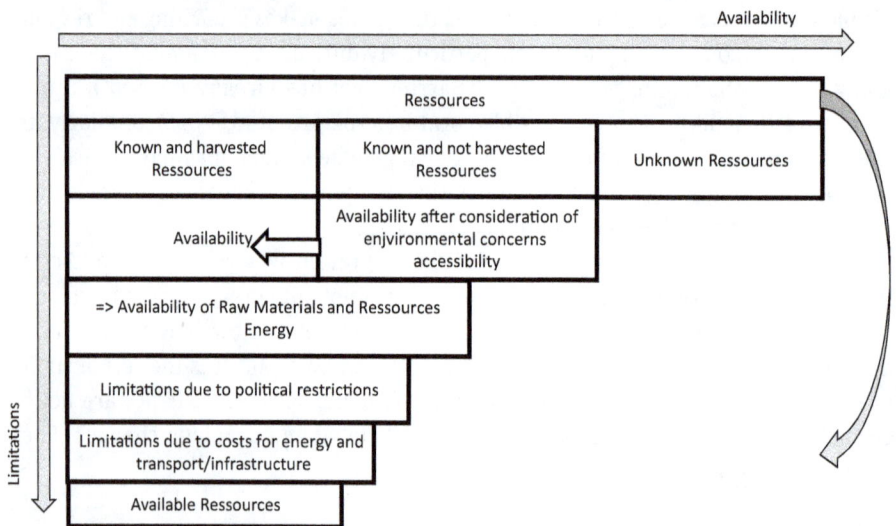

Figure 1.13: Limitation of the extractability of raw materials according to McKelvey et al. (1973). Illustration of the limitation of raw material extractability based on the McKelvey classification framework, which categorizes mineral resources according to their geological certainty and economic feasibility of extraction. The figure depicts how raw materials range from undiscovered and speculative resources to reserves that are economically extractable with current technology and market conditions. As resources move from inferred or indicated categories into economically viable reserves, their availability is constrained by factors such as ore grade, extraction costs, technological capabilities, and environmental regulations. This framework highlights the inherent limitations and challenges in expanding raw material supply, emphasizing that not all known deposits are immediately or economically accessible. The figure underscores the importance of resource efficiency, recycling, and sustainable management to address potential supply risks associated with finite and increasingly difficult-to-extract raw materials. The notion introduced in this figure is that of the resources available and due to several reasons, only a fraction becomes available at the point of need. Only some of these factors are linked to the raw material itself while others deal with the infrastructure required or with political forces involved.

1.10.1 Critical Raw Materials

To better assess resource availability, existing data and expert opinions are utilized. One example of this is the Critical Raw Materials List of the EU (see Table 1.1), which initially included about nine to ten raw materials (European Commission, 2008). In 2024, this list includes approximately 40 materials (European Commission, 2024).

Table 1.1: Critical minerals according to the assessment of the U.S. Geological Survey (USGS, 2022).

Critical mineral	Primary applications
Aluminum	Metallurgy; various industrial sectors
Antimony	Flame retardants; lead-acid batteries
Arsenic	Pesticides; semiconductors
Barite	Hydrocarbon (oil and gas) production
Beryllium	Aerospace industry; defense technology
Bismuth	Medical technology; metallurgy; nuclear research
Cerium	Catalysts; ceramics; glass; metallurgy; abrasives
Caesium	Research and development
Chromium	Metallurgy
Cobalt	Batteries; metallurgy
Dysprosium	Data storage; lasers; permanent magnets
Erbium	Fibreglass, glass coloring, lasers, optical amplifiers
Europium	Control rods in nuclear reactors
Fluorspar	Cement industry; industrial chemicals; metallurgy
Gadolinium	Medical imaging; metallurgy; permanent magnets
Gallium	Integrated circuits; optoelectronic devices
Germanium	Defense technologies; fiber optics
Graphite	Batteries; fuel cells; lubricants
Hafnium	Ceramics; nuclear control rods; metallurgy
Holmium	Lasers; nuclear reactors; permanent magnets
Indium	Liquid crystal displays (LCDs)
Iridium	Electrochemical coatings; catalysts
Lanthanum	Batteries; catalysts; ceramics; glass; metallurgy
Lithium	Batteries
Lutetium	Cancer therapy; electronics; medical imaging
Magnesium	Metallurgy
Manganese	Batteries; metallurgy
Neodymium	Catalysts; lasers; permanent magnets
Nickel	Batteries; metallurgy
Niobium	Metallurgy; alloy production
Palladium	Catalysts
Platinum	Catalysts
Praseodymium	Aerospace alloys, batteries, ceramics, dyes, and magnets
Rhodium	Catalysts; electronic components
Rubidium	Research and development
Ruthenium	Catalysts; electronic components; computer chips
Samarium	Cancer treatment, nuclear energy, and permanent magnets.

Table 1.1 (continued)

Critical mineral	Primary applications
Scandium	Ceramics; fuel cells; metallurgy
Tantalum	Capacitors; metallurgy
Tellurium	Metallurgy; solar cells; thermoelectric devices
Terbium	Fiber optics; lasers; permanent magnets; advanced devices
Thulium	Lasers; metallurgy
Titanium	Metallurgy; pigments
Tungsten	Metallurgy
Vanadium	Batteries; catalysts; metallurgy
Ytterbium	Catalysts; lasers; metallurgy; scintillators
Yttrium	Catalysts; ceramics; lasers; metallurgy
Tin	Metallurgy
Zinc	Metallurgy
Zirconium	Metallurgy; nuclear energy

In the United States, a system similar to the one available in Europe exists, in which the U.S. Geological Survey (USGS) classifies raw materials as "Scarce Materials" (U.S. Geological Survey, 2020b). This classification differs from that of the EU, as the USGS refers to materials whose availability is potentially restricted rather than "critical raw materials" (U.S. Geological Survey, 2020a).

An important factor in estimating resource availability is the "reach" of a resource, which is determined by the ratio of known reserves to annual consumption (U.S. Department of the Interior, 2024) (see Figure 1.14 below). Based on the data from the U.S. government (2024), it is possible to calculate how long certain raw materials will remain available. To perform this calculation, the mass of known reserves is divided by the annual consumption (U.S. Department of the Interior, 2024). However, these estimates have certain limitations.

On one hand, the calculation does not account for the impact of economic growth on annual consumption, which can make the estimate overly optimistic. On the other hand, political restrictions and unknown reserves are not considered, which could lead to an unrealistically positive assessment (U.S. Department of the Interior, 2024).

The calculations indicate that, for example, germanium and tellurium could be exhausted in about 10 years, and many other raw materials have a reach of only 10–35 years (U.S. Geological Survey, 2020). Note that this does not mean that these materials will become extinct, but rather that the current supply will no longer suffice to the degree used today. These findings are especially significant in light of political discussions around national security concerns. In the United States, President Donald Trump repeatedly emphasized the importance of raw materials for national security in his inaugural address (Trump, 2017). The USGS maintains a list that explicitly classifies materials as critical to U.S. national security (U.S. Geological Survey, 2020).

The limited availability of mineral raw materials is a significant economic issue.

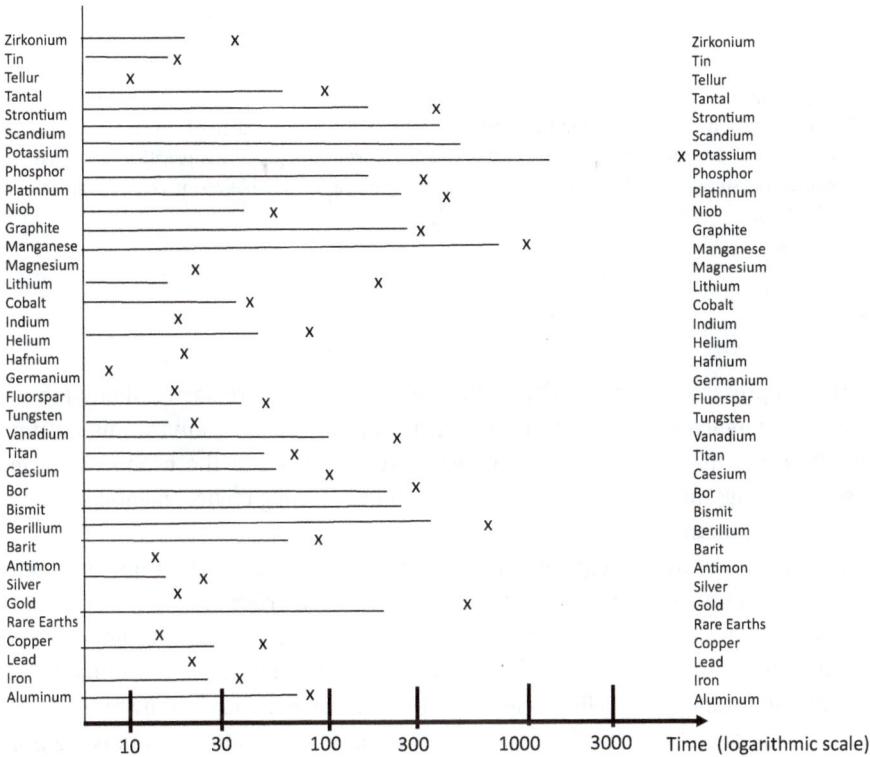

Figure 1.14: Temporal availability of materials on the critical raw materials list.
Temporal availability of materials (marked with an x in the figure) on the critical raw materials list according to the United States Geological Survey (USGS, 2022). This figure was calculated using the annual consumption data published by the United States, divided by the known resources according to the same source. The figure illustrates projected time horizons for the depletion or supply risk of various critical raw materials, taking into account current consumption rates, known reserves, geopolitical dependencies, and potential for recycling or substitution. Materials are categorized based on their expected availability over short-, medium-, and long-term timeframes, highlighting those that face imminent scarcity versus those with more stable supply prospects. The figure emphasizes the vulnerability of industrial and technological systems – particularly in sectors such as renewable energy, digital infrastructure, and defence – to disruptions in the supply of critical raw materials. It underscores the importance of strategic resource planning, diversification of supply chains, innovation in material substitution, and the development of circular economy approaches to enhance resource security over time.

According to recent calculations, iron will no longer be sufficiently available from primary sources in about 35–38 years. As a result, supply will only be possible through recycling (e.g. through urban mining) (Günther et al., 2020). This shift would have profound effects on many industrial processes, as the production of components and devices made from iron in its traditional form would no longer be feasible, fundamentally altering production methods:

Urban mining refers to the process of extracting valuable raw materials from urban waste or existing infrastructure. The goal is to recover materials such as metals, plastics, glass, or rare earth elements from products that have already been used (e.g. electronic waste, building demolition, or used machinery), rather than mining new resources from nature (Binnemans et al., 2013). Essentially, it involves "mining" raw materials from urban waste or discarded products, with an emphasis on recycling and reusing materials from the past. This practice helps reduce resource consumption, minimize waste, and promote more sustainable resource use (Binnemans et al., 2013).

1.10.2 Recycling and Circular Economy?

Recycling represents a key strategy within the context of a circular economy for reducing dependence on primary raw materials and minimizing environmental impacts. By recycling materials, raw materials are returned to the economic cycle, thereby reducing the need for new resources and alleviating environmental burdens (European Commission, 2014).

Advantages of Recycling: Recycling significantly contributes to resource conservation by reducing the demand for primary raw materials and alleviating pressure on natural reserves. For instance, recycling aluminium reduces energy consumption by up to 95% compared to primary production (European Aluminium, 2020). Additionally, recycling helps reduce environmental burdens by minimizing waste and decreasing landfill usage (Geyer et al., 2017). Another benefit is the prevention of environmental pollution caused by non-biodegradable substances such as plastics (Geyer et al., 2017). Furthermore, recycling processes are often (though not always) more energy-efficient than the production of new materials. For example, recycling steel causes only about one-third of the CO_2 emissions associated with primary production (World Steel Association, 2021).

Challenges of Recycling: Despite the numerous advantages, recycling also presents challenges. Aside from the energy and material consumption associated with recycling, one major issue is the loss of quality during recycling, particularly with plastics, which, after recycling, often cannot be used for high-quality applications due to deterioration in their chemical and mechanical properties (Hopewell et al., 2009). There are also technical and economic limitations, as not all raw materials can be recycled efficiently and cost-effectively. Complex composite materials and electronic products, for example, pose significant challenges in recovering valuable raw materials and are often expensive and technically demanding (Zeng et al., 2018). Another obstacle is insufficient collection and separation systems. A functioning circular economy requires well-developed infrastructure, which is lacking in many countries, leading to the loss of valuable resources (Bocken et al., 2016).

Circular economy is tightly linked to recycling.

A core principle of the circular economy is design for recycling, which involves designing products and materials from the outset to facilitate ease of recycling. This includes, for example, the use of single-type materials or detachable fastening elements (Stahel, 2016). That also enables the reuse and repair of products – rather than being discarded after use, products are repaired or repurposed, particularly in sectors such as electronics and machinery (Bressanelli et al., 2018). Another key principle is the integration of biological cycles, wherein biodegradable materials are returned to natural systems to replenish nutrients for new resources, as exemplified by compostable packaging or natural fibres (Braungart & McDonough, 2002).

Implementing a circular economy requires a profound systemic transformation not only of the product looked at but also involving adjustments along the entire value chain – from product design and logistics to the development of new business models (Kirchherr et al., 2018). Additionally, there are regulatory and economic barriers, as substantial initial investments and a lack of incentives for companies complicate the transition. Technological challenges primarily concern the recyclability of many materials, which are either difficult to recycle or require complex separation processes. These issues demand further innovations in materials science and recycling technologies (Parchomenko et al., 2019).

In summary, recycling within the framework of a circular economy represents a promising strategy for conserving resources and reducing environmental impacts. However, technical and infrastructural challenges must still be overcome to fully realize its potential – still recycling requires new materials and the use of energy. The circular economy is a holistic concept that goes beyond traditional recycling. It promotes sustainable resource use by aiming to eliminate waste, retain materials within the economic cycle for as long as possible, and design products for reintegration at the end of their life cycles (Ellen MacArthur Foundation, 2015).

Ensuring Resource Sustainability Through Circular Economy and Recycling

It is often assumed that recycling raw materials is a key strategy for ensuring resource sustainability. However, a study by François Grosse et al. (2011) demonstrates that recycling rates vary significantly depending on the type of material. Even under optimal conditions, recycling never achieves the theoretical maximum yield of 100%. In practice, the maximum recovery rate often falls short of three-quarters of its potential. In fact, recycling rates for many materials are considerably lower. Moreover, the study shows that as economic growth increases, resource consumption tends to rise not linearly but, in some cases, exponentially, as illustrated in the following figure. This implies that while recycling undoubtedly has positive effects, it cannot fully resolve the fundamental issue of limited raw material availability.

The availability of resources, commonly assessed in terms of the resource lifespan – i.e., the time period during which they will remain accessible – depends on a variety of factors, including:

- Availability of raw materials,
- Extraction and production rates,
- Recycling rates and the efficiency of the recycling process,
- New discoveries and technological advancements, particularly those aimed at improving efficiency across all stages of raw material supply, and
- Trends in demand for the respective raw materials.

Assuming a global recycling rate of 80% and considering a moderate annual economic growth rate of 1%, the findings of Grosse et al. (2011) suggest that raw material availability would be extended by only approximately 60 years (see Figure 1.15). This estimate underscores that recycling alone – even at high rates – cannot sufficiently ensure the long-term supply of raw materials.

In this context, it becomes evident that limiting economic growth and maximizing recycling rates are critical levers to avoid resource supply bottlenecks. Only through the combination of these measures can a collapse of the SES be effectively averted.

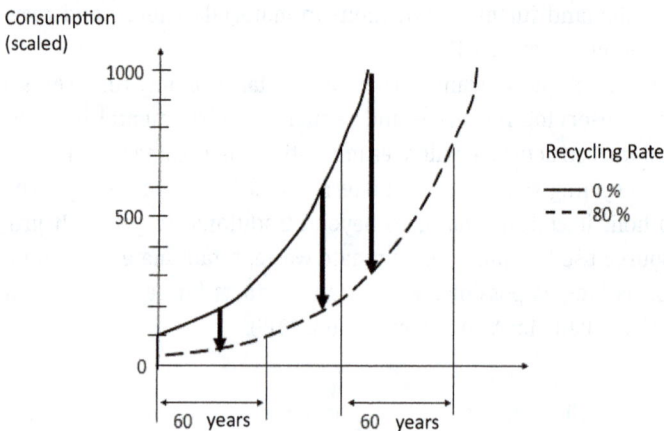

Figure 1.15: Development of consumption, assuming a growth rate of 1% and a global recycling rate of 80%, according to Grosse et al. (2011).
Growth goes along with increasing consumption. Recycling may counterbalance part of the effect caused by increasing consumption. This effect can, to a limited degree, extend the reach or raw materials at hands. This figure represents the effect as calculated by Grosse et al. under the assumption of a very high recycling rate of 80% and a very low economic growth rate of only 1%. Under those optimistic conditions, and according to that source, the effect of reduced resources can only be compensated for approximately 60 years.

1.10.3 Provision of Energy Resources

The considerations outlined above regarding the availability of mineral resources can be similarly applied to fossil energy carriers, with the additional aspect that, beyond their finite nature, their irreplaceable role in current industrial value creation must also be taken into account.

An often-considered hope is the assumption that sufficient fossil energy will remain available to maintain or even expand essential production processes, such as the manufacturing of fertilizers. However, recent reports from reputable sources – including the US Geological Survey (USGS), World Oil Resources, and British Petroleum (BP) – paint a different picture, as indicated in the table above (US Geological Survey, 2020; BP, 2023; World Oil Resources, 2019). The data platform Statista even estimates a lower availability of fossil fuels (Statista, 2021) (see Table 1.2 below). These figures underscore the absence of a realistic outlook for the unlimited availability of energy – particularly not for energy-intensive processes such as fertilizer production, which is, in turn, critical for agriculture.

Even if some projections were to be extended by 10–30 years, the fundamental issue remains unchanged:

Table 1.2: Estimated lifespan of key fossil energy carriers.

Energy resource	Availability of resource	Source
Gas	49 years	US Geological Survey (USGS) (2023) World Oil Resources; British Petroleum (BP) (2023)
Crude oil	52 years	US Geological Survey (USGS) (2023) World Oil Resources; International Energy Agency (IEA) 2023 – IEA Oil Market Report
Lignite	142 years	US Geological Survey (USGS) (2023), Mineral Commodity Summaries 2023: Coal; International Energy Agency (IEA) (2023), IEA Coal Market Report
Brown coal	400 years	US Geological Survey (USGS) (2023), Mineral Commodity Summaries 2023: Coal; International Energy Agency (IEA) (2023), IEA Coal Market Report; British Petroleum (BP) (2023); World Coal Association

According to the estimates from 2022 to 2023, the known reserves of fossil energy carriers will be available only for a limited period of time.

Energy sources are finite. Resources – whether food, fossil fuels, or mineral raw materials – are becoming increasingly scarce (Tilton, 2003). This scarcity has the potential to trigger geopolitical tensions and may even lead to armed conflicts (Kolb, 2015).

Socio-economic systems often anticipate these shortages even **before** they occur: rising prices, political tensions, and even military conflicts over declining resources are in many cases, already foreseeable (Klare, 2012). The consequences of resource overuse and growing demand are already evident, and eventually, the system will reach a tipping point. The question of whether a particular energy source should play a role in the future, or whether the focus should lie on reducing energy consumption, can be answered unequivocally: energy consumption must be reduced (Stern, 2004). Another aspect that must not be overlooked concerns the stability of energy supply: nuclear power supports centralized structures and large industrial organizations, whereas decentralized energy systems offer greater independence and resilience. The preceding arguments clearly support a decentralized energy system combined with a significant reduction in energy consumption.

1.10.4 Biogenic Raw Materials – Definition and Examples

In addition to traditional mineral and fossil resources, there are also biogenic resources. Biogenic resources are natural materials, distinct from fossil fuels, that originate from living organisms or their remains and are formed through biological processes (such as photosynthesis) in nature (Lange, 2012). They can be derived from plant or animal sources and are typically renewable, meaning that, compared to fossil resources, they can be – but not necessarily are – more sustainable, provided their use and harvest are in line with natural regeneration processes and the specific local conditions (Smeets et al., 2013). Biogenic resources are always considered in relation to local factors, such as climate, soil or water fertility, water supply, regional solar radiation, and resulting temperatures (Köhler et al., 2019).

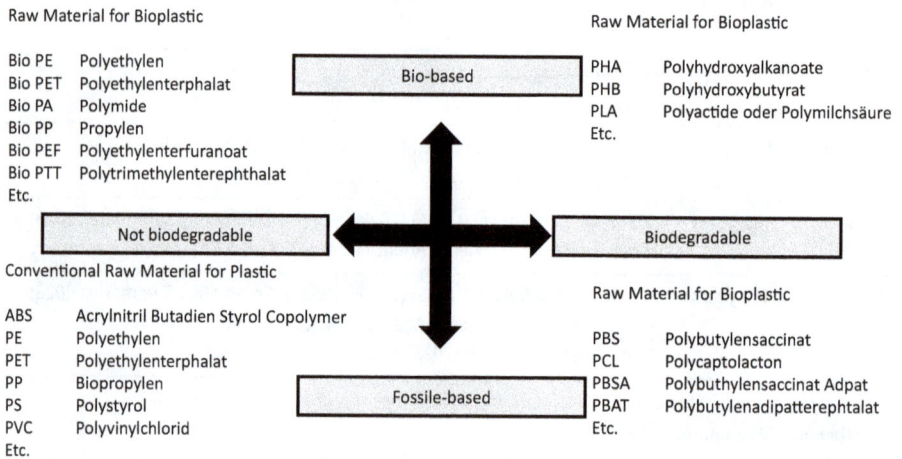

Raw Material for Bioplastic

		Raw Material for Bioplastic

Bio PE   Polyethylen
Bio PET   Polyethylenterphalat
Bio PA   Polymide
Bio PP   Propylen
Bio PEF   Polyethylenterfuranoat
Bio PTT   Polytrimethylenterephthalat
Etc.

Bio-based

PHA   Polyhydroxyalkanoate
PHB   Polyhydroxybutyrat
PLA   Polyactide oder Polymilchsäure
Etc.

Not biodegradable ⟷ **Biodegradable**

Conventional Raw Material for Plastic

Raw Material for Bioplastic

ABS   Acrylnitril Butadien Styrol Copolymer
PE   Polyethylen
PET   Polyethylenterphalat
PP   Biopropylen
PS   Polystyrol
PVC   Polyvinylchlorid
Etc.

Fossile-based

PBS   Polybutylensaccinat
PCL   Polycaptolacton
PBSA   Polybuthylensaccinat Adpat
PBAT   Polybutylenadipatterephtalat
Etc.

Figure 1.16: Classification of various chemicals from organic chemistry based on the raw materials they are derived from and their biodegradability.

Classification of selected organic chemicals according to the origin of their raw materials – whether fossil-based or bio-based – and their degree of biodegradability. The figure provides a comparative overview that maps various chemical compounds along two key axes: the source of their feedstocks (renewable vs. non-renewable) and their environmental fate, particularly their potential for natural degradation.

This classification highlights significant distinctions between conventional petrochemical-derived compounds, which often exhibit low biodegradability and persistence in the environment, and bio-based alternatives, which are typically designed for enhanced degradability and reduced ecological impact. By visualizing this spectrum, the figure underscores the relevance of green chemistry principles in developing sustainable materials and encourages the transition toward circular and bio-based chemical production pathways that minimize environmental burden and support long-term ecological balance.

As indicated in Figure 1.16, raw materials can be characterized by two parameters: their origin (biogenic vs. fossil) and their biodegradability (biological or non-biological).

Examples of biogenic raw materials are listed in Table 1.3.

Table 1.3: Examples of biogenic raw materials.

Biogenic raw materials	Sources	Areas of use
Wood	Trees or shrubs	Energy carrier, basis for cellulose and other chemical raw materials
Plant oils	Rapeseed oil, and soybean oil	Fuel, basis for medicinal and other chemical raw materials
Fibers	Cotton, hemp, and flax	Used in the textile and construction industries
Biomass	Wood pellets, agricultural, and household waste	Energy carrier, basis for chemical raw materials
Food raw materials	Grains, fruits, and vegetables	Energy carrier, basis for chemical raw materials
Animal products	Wool, leather, and beeswax	Basis for chemical raw materials

Details of the processing methods can be found in Chapter 4.

Biogenic or bio-based raw materials and the chemicals produced from them are considered an important component of a sustainable economy, as they generally have a lower environmental impact than fossil or mineral raw materials (Krausmann et al., 2009). Moreover, many biogenic raw materials can be reintroduced into the natural cycle through recycling or biological processes (e.g. composting) (Meadows et al., 1972). With appropriate product design, products made from such materials can be easily disposed of, as they decompose in nature or, more complexly, in an industrial composting process. Like other products, raw materials can be kept within the product cycle through recycling – see Figure 1.17. Therefore, the key to this aspect of sustainability lies in product design as well as in the consistent application of the right materials throughout the product lifecycle.

1.10.5 Distinction of Utilization Strategies for Biogenic Raw Materials

The use of biogenic raw materials in industry and agriculture can be divided into two main strategic lines, each following different objectives and conditions:

1. **Substitution of conventional (fossil) raw materials to alleviate the general raw material balance:** This strategy aims to replace fossil or mineral raw materials with renewable, biogenic materials. The focus is on ecological motives, such

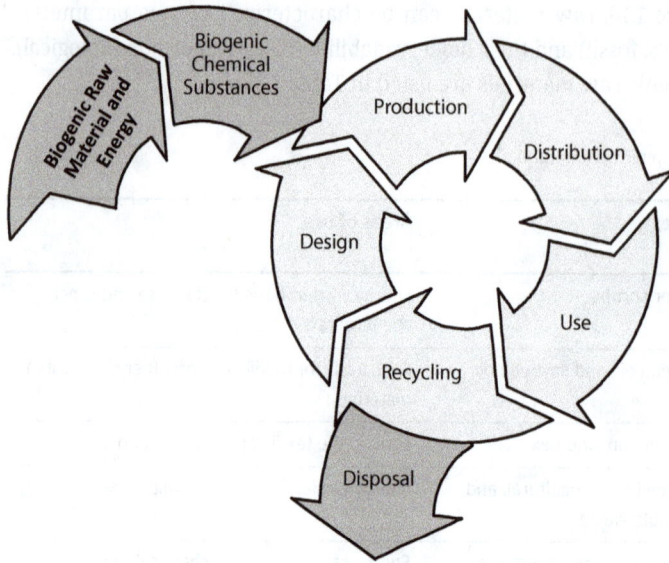

Figure 1.17: Biogenic raw materials as the foundation for product cycles.
Conceptual representation of biogenic raw materials serving as the foundational input for sustainable product cycles within a circular economy framework. The figure illustrates how renewable, bio-based resources – such as agricultural crops, forestry residues, and organic waste – can be transformed into a wide range of materials and products through biochemical, thermochemical, or mechanical processes. These biogenic inputs enable closed-loop systems where products are designed for reuse, biodegradation, or recycling, thereby reducing dependency on fossil resources and minimizing environmental impacts. The diagram emphasizes the regenerative nature of biogenic raw materials, highlighting their role in enabling carbon-neutral or carbon-negative cycles, supporting rural economies, and aligning industrial processes with ecological boundaries. By placing biogenic feedstocks at the core of product life cycles, the figure underscores their strategic importance in achieving long-term sustainability, resource efficiency, and climate goals.

as reducing CO_2 emissions, promoting closed material cycles, and reducing dependence on finite resources. Typical areas of application can be found in the bioeconomy, such as the production of bio-based plastics, paints, binders, or agricultural products, where petrochemical raw materials are replaced. The goal is to achieve long-term relief of global raw material consumption through regenerative, ideally regionally available resources.

2. **Substitution of so-called critical raw materials according to the USGS/EC classification:** In contrast to general raw material substitution, this strategy aims to specifically replace materials that are considered "critical" – either due to their economic significance, limited availability, or geopolitical risks (e.g. rare earth elements, cobalt, graphite, and lithium). Ensuring the security of industrial value chains is a primary focus here. The use of biogenic alternatives (e.g. lignin- or

cellulose-based materials as a potential replacement for graphite in batteries) is primarily driven by the wish to minimize systemic supply risks.

A list of raw materials and replacement strategies has been moved to the appendix, as its unwieldy length interfered with the flow of the main text. From this admittedly non-exhaustive and somewhat "snapshot-like" list, it is evident that research is already underway in many areas focused on replacing mineral resources and finding practical alternatives to mineral raw materials. Replacing each material, whether rare or not, with a biogenic one is a fundamental undertaking, and the research results do not indicate that the performance of the new materials matches that of the ones being replaced. Furthermore, especially in ongoing research, it remains unclear whether the results from the laboratory scale can be transferred to larger dimensions. However, even if this is only possible for some approaches, it already justifies a closer look. The purpose of the table is not to suggest that alternatives are already available for every raw material today, but rather to show that those who think in this direction are not at a disadvantage: there is sufficient evidence for the replacement of even rare minerals and elements with biogenic substances, and research in this direction is promising.

While the first of the two strategies mentioned above (substitution of conventional (fossil) raw materials to alleviate the general raw material balance) aims at a sustainable transformation of the entire raw material system, the second (substitution of so-called critical raw materials according to the USGS/EC classification) follows a risk-based resource strategy focused on resilience and supply security in key industries. Both approaches complement each other but require differentiated political, technological, and economic support mechanisms. The foundations of the second strategy have not yet been fully established. The development of biogenic raw materials to replace critical materials is still in its infancy but, as the above (and certainly overly extensive) table shows, provides grounds for some optimism.

Chapter 2
Agricultural Production Today – Availability, Fertility, and Dependence on Fertilizers

2.1 Overview

Erosion, monocultures, and soil sealing lead to a significant loss of soil fertility and pose a serious threat to food production (FAO, 2015; Montanarella et al., 2016). These developments not only affect agricultural productivity but also have far-reaching impacts on ecosystems, water balance, and climate (IPBES, 2018). To mitigate this, an integrated approach is required, encompassing both technological and political measures (UNEP, 2021). Agroecological practices such as crop rotation, minimal tillage, and the targeted use of organic fertilizers offer promising approaches to maintain soil fertility (FAO, 2018).

At the same time, political measures to reduce soil sealing and promote sustainable urban development are essential (UBA, 2020). In the long term, sustainable soil management will be crucial not only to ensure the productivity of agricultural land but also to preserve the ecological functions of soil (EEA, 2019). Research is increasingly focussing on innovative ways to minimize soil degradation and establish alternative management practices that are both ecologically and economically viable (Schulte et al., 2014). Climate change poses an additional threat to soil fertility by bringing rising temperatures, altered precipitation patterns, and extreme weather events (IPCC, 2022). Adaptation strategies such as sustainable soil management systems, innovative water technologies, and resilient agroecosystems are necessary to ensure long-term food security (Altieri & Nicholls, 2017).

Another challenge for global agriculture is the scarcity of phosphorus and potassium fertilizers. Phosphorus, which is limited in availability, and potassium, which is increasingly threatened by geopolitical uncertainties and high extraction costs, pose a long-term risk to agricultural productivity (Cordell et al., 2009; Scholz et al., 2013). Research is therefore focussing on recycling technologies, more efficient fertilizer application, and alternative raw material sources to reduce dependence on these finite resources (van Dijk et al., 2016).

In addition, the strong dependence of fertilizer production on fossil resources is increasingly identified as a significant factor for the future of agriculture. Nitrogen fertilizers, in particular, which are based on the energy-intensive Haber-Bosch process, as well as phosphorus and potassium fertilizers that require large amounts of fossil energy for their production, make agriculture vulnerable to rising energy prices and geopolitical uncertainties (IFA, 2022; Erisman et al., 2008). This calls for a sustainable transformation of fertilizer production to reduce reliance on fossil resources and to ensure food security (UNEP, 2021).

https://doi.org/10.1515/9783112218747-002

Fertilizer scarcity is increasingly being recognized as a threat to global agriculture. The reduced availability of synthetic fertilizers, especially in conventional agriculture, leads to declining yields and rising food prices (FAO, 2022). While wealthy countries are partly able to meet these challenges through technological innovations, developing countries are particularly affected. In the long term, alternative fertilization strategies, more efficient use concepts, and an expansion of circular economy practices are necessary to stabilize global yields and ensure food security (Smith et al., 2013).

Even today global food production is under pressure from climate change, resource scarcity, and geopolitical tensions, all of which impact crop yields and food prices (FAO, 2021; IPCC, 2022; World Bank, 2022). While technological innovations and political measures can help mitigate negative effects, long-term food security remains a key issue in global development policy (UNEP, 2021). A sustainable transformation of agriculture towards more resilient production systems will be necessary to cushion future crises and ensure the supply of a growing global population.

The challenges facing global food production particularly affect developing countries, which struggle with rising prices, food insecurity, and climate-related risks (FAO, IFAD, UNICEF, WFP & WHO, 2023). The agricultural sector is confronted with the need for structural changes and technological adaptation (OECD, 2022). Addressing these issues will require a combination of sustainable agricultural practices, technological innovation, and international political action to ensure global food security (Pretty et al., 2018).

Competition for agricultural resources is expected to intensify in the coming decades due to climate change, population growth, and geopolitical tensions (WWF, 2020; IPBES, 2019). Sustainable and equitable solutions are therefore essential to ensure global food security. A close integration of agroecological innovations, political measures, and fair resource distribution is crucial to prevent future food crises (Altieri & Nicholls, 2017).

2.2 Efficiency in Agriculture and Land Use Requirements

Agriculture has become increasingly efficient over the years, particularly in industrialized countries such as Germany.

Today, a single farmer in Germany can provide food for an average of around 135 people – a remarkable increase in productivity compared to the immediate post-war period, when the ratio was approximately 1:40 (BMEL, 2023). However, a similarly significant increase in efficiency is not to be expected in the future. In a densely populated country like Germany, expanding agricultural land is hardly feasible: already about half of the country's total land area is used for agriculture, around 30% is covered by forests, and approximately 14% is designated for urban use (UBA, 2022). Conservation organizations warn that one-third of Germany's land area should remain undeveloped to preserve biodiversity (WWF, 2021).

Furthermore, Germany currently requires about 1.3 times its own agricultural land area to meet domestic food demand (Notepad, 2020). This suggests that ensuring local food self-sufficiency will become increasingly difficult – if not impossible – if the German population grows.

In addition to the **absolute area**, soil quality is also a critical factor. The **soil's ability to supply sufficient nutrients** is a complex process that has been studied since the nineteenth century and can be described by Liebig's law of the minimum and Mitscherlich's law of diminishing returns (Fink, 2002).

2.2.1 Nutrients for Plants

Plants require a wide range of nutrients that are crucial for their development and growth. These nutrients can be categorized into different groups, with each group playing a specific role in the plant's metabolism and physiological processes. The main nutrient categories include macronutrients, secondary nutrients, and micronutrients, which are required in varying quantities but are all equally essential for the healthy growth and optimal development of plants (Marschner, 2012; Mengel & Kirkby, 2001).

Macronutrients are the key components for plant growth and are needed in larger quantities. Three of these macronutrients play particularly important roles:

Nitrogen (N) is a fundamental component of amino acids and proteins and has a direct effect on the vegetative growth of the plant. In particular, it promotes the growth of leaves and shoots, as well as the formation of chlorophyll, the green pigment required for photosynthesis. Nitrogen deficiency is often manifested as a general growth halt or yellowing of the leaves, which impairs photosynthetic performance (Taiz et al., 2015).

Phosphorus (P) is essential for the development of the root system as well as for the formation of flowers and fruits. Phosphorus also plays an important role in energy transfer within the plant, particularly in relation to adenosine triphosphate (ATP), which is required for many biochemical reactions. A phosphorus deficiency often results in weak root growth and poor fruit formation (Havlin et al., 2014).

Potassium (K) contributes to the regulation of the plant's water balance and supports disease resistance as well as cell wall stability. Potassium acts as an activator of numerous enzymes and plays a key role in osmoregulation, which is the plant's ability to efficiently transport and store water and nutrients. A potassium deficiency can lead to reduced resilience against environmental stress such as drought or diseases (Marschner, 2012).

Secondary nutrients are also required in larger quantities, but their significance is less comprehensive than that of the macronutrients. These nutrients include:

Magnesium (Mg) serves as an essential component of chlorophyll and thus plays a key role in photosynthesis. Magnesium deficiency often leads to a reduction in photosynthesis efficiency and a general weakening of the plant (Taiz et al., 2015).

Calcium (Ca) is important for the stability of cell walls and promotes root development. It is also involved in signal transmission within the plant and supports defence mechanisms against diseases. A calcium deficiency can weaken the cell structure and impair growth (Havlin et al., 2014).

Sulphur (S) is an essential component of amino acids and enzymes and plays an important role in protein metabolism. Without adequate sulphur, the plant cannot efficiently synthesize its proteins, leading to growth disturbances (Marschner, 2012).

Micronutrients or trace elements are required in very small amounts, yet they are no less important for the plant. The key micronutrients include iron (Fe), zinc (Zn), copper (Cu), boron (B), and manganese (Mn). These elements are crucial for a variety of enzymatic reactions and metabolic processes necessary for the growth and development of the plant. A deficiency of micronutrients can lead to specific symptoms such as growth disturbances, discoloration of leaves, or reduced resistance to diseases. In particular, iron is required for the synthesis of chlorophyll, and a deficiency can lead to chlorosis (yellowing of the leaves), while zinc and copper are important for numerous enzymes in metabolic processes (Mengel & Kirkby, 2001; Taiz et al., 2015).

Plants require, in addition to specific environmental conditions, a complex mixture of macro-, secondary, and micronutrients to optimize their growth, development, and ability to self-regulate. A balanced ratio of these nutrients is essential to ensure efficient nutrient uptake and utilization, as a deficiency in any single nutrient can negatively impact the overall growth and health of the plant.

2.2.2 Liebig's Law of the Minimum

The "law of the minimum," observed and formulated by Justus von Liebig in the nineteenth century, states that the growth of a plant is not determined by the total amount of nutrients available but by the nutrient whose concentration is the lowest in relation to the plant's needs (Liebig, 1855/2004). This theory is often illustrated by the image of a "barrel with staves of varying heights." The shortest stave represents the limiting nutrient and determines the barrel's maximum capacity to hold water (symbolic of plant growth) (see Figure 2.1). For agriculture, this means that to maximize yields, the most limiting nutrient must be provided in the right amount. The law of the minimum applies not only to nutrients, as originally formulated, but also to other growth factors such as light, temperature, and water (Schröder et al., 2018). This insight is crucial for agricultural practice, especially in fertilization.

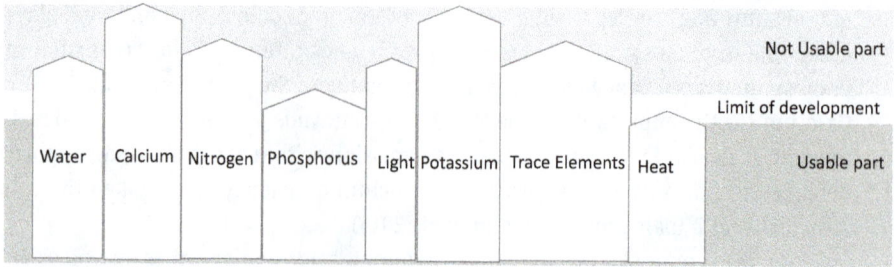

Figure 2.1: Liebig's law of the minimum.
Illustration of Liebig's law of the minimum, a foundational principle in ecological and agricultural sciences, which states that the growth and productivity of an organism (such as a plant) are limited not by the total amount of available resources but by the scarcest essential resource (the "limiting factor"). The figure typically visualizes this concept using a barrel analogy, where each stave represents a different nutrient or environmental factor, and the shortest stave determines the maximum level to which the barrel (symbolizing growth) can be filled. This depiction emphasizes that even if all other resources are abundant, the deficiency of a single critical input – such as nitrogen, phosphorus, water, or light – can restrict overall growth. The concept has broad relevance across disciplines, from agriculture and ecology to industrial processes and resource management, where system performance is similarly constrained by the most limiting factor. The figure underscores the importance of balanced resource management to optimize productivity and sustainability.

Plant growth is positively influenced by new resources or additions only as long as they increase the portion that is available for the plant. Beyond the development limit, any further additions become unusable and do not lead to the desired result of continued plant growth.

2.2.3 Mitscherlich's Law and Diminishing Returns in Yield

The Mitscherlich law describes the relationship between the amount of fertilizer applied and the resulting yield increase. Initially, an increase in nutrient supply leads to a significant boost in yield. However, with increasing fertilization, the additional yield increase diminishes and eventually levels off as shown in Figure 2.2. Over-fertilization leads to saturation, where additional nutrients no longer have positive effects and may even become toxic. This highlights the need for sustainable fertilization to focus on the optimal nutrient level in order to maximize yield without harming the environment (Mitscherlich, 1909; Fink, 2002).

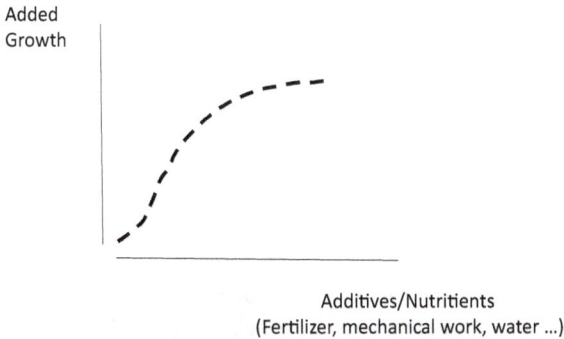

Figure 2.2: Mitscherlich's law.
With increasing fertilizer application or additional mechanical work, it is only within a certain range that the yield proportionally increases. Beyond that window, further increases in fertilizer or intensified work do not lead to significant returns any more.

2.2.4 Synthesis of Liebig's law, Mitscherlich's Law, and the Model of Adaptive Cycles

The synthesis of Liebig's law and Mitscherlich's law is crucial for efficient and sustainable agriculture. While Liebig's law describes the limiting nutrient as a factor for growth, Mitscherlich's law highlights the limits beyond which additional fertilization no longer leads to further yield increases (Mitscherlich, 1909; Taiz et al., 2015). Both principles are essential for the sustainable use of resources, as they enable targeted and resource-efficient management practices (Fink, 2002).

Both Liebig's law of minimum and Mitscherlich's law of diminishing returns postulate a relationship between nutrient availability and plant growth that begins at a zero point. This starting point implies that a plant will show no growth without the limiting nutrients. However, such an assumption is only valid under idealized conditions and primarily applies to a few, highly resilient plant species (Taiz et al., 2015). Both concepts implicitly assume universal applicability across different plant species and soil types, without considering the specific ecological requirements or necessary soil biology. In practice, however, it becomes apparent that many plant species engage in complex interactions with their environment, and specific ecological conditions must be met before they can successfully establish themselves in the cultivation area. This includes a preparatory phase of soil development, during which not only the availability of mineral nutrients but also the presence of microorganisms, mycorrhiza fungi, and other plant and animal species play a crucial role (Wardle et al., 2004). Only the interaction of these biotic components enables the formation of a functional ecological niche within which the target plants can establish themselves. This process of establishing an ecological network requires time and is referred to as the reorganization phase within the framework of the adaptive cycles model (Gunderson & Holling, 2002).

The neglect of these dynamic processes in classical agronomic growth models represents a significant limitation in their applicability to ecologically complex systems.

2.2.5 Fertilizers in Modern Agriculture

Fertilizer is indispensable in modern agriculture, as it ensures the nutrient supply necessary for plant growth. Without adequate nutrient availability, plants would not thrive optimally, negatively affecting both the quality and quantity of the harvest (Havlin et al., 2014). In natural ecosystems, decomposition processes recycle nutrients back into the soil. In agriculture, however, harvesting leads to a loss of biomass and nutrients, making continuous fertilization necessary (Marschner, 2012).

As shown in Figure 2.3, another issue is the loss of nutrients through erosion, leaching, and soil degradation. Sandy soils, in particular, are susceptible to nitrogen and potassium losses, which are accelerated by rain or irrigation (Blume et al., 2016). Targeted fertilization helps counteract these losses and maintain soil fertility. Additionally, organic fertilization, such as compost or manure, contributes to improving soil structure and fertility (Fink, 2002; Taiz et al., 2015).

However, the production of fertilizers is energy-intensive and requires natural resources, particularly phosphorus. Phosphorus is a limited resource, and its reserves could be exhausted in the coming decades (Cordell & White, 2014). This poses a significant challenge for agriculture, as phosphorus is essential for fertilizer production. Global phosphorus demand is influenced by geopolitical and economic factors, leading to price fluctuations and limited availability (van Dijk et al., 2016).

2.2.6 Forecast: Increasing Pressure on Agricultural Supply

Climate change, reduced biodiversity, and other passing planetary boundaries represent an escalating threat to agricultural production. Especially altered precipitation patterns, extreme weather events, and rising temperatures negatively impact crop yields and reduce available agricultural land (IPCC, 2022). Additionally, soil degradation caused by intensive farming and improper land management leads to a decline in soil fertility and, consequently, lower crop yields (FAO, 2015).

Furthermore, biodiversity loss adversely affects agriculture, as pollinator insects such as bees and butterflies are essential for many crops (Potts et al., 2016). Overall, we face an increasing challenge to ensure future food production. Efficient management of agricultural land, nutrients, and fertilizers, as well as adaptation to climate change, are crucial for securing global food security and determine the size of populations that can be fed with plant-based products (FAO, 2017). Even independent of the lack of artificial fertilizers –

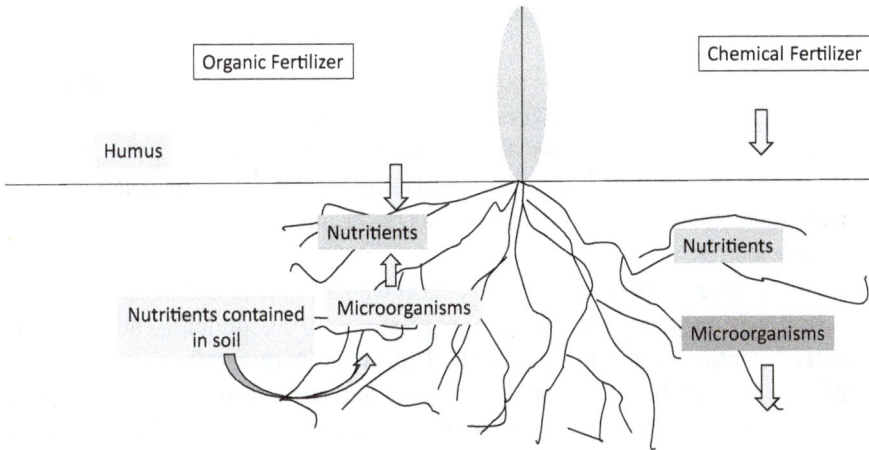

Figure 2.3: Pathways of nutrients into plant roots.
Nutrients enter plant roots. For that nutrient uptake intake organic matter of soil – particularly humus – plays a crucial role. The diagram shows how nutrients from chemical fertilizers and organic sources move through the soil profile, interact with soil particles, and are absorbed by plant roots through processes such as diffusion, mass flow, and root interception. While chemical fertilizers can supply essential nutrients like nitrogen, phosphorus, and potassium, their effectiveness is significantly enhanced when humus is present. Acting as a biologically active organic layer, humus improves soil structure, increases water retention, enhances microbial activity, and helps chelate nutrients, making them more bioavailable to plant roots. The figure highlights that without sufficient humus, synthetic fertilizers may remain in forms that are less accessible or may leach into groundwater, reducing both efficiency and environmental sustainability. While chemical fertilizers are often not absorbed without humus, fertilizers are more easily taken up when humus, as an organic layer, covers the surface and processes the fertilizers. This visual representation underscores the importance of maintaining soil health through organic matter management to ensure efficient nutrient cycling and sustainable crop production.

given the growing strain on agricultural systems due to biodiversity losses, food supply is likely to come under pressure in the long term (Rockström et al., 2020).

2.3 Soil Fertility

2.3.1 Monocultures and Soil Sealing

Soil fertility, understood as the ability of soil to provide essential nutrients to plants in the right amounts and proportions for healthy growth and high yields, **is one of the fundamental prerequisites for agricultural production and the sustainable assurance of global food security.** It is determined by a complex interplay of physical, chemical, and biological factors, including nutrient availability, water retention capacity, soil structure, and the activity of microorganisms and soil fauna (Blume et al., 2016; Lal, 2015):

Key factors influencing soil fertility are:

1. Nutrient Content:

   The presence of essential plant nutrients such as nitrogen (N), phosphorus (P), potassium (K), calcium (Ca), magnesium (Mg), sulphur (S), and micronutrients like iron (Fe), zinc (Zn), and copper (Cu) is fundamental to soil fertility. The availability and balance of these nutrients are determined by soil composition, organic matter, and management practices such as fertilization.

2. Organic Matter:

   Organic matter (the decomposed remains of plants, animals, and microorganisms, as well as substances they produce) is essential for maintaining soil structure, water retention, and nutrient supply. It provides a reservoir of nutrients for plants, promotes the activity of beneficial soil organisms, and improves soil aggregation. Decomposing organic matter releases nutrients in forms that plants can absorb, which is vital for long-term fertility.

3. Soil pH:

   The pH level of the soil is a key factor that influences nutrient availability. Most nutrients are readily available to plants in soils with a pH between 6 and 7. Extreme pH levels (either too acidic or too alkaline) can reduce the solubility of certain nutrients, making them unavailable to plants.

4. Soil Structure:

   Soil structure refers to the arrangement of soil particles (sand, silt, and clay) and the spaces between them. Good soil structure allows for the adequate movement of air, water, and nutrients through the soil. Well-structured soils have a balance of pore spaces that promote root growth, water infiltration, and drainage, which are essential for plant health.

5. Soil Microorganisms:

   Soil fertility (the ability of soil to support plant growth by providing all the essential nutrients and conditions that plants need to grow and produce crops) is strongly influenced by the biological activity of microorganisms, such as bacteria, fungi, and earthworms, that contribute to nutrient cycling, organic matter decomposition, and soil aeration. Symbiotic relationships between plants and soil organisms, such as mycorrhiza fungi, also enhance nutrient uptake, especially for phosphorus.

Soil fertility can degrade over time due to various factors, particularly in intensive agricultural systems. Key causes of fertility loss include:

Erosion: Soil erosion, caused by wind or water, removes the topsoil, which contains the highest concentration of nutrients and organic matter. This leads to a reduction in soil fertility and the depletion of vital nutrients.

Overuse of Fertilizers: Excessive or improper application of chemical fertilizers can degrade soil fertility by disrupting the natural nutrient cycles, causing nutrient imbalances, and leading to pollution of surrounding water bodies. Over-fertilization may also harm soil organisms and reduce soil biodiversity.

Soil Acidification: The excessive use of nitrogen-based fertilizers can lead to soil acidification, which reduces the availability of certain nutrients, such as calcium and magnesium, and harms soil organisms.

Salinization: Irrigation with water that contains high levels of salts can cause soil salinization, which reduces the ability of plants to absorb water and nutrients. Salinized soils are often less productive and require specific management practices to restore fertility.

Over the past decades, significant global declines in soil quality have been observed, driven by factors such as erosion, monoculture practices, and soil sealing (FAO, 2015). These processes not only pose a serious threat to agricultural productivity but also have far-reaching ecological and climatic consequences (IPBES, 2018).

2.3.2 Erosion as a Major Cause of Soil Fertility Loss

Soil erosion is a natural process caused by wind or water. In the natural environment, this process usually occurs slowly and is regulated by vegetation, root systems, and soil organisms (Morgan, 2005). However, human interventions such as intensive agriculture, deforestation, or inappropriate soil management practices significantly accelerate erosion. Particularly problematic is the loss of the topsoil (humus layer) as sketched in Figure 2.4, as this layer contains the highest concentration of nutrients and organic matter, making it essential for plant growth (Lal, 2001).

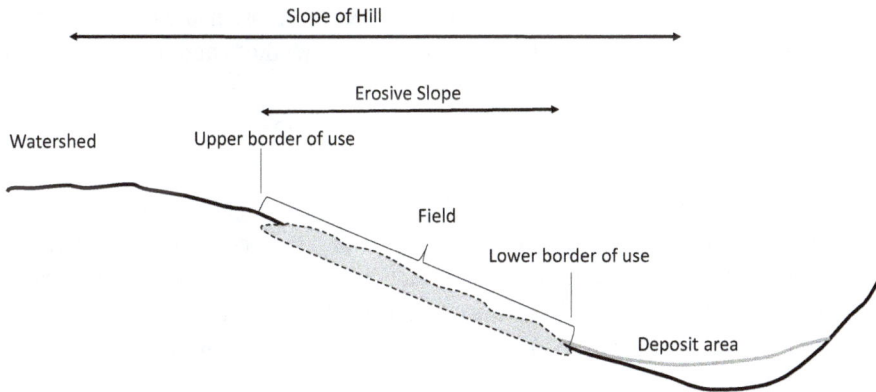

Figure 2.4: Soil erosion on erosive slope areas.
Recent studies indicate that approximately 24 billion tonnes of fertile soil are lost annually worldwide due to erosion. According to the FAO, the global average rate of soil loss is between 10 and 20 tonnes per hectare per year, while natural soil formation is only about 0.5–1 tonne per hectare per year (FAO, 2015). Regions particularly affected are those with intensive agriculture, sloping terrain, or arid climates, where vegetation cover is limited.

The scientific study of soil erosion focusses on a variety of strategies to mitigate erosion-related damage, with different agricultural and forestry practices being employed to stabilize soil structure and prevent erosion (Montgomery, 2007). Key approaches include conservation tillage, reforestation, agroforestry, and specific soil protection measures on sloping terrain.

Conservation tillage aims to design soil cultivation in such a way that the soil cover remains as undisturbed as possible. This is achieved by avoiding deep ploughing, which can lead to the destabilization of soil structure. Instead, methods such as mulching or direct seeding are used, where a soil cover of organic material or directly sown plants prevents water infiltration and erosion while simultaneously maintaining soil fertility (Derpsch et al., 2010).

Reforestation and agroforestry integrate trees and shrubs into agricultural systems, thus contributing to soil stabilization. Trees and shrubs not only stabilize the soil through their root systems but also regulate the water balance and improve wind protection. These measures promote long-term soil fertility and reduce erosion processes, especially in regions prone to erosion due to intensive agriculture or natural climatic conditions (Jose, 2009).

Terracing and other soil protection measures are primarily employed on slopes, where water runoff and gravity promote soil erosion. The construction of terraces or the establishment of contour planting helps to mitigate the steepness of the slope and slow down water flow. This leads to better infiltration of rainwater and prevents large amounts of water from eroding the soil in a short period (Nyberg et al., 2012). These measures not only help prevent erosion but also promote sustainable agricultural use in topographically challenging regions.

These approaches offer promising ways to reduce soil erosion and ensure the sustainable use of soils. The combination of agroecological methods and practical protection measures represents an important strategy to prevent the long-term loss of fertile soils and secure soil quality (Montgomery, 2007; IPBES, 2018).

Monocultures and Their Impact on Soil Quality

Another significant issue for soil fertility is the widespread cultivation of monocultures. Monocultures are agricultural systems in which the same plant species is cultivated over several cropping cycles. This practice is often preferred for economic reasons, as it allows for higher short-term yields and more efficient use of machinery (Altieri, 1999).

However, scientific studies indicate that long-term monoculture farming has considerable negative effects on soil quality:

One of these effects is nutrient depletion. Since each plant species absorbs specific nutrients from the soil in varying amounts, a one-sided management approach leads to the excessive extraction of certain nutrients while others remain in high concentrations. Over time, this can result in nutrient imbalances, which are often insufficiently addressed through fertilization (Lal, 2006).

Equally problematic is the loss of soil biology and biodiversity. Monocultures reduce the diversity of organisms living in the soil, particularly microorganisms, fungi, and earthworms, which are essential for humus formation and nutrient cycling (Tautges et al., 2019). Studies have shown that in healthy, biodiverse soils, symbiotic relationships between plant roots and mycorrhiza fungi are established, enhancing nutrient uptake (Smith & Read, 2008). However, in monocultures, these synergies are lost, reducing the nutrient efficiency of plants.

To improve soil quality and prevent the negative consequences of monocultures, scientists recommend various measures including:

- **Crop Rotation and Intercropping**: Alternating cropping cycles with different plant species to regenerate soil nutrients (Döring et al., 2015).
- **Cover Crop Cultivation**: The use of legumes or other nitrogen-fixing plants to enrich the soil with essential nutrients (Drinkwater et al., 1998).
- **Integrated Agricultural Approaches**: Combining livestock farming and crop cultivation to close nutrient cycles and provide organic fertilization (Gattinger et al., 2012).

Soil Sealing and Its Impact on Soil Fertility

A third important factor contributing to the decline in soil fertility is the increasing soil sealing caused by urbanization and infrastructure development. Soil sealing refers to the covering of soil with concrete, asphalt, or other impermeable materials, which disrupts natural soil functions such as water retention, nutrient exchange, and biological activity (Scalenghe & Marsan, 2009).

Soil sealing refers to the process in which the natural soil surface is covered with impermeable materials such as asphalt, concrete, or bricks, significantly restricting the natural exchange of water and air within the soil (Blöschl et al., 2005). This process is particularly widespread in urban and industrial areas and has significant effects on soil quality, the environment, and agriculture.

Causes and Consequences of Soil Sealing

The primary cause of soil sealing is the increasing urbanization and industrialization. The expansion of roads, residential areas, commercial zones, and infrastructure projects leads to the coverage of natural surfaces with impermeable materials. Causes for soil sealing (Schröder et al., 2010) include the construction of parking lots, storage areas, sports facilities, and other large-scale projects that result into consequences such as

1. **Reduction in Soil Fertility:** Soil sealing leads to the destruction of the natural soil structure. The soil loses its ability to retain water and supply nutrients, which results in a decrease in fertility. This sealing can significantly affect agricultural productivity in the affected areas, as the soil is no longer suitable for plant growth (FAO, 2015).

2. **Water Runoff and Flooding:** Sealed surfaces prevent water from infiltrating the soil, leading to increased surface runoff. This results in intensified erosion and heightened flood risks, as rainwater can no longer be absorbed into the ground but instead flows directly into rivers and streams. This puts additional pressure on drainage systems and can lead to flooding (Blöschl et al., 2005).
3. **Loss of Biodiversity:** Soil sealing destroys habitats for many plant and animal species. Especially for soil-dwelling organisms such as microbes, worms, and insects, the loss of unsealed soil is detrimental. These organisms play essential roles in nutrient cycling and soil fertility (Wardle et al., 2004).
4. **Increased Air Temperatures and "Heat Island Effect":** In urbanized areas, soil sealing leads to higher surface temperatures. Sealed surfaces absorb and store heat, contributing to the formation of urban heat islands, where temperatures are significantly higher than in surrounding, unsealed areas. This increases the demand for cooling energy and can negatively affect the quality of life (Santamouris, 2014).

Mitigation measures for soil sealing would naturally include preventing of causes and, in general

Green Infrastructure: Promoting green infrastructure, such as green roofs, green facades, and the creation of green spaces in urban areas, can help mitigate the negative effects of soil sealing. These measures promote water absorption, reduce surface runoff, and improve air quality (Faulhaber et al., 2017).

Increasing Soil Permeability: The use of permeable materials for the construction of roads, walkways, and parking lots can help restore the soil's natural water retention capacity. Permeable surfaces, such as grass pavers or porous concrete, allow water to infiltrate the soil, thus reducing the risk of flooding and erosion (Schröder et al., 2010).

Soil Renaturation: In areas already heavily sealed, renaturation measures can help restore soil quality. This can be achieved by creating green spaces, restoring wetlands, or planting trees and vegetation that stabilize the soil and promote biodiversity (McKinney, 2008, Gill et al., 2007).

Sustainable Urban Planning: Promoting "green" cities with more permeable surfaces and urban green spaces (Kabisch et al., 2016).

Permeable Materials: Using water-permeable paving stones and surfaces to reduce sealing effects (Pratt et al., 2007).

Soil sealing presents one of the significant threats (see Figure 2.5) to the environment and agricultural production by deteriorating soil quality, disrupting water cycles, and endangering biodiversity. Sustainable urban planning and targeted management of sealed areas are essential to counteract the negative consequences of soil sealing. Innovative methods or strategies such as green infrastructure, permeable surface materials, and soil renaturation offer promising approaches to mitigate the impacts of soil sealing and strengthen the resilience of urban and agricultural ecosystems.

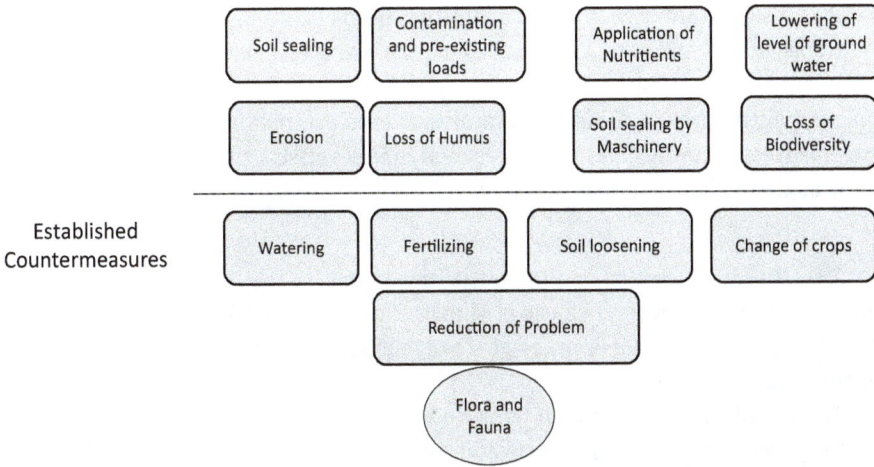

Figure 2.5: Some mechanisms of influence on plant and ultimately on animal life.
Various mechanisms through which environmental, biological, and anthropogenic factors influence plant life, and how these changes cascade through ecosystems to ultimately affect animal life. The diagram outlines both direct and indirect pathways, including alterations in climate conditions (e.g. temperature, precipitation, and CO_2 levels), soil composition, nutrient cycles, and pollution exposure, all of which impact plant growth, health, and distribution. Changes in plant communities – such as shifts in species composition, biomass production, or flowering and fruiting patterns – subsequently influence herbivores and higher trophic levels by modifying food availability, habitat structure, and ecosystem stability. The figure also includes human-driven factors such as land use change, deforestation, introduction of invasive species, and the use of agrochemicals, which further disrupt plant-animal interactions. These interconnected mechanisms underscore the delicate balance within ecosystems and highlight the importance of plant health as a foundational element for sustaining animal life and broader ecological integrity.

According to recent estimates, approximately 4.2 million hectares of soil are sealed annually worldwide, which not only reduces valuable agricultural land but also brings hydrological and climatic consequences (FAO, 2015). Sealed soils cannot absorb water, which limits groundwater recharge and increases flood risk. Additionally, the reduction of vegetative areas leads to a decrease in carbon storage, resulting in increased CO_2 emissions (Seto et al., 2012).

2.3.3 Impact of Climate Change on Agriculture

Soil Fertility and Climate Change
Soil fertility is a crucial prerequisite for agricultural productivity and the long-term security of global food supply. It is determined by a variety of physical, chemical, and biological factors, including organic matter content, nutrient availability, soil structure, as well as the activity of microorganisms and soil organisms (Lal, 2006). In recent

decades, a significant decline in soil quality has been observed, accelerated by various anthropogenic factors such as intensive land use, erosion, and monocultures (Scharlemann et al., 2014). However, an increasingly relevant influencing factor is climate change, which, through rising temperatures, altered precipitation patterns, and more extreme weather events, has both direct and indirect effects on soil fertility and agricultural productivity (Smith et al., 2016).

Temperature Increase and Its Impact on Soil Fertility

One of the most immediate effects of climate change on soils is the rise in temperature. Higher average temperatures affect biological, chemical, and physical soil processes in various ways:

1. **Increased Decomposition of Organic Matter**:
   Organic matter (humus) is essential for soil fertility, as it stores nutrients, improves soil structure, and increases water retention capacity (Jobbágy & Jackson, 2000). Higher temperatures intensify microbial activity in the soil, leading to a faster breakdown of organic matter. This results in increased CO_2 emissions, which, in turn, enhance the greenhouse effect – a feedback mechanism that is widely discussed in the scientific literature (Lal, 2004).

2. **Changes in Nutrient Availability**:
   The rise in temperature can influence the mobility and availability of plant nutrients. For example, nitrogen mineralization is accelerated at higher temperatures, which increases nitrogen availability in the short term but may lead to soil depletion in the long term (Sullivan et al., 2015). At the same time, high temperatures can enhance the fixation of phosphorus in the soil, making it more difficult for plants to absorb (Frossard et al., 2000).

3. **Impact on Soil Life**:
   Soil biodiversity also plays a key role in soil fertility, as soil organisms are involved in organic matter decomposition, nutrient mobilization, and soil structuring (Bardgett et al., 2005). Temperature increases can disrupt the balance between various microorganisms and soil organisms. Studies show that some beneficial soil bacteria and fungi are sensitive to heat, while pathogenic microorganisms may benefit from rising temperatures (van der Putten et al., 2010).

Altered Precipitation Patterns and Their Impact on Soil Quality

In addition to the rise in ambient temperature, changes in precipitation patterns are also a consequence of climate change, with significant consequences for soil fertility (IPCC, 2014).

– **Increase in Extreme Events**:
   Scientific studies show that climate change is increasing the frequency and intensity of extreme weather events such as heavy rainfall and droughts. Heavy rainfall can lead to increased soil erosion, particularly in agricultural areas with intensive soil

cultivation or low vegetation cover (Lal, 2001). Erosion contributes to the depletion of topsoil, causing valuable nutrients and organic matter to be lost. Drought periods, on the other hand, reduce soil moisture and can lead to degradation of soil structures, further decreasing the soil's water retention capacity (Trnka et al., 2011).

– **Deterioration of Soil Structure**:
Alternating phases of extreme rainfall and drought can significantly impact the physical properties of soil. A common consequence is soil compaction, especially in agricultural areas with high machinery use. Compacted soils have a lower water infiltration rate, which increases the risk of surface runoff and erosion (Koch et al., 2014). Additionally, plant root systems are affected, as there is less oxygen available in the soil.

– **Altered Groundwater Availability**:
Long-term changes in precipitation affect groundwater recharge and, consequently, the availability of water for agriculture. In many regions, particularly in arid and semi-arid areas, it is projected that water scarcity will further increase due to climate change (Oki & Kanae, 2006). This not only has direct implications for irrigated agriculture but can also impact soil fertility, as dry soils are more vulnerable to erosion and salt accumulation.

Arid and Semi-arid Areas

Arid and semi-arid areas are regions characterized by low precipitation and often high evaporation, which significantly influence agricultural use and ecosystems (Reynolds et al., 2007):

– **Arid Areas (or Desert Regions)**:
Arid areas are regions where precipitation is so low that it cannot meet the evaporation and water requirements of plants. In such areas, precipitation typically amounts to less than 250 mm per year. The plant and animal life in arid areas is adapted to extreme conditions and often consists of drought-tolerant plants such as cacti or thorny shrubs as well as animals that can survive on minimal water (Zohary, 2017). Examples of arid areas include the Sahara in Africa, the Atacama Desert in South America, and the Arizona Desert in the USA.

– **Semi-arid Areas (or Semi-dry Areas)**:
Semi-arid areas lie between arid regions and temperate climate zones, receiving more precipitation than desert areas but still less than in wetter regions. Precipitation in semi-arid areas typically ranges from 250 to 500 mm per year. These areas experience seasonal fluctuations in rainfall, often with a distinct wet season and dry season. Plants and animals in semi-arid areas are adapted to periodic droughts and can cope with extreme temperature fluctuations (Dregne, 2002). Examples of semi-arid areas include parts of the southern USA, the Sahel region in Africa, and the Indian subcontinent.

Both types of regions are particularly vulnerable to soil erosion and desertification, as the soil often has little vegetation cover to protect it from erosion (Sivakumar, 2005). Agriculture in these areas often requires adapted techniques such as irrigation or the use of drought-resistant plants. Maintaining soil fertility and promoting sustainable agricultural practices are crucial for stabilizing the ecosystems in these areas and ensuring the livelihoods of the people living there (FAO, 2011).

Soil Salinization and Soil Acidification as Consequences of Climate Change

Another aspect that is gaining increasing importance in scientific discussions is the intensification of soil salinization and acidification as a result of climate change (Rengasamy, 2006; Schjønning et al., 2015).

Soil Salinization: In arid regions, increased evaporation due to rising temperatures can cause dissolved salts to be transported to the soil surface (Munns & Tester, 2008). This is particularly problematic in irrigated agricultural systems, as evaporation further increases the salt concentration in the soil and leads to long-term degradation of soil fertility. Scientific studies show that today, around one-fifth of irrigated agricultural land worldwide is affected by soil salinization – a figure that could rise with increasing water scarcity and the acidification of agricultural land (FAO, 2011; Munns & Tester, 2008).

Soil Acidification: Climate change can also exacerbate processes of soil acidification. The increased concentration of CO_2 in the atmosphere can lead to a heightened dissolution of carbonic acid in the soil, which lowers the pH. This affects the availability of essential nutrients such as calcium and magnesium and can promote the mobilization of toxic elements like aluminium, thereby inhibiting plant growth (Zhao et al., 2015).

2.3.4 Strategies for Adapting Agriculture to Climate Change

In light of the pressing challenges posed by climate change, intensive research is being conducted on strategies to adapt agricultural practices, with the goal of enhancing the resilience of agricultural systems and ensuring their long-term sustainability (Lipper et al., 2014). To ensure food security and sustainable agricultural practices, various adaptation strategies are being implemented:

1. **Diversification and Genetic Improvement of Crops and Livestock of Crops and Livestock**

   Introducing more diverse and more resilient crop species and livestock breeds can enhance resilience to climate-related risks. By selecting varieties that are better adapted to local climate conditions (e.g. drought-resistant crops or heat-tolerant livestock), farmers can reduce the vulnerability of their operations. Also genetic improvement can provide higher yields under changing climate conditions. This includes developing traits such as drought tolerance, disease resistance, and early maturation.

2. **Improved Water Management**

   Efficient irrigation systems, rainwater harvesting, and the use of drought-resistant crops can help mitigate the impacts of water scarcity. Modern technologies such as precision irrigation and moisture sensors allow farmers to optimize water use and reduce waste. The use of efficient irrigation technologies, such as drip irrigation or water-saving cultivation methods, is crucial. These technologies enable precise and resource-efficient irrigation, minimizing water loss and optimizing water use in agriculture (Rockström et al., 2009).

3. **Soil Conservation and Fertility Management**

   Implementing soil conservation practices, such as no-till farming, cover cropping, and agroforestry, helps prevent soil erosion, improve soil fertility, and maintain soil moisture. These practices also reduce the carbon footprint of farming. These practices help conserve soil moisture by reducing evaporation and leaving a protective layer on the soil surface, which also reduces erosion (Lal, 2015).

   In this context, conservation tillage methods such as mulch seeding or direct seeding play an important role. By avoiding deep ploughing, the soil structure remains stable, and biological activity in the soil is promoted, which supports long-term soil fertility.

4. **Climate-Smart Agriculture**

   This integrated approach involves using technologies, practices, and policies that enhance productivity and environmental sustainability. For example, precision farming, agroecological practices, and the use of biotechnology for improved pest and disease management are components of climate-smart agriculture.

   Another critical approach is the promotion of humus-rich soils. The targeted use of organic fertilizers and the integration of agroforestry systems can help increase the humus content in the soil. Humus-rich soils have a higher water retention capacity and provide better nutrient availability for plants (Smith et al., 2012). These properties enhance the soil's resilience to climatic changes, particularly during periods of drought or extreme rainfall events.

5. **Early Warning Systems and Climate Forecasting**

   Access to accurate climate forecasts and early warning systems for extreme weather events (e.g. storms, floods, and droughts) allows farmers to plan better and implement precautionary measures, minimizing damage to crops and infrastructure.

6. **Agroecological Approaches**

   Agroecology emphasizes the use of ecological principles in farming systems. By working with nature, such as enhancing biodiversity and using organic fertilizers, agroecological approaches help build resilience to climate change while also improving food security.

7. **Supportive Policies and Financial Instruments**
   Governments and international organizations can provide subsidies, insurance schemes, and funding for research to promote the adoption of climate-resilient agricultural practices. Policies that support knowledge exchange and farmer education are also crucial for long-term adaptation.

Measures such as these demonstrate that a holistic adaptation strategy is both necessary and feasible to adjust agriculture to the challenges of climate change. By combining innovative techniques with the promotion of ecologically sustainable practices, agriculture can become more resilient to climatic fluctuations while maintaining its long-term productivity.

2.4 Fertilizer Scarcity

Fertilizer scarcity has become an increasingly pressing issue in global agriculture. Fertilizers, particularly nitrogen, phosphorus, and potassium, are essential for maintaining soil fertility and supporting crop yields. However, various factors, such as geopolitical tensions, resource depletion, rising energy costs, and environmental concerns, are contributing to the limited availability and increasing costs of fertilizers.

1. **Causes of Fertilizer Scarcity:**
   - **Resource Depletion:** The extraction of natural resources needed for fertilizer production, particularly phosphorus, is limited. Phosphorus, for example, is a finite resource, and its reserves are concentrated in only a few countries, which creates vulnerability to supply disruptions.
   - **Energy Costs:** The production of nitrogen fertilizers, particularly through the Haber-Bosch process, is energy-intensive. Rising global energy prices, due to factors such as oil shortages or increased demand, make fertilizer production more expensive.
   - **Geopolitical Tensions and Trade Disruptions:** Trade restrictions, sanctions, and conflicts between fertilizer-producing countries and others can lead to supply chain disruptions, exacerbating the scarcity of fertilizers.
   - **Environmental Concerns:** Increasing concerns over the environmental impact of excessive fertilizer use, such as water pollution and greenhouse gas emissions, have led to regulations and a push for more sustainable agricultural practices. These efforts can limit the availability of conventional fertilizers.

2. **Implications for Agriculture:**
   Fertilizer scarcity can severely impact global food production. Reduced access to fertilizers may result in lower crop yields, reduced soil fertility, and poorer quality produce, which can contribute to food insecurity, particularly in regions dependent on intensive farming systems. Farmers may be forced to reduce fertilizer

applications, use alternative (often less effective) methods, or shift to lower-input agricultural practices, potentially leading to decreased productivity.

3. **Strategies for Mitigation:**
   - **Efficient Fertilizer Use:** The adoption of precision farming techniques, such as nutrient management systems and targeted application technologies, can help optimize fertilizer use, reducing waste and minimizing environmental impacts.
   - **Alternative Fertilizers:** Developing and utilizing organic fertilizers, recycled nutrients, and biological fertilizers (e.g. nitrogen-fixing bacteria) can provide alternatives to conventional chemical fertilizers. Additionally, promoting agroecological practices, such as crop rotation and agroforestry, can help maintain soil fertility without heavy reliance on chemical fertilizers.
   - **Sustainable Farming Practices:** Encouraging practices like composting, reduced tillage, and cover cropping can naturally improve soil fertility and reduce the need for synthetic fertilizers. These approaches also enhance soil structure and water retention, improving agricultural resilience to climate change.

Addressing fertilizer scarcity requires a multifaceted approach involving technological innovations, policy interventions, and sustainable agricultural practices to ensure the future of global food production while minimizing environmental impacts.

2.4.1 Fossil Resources for Fertilizer Production

Global agriculture heavily relies on synthetic fertilizers to maximize crop yields and feed the growing world population (Tilman et al., 2002). A significant proportion of these fertilizers is derived from fossil resources. Particularly natural gas, coal, and phosphate rock serve as essential raw materials for the production of nitrogen, phosphorus, and potassium fertilizers (Van Vuuren et al., 2010). This dependence on finite resources poses an increasing challenge to sustainable agriculture and global food security (Foley et al., 2011). Scientific debates focus on the availability of these raw materials, the implications of rising energy costs, and potential alternatives to reduce reliance on fossil resources (Sutton et al., 2013).

The industrial production of fertilizers is largely based on three primary nutrients: nitrogen (N), phosphorus (P), and potassium (K) (Cordell et al., 2009). While phosphorus and potassium fertilizers are extracted from mineral deposits, the production of nitrogen fertilizers is particularly energy-intensive and heavily dependent on fossil resources (Sutton et al., 2013).

2.4.2 Scarcity of Phosphorus and Potassium Fertilizers

The availability of essential fertilizers is critical to the productivity of global agriculture. While nitrogen, phosphorus, and potassium are considered the three primary nutrients for plant growth, increasing attention is being paid to the emerging scarcity of phosphorus and potassium fertilizers in scientific and policy discussions. These two nutrients are indispensable for plant development, as they influence root formation, flower and fruit development, and the overall resilience of crops to stress factors (Cordell & White, 2014). However, due to limited natural reserves and growing geopolitical uncertainties, the supply of phosphorus and potassium is becoming increasingly insecure (Elser & Bennett, 2011). The consequences of this development are far-reaching, affecting not only agricultural production but also global food security, economic stability, and ecological sustainability (Bennett et al., 2001).

Phosphorus

Phosphorus is relatively abundant in the Earth's crust, yet it is only available in agriculturally usable forms to a limited extent (Cordell & White, 2014). The most significant phosphate reserves are concentrated in a few countries – primarily Morocco, China, the USA, and Russia. Morocco alone holds over 70% of the world's known phosphate reserves (Gates et al., 2015). This geopolitical concentration creates substantial dependencies on the global market and heightens the risk of supply disruptions, particularly during crises or trade conflicts (Gilbert, 2009). Another pressing issue is that phosphate deposits are often contaminated with heavy metals such as cadmium and uranium (Cordell et al., 2009). Purifying these raw materials is both costly and technically demanding, which is why the scientific debate increasingly focusses on how to develop sustainable alternatives to conventional phosphate extraction (Bennett et al., 2001).

Phosphorus is an essential macronutrient for plants, required for DNA synthesis, energy transfer in the form of ATP, and numerous other cellular processes in both plants and animals (Elser & Bennett, 2011). In agriculture, phosphorus is primarily applied in the form of phosphate fertilizers – such as superphosphate or diammonium phosphate – which are derived from mineral phosphate rock (Scholz et al., 2014).

Availability and Scarcity of Phosphorus

The increasing scarcity of phosphorus poses potentially far-reaching consequences for global agriculture:

> Phosphate fertilizers are already subject to significant price volatility. A decline in production or the imposition of export restrictions by major producer countries could drive prices even higher, thereby increasing the production costs for farmers (Heffer, 2013). Phosphorus deficiency in soils leads to reduced root development and impaired nutrient uptake, which can severely affect the

yields of staple crops such as wheat, maize, and rice (Bennett et al., 2001). This is especially critical in phosphorus-deficient soils, which are widespread in regions such as Africa and Asia, where fertilizer scarcity could directly threaten food security (Sattari et al., 2012).

The use of phosphate fertilizers also has environmental side effects. Inefficient phosphorus application contributes to the eutrophication of water bodies, as excess phosphorus is washed into rivers and lakes, stimulating the excessive growth of algae and cyanobacteria (Carpenter et al., 1998). As a result, scientific studies emphasize the urgent need to use phosphorus more efficiently and to establish circular economy models that enhance phosphorus recovery and recycling (Cordell et al., 2009).

To address the limited availability of mineral phosphorus, agricultural science is increasingly focussed on strategies to reduce dependence on finite phosphorus reserves and to promote more sustainable use of this critical resource. One promising approach is the recycling of phosphorus from wastewater and sewage sludge (Bennett et al., 2001). Various methods for recovering phosphorus from these waste streams are currently under investigation. Among the most widely studied is the precipitation of struvite (magnesium ammonium phosphate), which can be used as a fertilizer (Kämpf et al., 2018). This method provides a viable way to recycle phosphorus from wastewater while also offering an environmentally friendly alternative to synthetic fertilizers.

Another important strategy is to reduce overall phosphorus use – particularly through more efficient fertilization technologies. Precision agriculture enables targeted application of phosphate fertilizers, minimizing nutrient losses. By employing modern technologies such as GPS-guided spreading systems and sensor-based monitoring, phosphorus consumption can be significantly reduced while improving nutrient uptake by crops (Basso et al., 2016). These techniques not only enhance efficiency but also help mitigate environmental damage caused by nutrient over-application (Basso et al., 2016).

Another highly promising approach is the breeding of phosphorus-efficient crops. The development of crop varieties capable of absorbing and utilizing phosphorus more efficiently from the soil is an active area of research (Zhao et al., 2018). Such crops could significantly reduce the need for mineral phosphorus fertilizers while also improving productivity in phosphorus-deficient soils. Through targeted breeding efforts aimed at enhancing the ability of plants to thrive under low-phosphorus conditions, it is possible to move towards a more sustainable form of agricultural production in the long term (Lynch, 2011).

Overall, it becomes clear that reducing dependence on mineral phosphorus through a combination of innovative recycling technologies, more efficient fertilization methods, and the development of better-adapted crop varieties represents a promising strategy for ensuring the sustainable use of phosphorus and securing the long-term viability of agriculture.

Potassium

Potassium is another essential nutrient that plays a critical role in plant water regulation, photosynthesis, nutrient transport, and protein synthesis (Marschner, 2012). Potassium fertilizers are primarily derived from potash salts – such as potassium chloride and potassium sulphate – which are extracted from underground mineral deposits (Pampolino et al., 2014).

Similar to phosphorus, potassium reserves are unevenly distributed across the globe. The largest deposits are located in Canada, Russia, Belarus, and Germany (FAO, 2014). Because Russia and Belarus are among the world's leading exporters of potassium fertilizers, geopolitical tensions and sanctions in recent years have led to considerable market instability (Schmitt et al., 2017). Additionally, the extraction of potassium minerals is often challenging, as deposits are frequently located at significant depths (White & Brown, 2017). This results in rising production costs and increases the risk of supply shortages.

> The scarcity of potassium has significant implications for agriculture, as potassium is an essential nutrient that strongly influences plant health, development, and yield potential (Marschner, 2012). A deficiency in potassium can lead to substantial reductions in both yield and quality. Potassium-deficient plants often exhibit reduced disease resistance, weaker structural integrity, and diminished fruit development capacity (Sakaki et al., 2012). Crops such as potatoes, sugar beets, and various fruit species are particularly sensitive to potassium shortages, which can result in lower yields and diminished product quality (Bauer & Döring, 2014).
>
> Another adverse effect of potassium scarcity is increased soil salinization. Excessive use of potassium chloride fertilizers contributes to salt accumulation in the soil, which can reduce soil fertility over time (Müller et al., 2015). High salt concentrations in the soil impair water uptake by plants and damage soil structure, ultimately leading to a decline in agricultural productivity (Rengel, 2015).
>
> In addition, potassium scarcity has economic consequences, particularly for smallholder farms in developing countries. Rising prices for potassium fertilizers – driven by limited resource availability – increase production costs for farmers. These higher costs can lead to greater reliance on expensive imports and threaten the economic sustainability of small-scale agricultural operations (Fischer et al., 2013).

To address potassium scarcity, various strategies are being researched to secure potassium supply. One approach involves exploring new sources of potassium. These include alternative sources such as seawater desalination, where potassium could potentially be extracted as a by-product from desalted seawater, as well as the use of volcanic ash, which contains potassium in the form of bound minerals (González & Gutiérrez, 2017). These alternative sources could contribute to ensuring a long-term potassium supply. As with all measures, the primary focus in light of scarcity is on conservation. Accordingly, the development of more efficient fertilizer applications deserves significant attention. By combining potassium with other nutrients, nutrient uptake efficiency could be improved, potentially reducing potassium demand and the amount of fertilizer required (Cakmak, 2010).

In addition, the use of biological alternatives is also being explored. Certain microorganisms are capable of releasing potassium from insoluble soil deposits and making it available to plants. These biological solutions could provide a sustainable and environmentally friendly method for mobilizing potassium in agricultural soils in the long term (Zhu et al., 2014).

Overall, it is crucial to develop innovative solutions to secure potassium supply in agriculture and minimize the negative impacts of potassium scarcity on yields and soil fertility. This requires close collaboration between science, agriculture, and policy to promote sustainable and efficient approaches (Schmitt et al., 2017).

2.4.3 Scarcity of Nitrogen Fertilizers

Nitrogen is a vital nutrient for plants, as it is essential for protein synthesis and growth (Leach et al., 2012). Although nitrogen is abundant in the atmosphere, its molecular form (N_2) is not accessible to plants. Therefore, it must be converted into a plant-available form through energy-intensive chemical processes (Erisman et al., 2008).

The primary industrial method for producing nitrogen fertilizers is the Haber-Bosch process, in which ammonia (NH_3) is synthesized from hydrogen (H_2) and nitrogen (N_2). Hydrogen is predominantly obtained via steam reforming of natural gas (methane, CH_4), making the production of nitrogen fertilizers highly dependent on fossil energy sources (Sutton et al., 2013).

Impact of Rising Natural Gas Prices on Fertilizer Production
Since natural gas accounts for approximately 70–80% of the production costs of nitrogen fertilizers, fluctuations in energy markets have a direct impact on fertilizer availability and pricing (Van der Zee et al., 2012). The energy crisis in Europe during 2021/2022 demonstrated that rising natural gas prices led to significant reductions in the production of ammonia and urea fertilizers. During periods of high energy costs, fertilizer production can become economically unviable for manufacturers, thereby increasing agricultural expenses and potentially threatening global food security (Olesen et al., 2021).

Another major concern is the CO_2 emissions associated with nitrogen fertilizer production, which amount to approximately 450 million tonnes of CO_2 per year (Bouwman et al., 2002) – equivalent to about 2% of global greenhouse gas emissions. Consequently, the search for alternative production methods, such as the use of green hydrogen derived from renewable energy sources, has become an area of intense research (Sutton et al., 2013).

Interdependence Between Mining and Fossil Resources
Unlike nitrogen fertilizers, phosphorus and potassium fertilizers are not directly dependent on fossil fuels. However, their extraction and processing require substantial amounts of energy, which is often derived from fossil sources (Cordell et al., 2009):
1. **Phosphorus fertilizers** are produced from phosphate rock, which is mined on a large scale and subsequently processed chemically. This processing requires sulphuric acid, which is commonly obtained from sulphur – a by-product of petroleum refining (Dawson & Hilton, 2011).

2. **Potassium fertilizers** are derived from potash deposits, which are accessed through energy-intensive mining operations. The transportation and processing of these raw materials are frequently associated with high CO_2 emissions (Sutton et al., 2013).

Both types of fertilizers are also heavily influenced by geopolitical factors, as their primary reserves are concentrated in a few countries (e.g. Morocco for phosphorus; Russia and Canada for potassium) (Cordell et al., 2009).

Consequences of Fossil Fuel Dependency in Fertilizer Production

The close linkage between fertilizer production and fossil fuels has far-reaching implications for agriculture, the environment, and the global economy. In particular, the volatility of prices and the security of fertilizer supply pose major challenges – challenges that are exacerbated by the heavy reliance on fossil fuels and their fluctuating prices (Van der Zee et al., 2012).

Price Volatility and Supply Security

Rising energy prices directly lead to higher production costs, which burden agriculture and can drive up food prices (Bouwman et al., 2002; Sutton et al., 2013). Countries that are heavily dependent on fertilizer imports are particularly vulnerable to price fluctuations and supply shortages. These countries are forced to adapt to international markets, where fertilizer prices can fluctuate, potentially destabilizing their agricultural production (Cordell et al., 2009). Additionally, trade conflicts and geopolitical tensions can severely impact fertilizer supply. A notable example of this was Russia's export restriction in 2022, which significantly limited global access to fertilizers, particularly phosphate and potassium, jeopardizing supply security in many countries (Euractiv, 2022).

2.5 Consequences

2.5.1 Environmental Impact

In addition to the economic implications, fertilizer production also has significant environmental impacts. The high energy consumption involved in fertilizer production contributes substantially to global CO_2 emissions, exacerbating climate change (Smil, 2001). Furthermore, over-fertilization of agricultural land, often resulting from the excessive use of chemical fertilizers, leads to severe environmental issues such as water eutrophication and soil acidification (Sutton et al., 2013). These processes degrade soil fertility and threaten water quality (Bouwman et al., 2002). The long-term reliance on fossil resources for fertilizer production also contradicts the goals of sustainable agriculture, which focusses on resource conservation and minimizing environmental impacts (Cordell et al., 2009).

Long-Term Resource Scarcity

Another pressing issue associated with the fossil fuel dependence of fertilizer production is the long-term resource scarcity, previously discussed in relation to the industry. In particular, phosphorus reserves are limited and could become increasingly difficult to access in the coming decades (Cordell et al., 2009). Phosphorus is an irreplaceable component of fertilizers, and its inexorable depletion poses a serious threat to long-term agricultural production (Smil, 2000). A similar situation applies to nitrogen production, which is heavily dependent on the availability and price fluctuations of natural gas (Sutton et al., 2013). Given the geopolitical uncertainties and the future depletion of fossil resources, this could lead to further price increases and supply issues, which would place a sustainable burden on agriculture (Bouwman et al., 2002).

The close link between fertilizer production and fossil resources not only presents economic and environmental challenges but also threatens the long-term sustainability of agricultural production. To ensure a more sustainable and resilient agriculture, alternative fertilizer sources and new, resource-efficient production methods are urgently needed.

Strategies for Reducing Fossil Dependency in Fertilizer Production

The development of alternatives to conventional fertilizer production aims to reduce dependence on fossil resources while simultaneously minimizing the environmental impacts of fertilizer manufacturing (Erisman et al., 2008). Various approaches are being explored that could promote both resource conservation and the reduction of CO_2 emissions in agriculture:

- A promising approach is the use of green hydrogen for ammonia production. Transitioning to electrolytically produced hydrogen from renewable energy sources – using so-called Power-to-X technologies – could drastically reduce the CO_2-emissions associated with nitrogen production (Häfner et al., 2021). This process holds the potential to replace fossil fuels, currently used in ammonia production, with clean, renewable sources. Pilot projects for green ammonia production are already in planning, particularly in countries with high availability of renewable energy (Erisman et al., 2008).
- Another approach is the recycling of nutrients from waste. In particular, the recovery of phosphorus from sewage sludge, animal manure, or food waste could help reduce dependence on conventional mined phosphorus, whose reserves are limited and becoming increasingly difficult to access (Cordell et al., 2009). Moreover, the development of biological processes for nitrogen fixation by microor-

ganisms is being intensively researched. These processes could provide an environmentally friendly alternative to synthetic nitrogen fertilization in the long term, utilizing the natural ability of microbes to convert atmospheric nitrogen into plant-available nitrogen (Galloway et al., 2008).

–   Finally, more efficient fertilization strategies are being developed to optimize nutrient use in agriculture and minimize losses. Precision farming and targeted fertilization techniques are promising methods that use sensor technology and GPS-controlled application to match fertilizer use precisely to plant needs (Sutton et al., 2013). These technologies allow for significant reductions in nutrient losses and contribute to resource conservation. In addition, research into new fertilizer forms, such as slow-release fertilizers or biological alternatives, is increasingly being intensified to ensure a more sustainable and efficient supply of fertilizers in the long term (Bouwman et al., 2002).

The development of alternative methods for fertilizer production and application is a major goal of modern agricultural research. These approaches not only offer potential to reduce dependence on fossil resources but also contribute to mitigating the environmental impacts of agriculture and ensuring the sustainability of food production. However, as shown below, although promising the conflict demand for raw materials will cause significant price increases which may make production too expensive.

2.5.2 Impact on Agricultural Yields

Fertilizer scarcity represents a significant challenge for global agriculture and food security. Synthetic fertilizers, which provide essential nutrients such as nitrogen (N), phosphorus (P), and potassium (K), play a critical role in modern agriculture by enabling high yields and compensating for natural soil depletion processes (Smil, 2001). Increasing scarcity of these fertilizers can, therefore, have severe consequences on agricultural yields, which could directly impact food prices, global markets, and social stability (Sutton et al., 2013). The scientific discussion focusses on the causes of fertilizer scarcity, its short- and long-term impacts on various agricultural systems, and potential ways to mitigate its negative effects (Cordell et al., 2009).

Influencing Factors on Crop Yields

Crop yields result from a complex interplay of numerous factors both natural and anthropogenic in origin. These factors affect agricultural productivity and have far-reaching implications for global food supply and the economy.

Climate change plays an increasingly significant role in impacting agricultural yields. Rising average temperatures, changing precipitation patterns, and the increase in extreme weather events such as droughts, floods, and storms pose a serious threat to agricultural production worldwide (IPCC, 2022). An analysis by Zhao et al. (2017) reveals that climate change has already had significant effects on the yields of staple crops like wheat, maize, and rice, particularly due to the frequency and intensity of heatwaves during growth phases, which can cause yield losses of up to 10–20%. In tropical regions, where many crops are already grown at the upper limit of their temperature tolerance, yield losses due to rising temperatures are particularly severe (Lobell et al., 2011).

The availability of essential nutrients such as nitrogen (N), phosphorus (P), and potassium (K) is another key factor for high crop yields. The use of mineral fertilizers has contributed significantly to yield increases over the past decades (Bouwman et al., 2002). Many agricultural soils, due to intensive cultivation, no longer contain sufficient natural nutrient reserves (Bouwman et al., 2002). In particular, nitrogen fertilization has significantly contributed to increasing crop yields in recent decades (FAO, 2021). According to FAO (2021), the use of mineral fertilizers is responsible for approximately 50% of global agricultural yield growth. However, inadequate fertilizer supply can lead to a range of negative effects on plant growth and productivity (Zhang et al., 2020). Moreover, the scarcity of these fertilizers, caused by rising energy prices, geopolitical conflicts such as the Ukraine war, and environmental regulations, has led farmers in many regions to reduce their fertilizer applications (FAO, 2021). This directly impacts yields and presents a growing risk to agriculture.

The long-term maintenance of soil fertility is crucial for sustainable yields. However, intensive agriculture, monocultures, and soil degradation are leading to a global decline in productivity. A study by Montgomery (2007) shows that approximately 24 billion tonnes of fertile soil are lost annually due to erosion, which significantly affects agricultural productivity, especially in Africa and Asia. Measures such as conservation agriculture and agroforestry systems could partially reverse this trend and contribute to yield stabilization (Pretty, 2008), though these approaches come with additional costs and technological changes:

- **Nitrogen Deficiency:** Nitrogen deficiency leads to stunted growth, reduced leaf development, and decreased photosynthetic capacity, directly affecting yields (Bouwman et al., 2002).
- **Phosphorus Deficiency:** Limited access to phosphorus results in weak root systems, reduced water and nutrient uptake, and delayed fruit formation (Cordell et al., 2009).
- **Potassium Deficiency:** A potassium deficiency can increase susceptibility to diseases and reduce the shelf life of harvested products (Sutton et al., 2013).

Scientific studies have shown that a reduction in fertilizer inputs by 30–50% can lead to yield losses of 20–40% in many agricultural systems (Zhang et al., 2020). Crops with high nutrient requirements, such as maize, wheat, and rice, which account for a large portion of global food production, are particularly affected (Sutton et al., 2013).

More than 40% of global food production depends on irrigated land (FAO, 2020). However, increasing water scarcity, exacerbated by overuse, groundwater depletion, and climate change, such as the exceeding of planetary boundaries, limits the availability of water for agriculture, particularly in regions like the Middle East, North Africa, and parts of India (see Figure 2.6). Projections suggest that by 2050, water stress could lead to yield losses of up to 30% in particularly affected regions (Rockström et al., 2017).

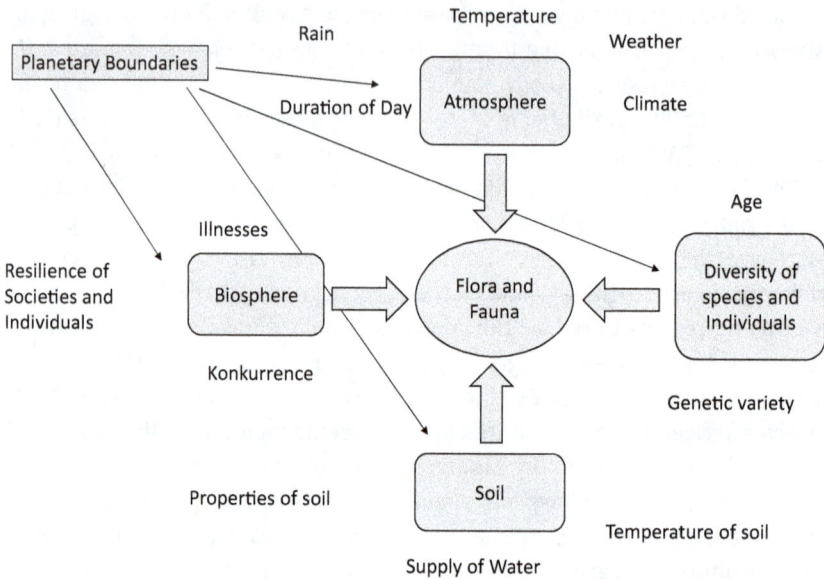

Figure 2.6: Indirect influencing factors of reaching the planetary boundaries on crop yields. The attainment of planetary boundaries has both direct and indirect impacts on flora and fauna, with these effects being complex and, due to their intricacy, sometimes difficult to quantify. The planetary boundaries framework defines critical thresholds for Earth system processes – such as climate change, biodiversity loss, land-system change, and biogeochemical flows – that, if crossed, risk destabilizing the global environment. This figure highlights how exceeding or approaching these boundaries can indirectly affect flora and fauna by altering ecosystem services, soil, atmosphere, and biodiversity dynamics, which are essential for sustainable agricultural production. The intricate relationships between these environmental stressors and biological systems lead to multifaceted impacts on flora and fauna, complicating the accurate quantification and prediction of outcomes. Understanding these indirect influences is crucial for developing adaptive agricultural practices and policies that support food security while maintaining the resilience of the planet's ecosystems.

Regional Differences in the Impact of Fertilizer Scarcity

The consequences of limited fertilizer availability vary significantly across regions and are largely influenced by the specific agricultural systems, soil conditions, and political frameworks in place (Smil, 2000):

- **Industrial Agriculture in Developed Countries**: In regions with intensively farmed land, such as Europe, North America, and China, modern precision agriculture and more efficient fertilization strategies could help partially mitigate fertilizer shortages. However, yield losses are still expected, as many soils, due to decades of overuse, may not provide enough nutrients for high yields without external fertilization (Bouwman et al., 2002).
- **Small-Scale Farming in Developing Countries**: Countries with limited access to synthetic fertilizers, particularly in sub-Saharan Africa, Southeast Asia, and parts of South America, are especially affected. Small-scale farmers often lack the financial resources to purchase more expensive fertilizers, leading to increased yield losses (FAO, 2021). The FAO warns that ongoing fertilizer shortages in many developing countries could lead to increasing food insecurity (FAO, 2021).
- **Special Case: Organic Farming**: Ecological agricultural systems that do not rely on synthetic fertilizers are less directly impacted by the current fertilizer scarcity. However, long-term challenges are expected, as organic fertilizers are often not available in sufficient quantities to sustain large-scale agricultural systems (Galloway et al., 2008). In particular, the shift to organic fertilization strategies could lead to yield declines in some cases, as these fertilizers act more slowly and less predictably in terms of nutrient availability than synthetic alternatives (Galloway et al., 2008).

Fertilizer scarcity represents a far-reaching and growing challenge for global agriculture, with its impacts varying across regions in terms of both crop yields and food security (Sutton et al., 2013). To mitigate these negative effects, not only efficient fertilization and innovative fertilizer technologies are needed but also comprehensive political and economic measures that can ensure access to nutrients, particularly in developing countries (Cordell et al., 2009).

The impacts of the looming fertilizer scarcity extend beyond agricultural production, affecting global markets and economies as well (Sutton et al., 2013). The shortage of fertilizers not only leads to reduced crop yields but also has far-reaching economic (see Section 2.5.3) and societal consequences (Cordell et al., 2009).

Social and Political Instability

Another significant aspect is the social and political instability that can be triggered by rising food prices. In countries with a high dependence on food imports, such as many developing nations, inadequate food supplies and rising prices can lead to social unrest (Headey et al., 2014). This was evident during the global food crises of 2008 and 2011, when the prices of key agricultural products surged sharply (Gilbert, 2010). In response to these developments, governments may be forced to increase subsidies for fertilizers or food imports, which would lead to substantial fiscal challenges and further strain national finances (Headey et al., 2014).

Changes and Strategies to Mitigate Negative Impacts

The ongoing fertilizer shortage could lead to long-term structural changes in agricultural production. Farmers may be forced to switch to less nutrient-intensive crops or adjust their crop rotations to compensate for nutrient deficiencies (Zhang et al., 2020). Additionally, a stronger promotion of alternative farming methods, such as agroecological systems or intercropping, could contribute to reducing the long-term need for synthetic fertilizers (Altieri, 2002). These methods focus on fostering more sustainable agriculture, which relies less on external inputs while simultaneously enhancing soil fertility (Gliessman, 2007).

Given the far-reaching consequences of fertilizer shortages, various strategies are being discussed to reduce dependence on synthetic fertilizers and stabilize agricultural yields:

1. **More Efficient Use and Prudent Application of Fertilizers**
   Precision agriculture represents a key approach to applying fertilizers in a targeted and efficient manner (McBratney et al., 2005). By utilizing modern technologies, such as GPS-guided fertilizer application and soil analysis, farmers can more accurately determine the nutrient requirements of plants and minimize losses (Liu et al., 2020). Additionally, new fertilization strategies are being explored, such as long-term fertilizers or soil biological methods for nitrogen fixation, which could ensure a sustainable nutrient supply in the long term (Bouwman et al., 2002).

2. **Promoting the Circular Economy in Agriculture**
   Another promising approach is the expansion of the circular economy in agriculture. This primarily involves nutrient recovery, such as the recycling of phosphorus from sewage sludge (Cordell et al., 2009). The increased use of organic fertilizers, such as compost or manure, also holds potential to reduce the reliance on synthetic fertilizers while simultaneously improving soil quality (Galloway et al., 2008).

3. **Alternative Fertilizer Sources**
   Research is increasingly focussed on alternative fertilizer sources. In particular, biological fertilizers based on microorganisms could offer an environmentally friendly and sustainable alternative (Liu et al., 2020). Additionally, algae and biochar are being investigated as potential long-term nutrient sources for the soil, which not only provide nutrients but could also contribute to soil improvement (Xu et al., 2017).

4. **International Cooperation and Political Measures**
   To mitigate the negative impacts of fertilizer scarcity, international cooperation and political measures are essential. Increased research into reducing fertilizer dependence, as well as strengthening resilience in global supply chains, could contribute to enhancing long-term supply security (FAO, 2021). International partnerships, particularly in developing countries, could help by exchanging technologies and knowledge to improve efficiency and resource conservation (Sutton et al., 2013).

Overall, addressing fertilizer scarcity requires a comprehensive approach that includes both technological innovations and political and economic measures to minimize the negative impacts on agriculture and global food security (Cordell et al., 2009).

2.5.3 Consequences on Crop Prices

Global food production is a highly complex system influenced by a variety of ecological, economic, and political factors (Foley et al., 2011). In recent years, numerous challenges to agricultural production have intensified, including climate change, geopolitical crises, resource scarcity, and rising operating costs (Porter et al., 2014). Particularly, the availability and cost of inputs such as fertilizers, water, and energy have direct impacts on crop yields and food prices (Tilman et al., 2011). Scientific studies and reports from international organizations like the Food and Agriculture Organization (FAO) and the World Bank indicate that these developments have long-term consequences for global food security (World Bank, 2022; FAO, 2021).

Rising Food Prices as a Result of Declining Yields
The combination of reduced crop yields, rising production costs, and geopolitical uncertainties has significant effects on global food markets, leading to substantial price increases (FAO, 2021). Lower crop yields result in a reduced food supply, which in turn drives up food prices. According to projections by the World Bank, a sustained fertilizer shortage, combined with the impacts of climate change, could increase global food prices by up to 20% (World Bank, 2022). Particularly vulnerable are poorer households, which spend a significant portion of their income on food, are susceptible to these price hikes, potentially exacerbating poverty and food insecurity (FAO, 2021). Agricultural market pricing is determined by the interplay of supply and demand. Declining yields reduce supply, while global demand, especially in growing economies like China and India, continues to rise (Trostle, 2010). Furthermore, speculation in commodity markets contributes to price volatility. A scientific analysis by Clapp and Isakson (2018) indicates that financial investments in agricultural commodities significantly contributed to price increases during the 2008 food crisis. According to World

Bank data, prices for wheat, maize, and rice have increased by an average of 30–50% since 2020, primarily due to climatic factors, rising energy costs, and higher fertilizer prices (World Bank, 2022). Historical comparisons show that similar price increases in the past have led to social unrest, as seen during the food crises of 2008 and 2011 (FAO, 2021). Rising food prices have led to an increase in food insecurity in many countries.

Low-income households are particularly affected by rising food prices, as a significant portion of their budget is allocated to food (FAO, 2021). In developing countries, price inflation exacerbates food scarcity and can further intensify malnutrition. In some regions, such as India, export restrictions on certain agricultural products, like wheat, have already been implemented, further driving price increases on global markets (World Bank, 2022).

Overall, the increasing uncertainty regarding agricultural yields highlights that climate change, fertilizer shortages, soil quality, and water availability are key factors for future food production (Rockström et al., 2017). The interactions between these factors are driving rising food prices, which significantly burden both global markets and socially vulnerable populations.

2.5.4 Consequences for Developing Countries and the Agricultural Industry

The global food production system is facing profound challenges, which are particularly evident in developing countries and the agricultural industry. Factors such as climate change, resource scarcity, geopolitical uncertainties, and rising production costs have significant impacts on agriculture and food security. Studies show that countries with low-economic resilience and high-import dependence are particularly affected by these developments (FAO, 2022). At the same time, the agricultural industry must respond to structural changes in order to remain competitive in the long term.

Impacts on Developing Countries

Developing countries are affected by the current challenges in global food production in several ways. Particularly severe are the rising food prices, food insecurity, and structural deficits in agricultural production (see also Figure 2.7):

1. **Food Security and Rising Food Prices**
   Price trends on global agricultural markets have direct implications for food availability in developing countries. According to a World Bank study (2023), prices for wheat, maize, and rice have increased by up to 50% in recent years. As a result, low-income households are spending an increasing portion of their budgets on staple foods, exacerbating hunger and malnutrition.

- **Dependence on Imports**: Many developing countries are reliant on imports of grains, oilseeds, and fertilizers. Import costs have significantly increased due to rising energy prices and geopolitical crises. Countries in Sub-Saharan Africa, which import large quantities of wheat from Ukraine and Russia, are facing supply shortages and rising prices (FAO, 2023).
- **Inflation and Economic Instability**: Price hikes affect the entire economy, leading to inflation and social tensions. In countries such as Egypt and Pakistan, protests against rising food prices have already intensified.

2. **Impact on Smallholder Farmers and Rural Development**

The majority of agricultural production in developing countries is carried out by smallholder farmers, who often have limited access to modern technologies, financial resources, and markets:

- **Increased Input Costs:** The rising prices of seeds, fertilizers, and energy place a particularly heavy burden on smallholder farmers. Many can no longer afford the necessary investments in their farms, leading to reduced yields (IFPRI, 2022).
- **Decreasing Productivity:** Due to poor soil quality, inadequate infrastructure, and limited access to irrigation technologies, agricultural yields in many developing countries are already lower than in industrialized nations. The additional pressures from climate change and resource scarcity exacerbate this problem.
- **Rural Exodus and Social Instability:** The economic hopelessness faced by many farmers increasingly leads to migration to urban areas, intensifying urbanization and the social challenges in metropolitan areas (UNDP, 2021).

3. **Climate Change and Agricultural Vulnerability**

Developing countries are particularly vulnerable to the impacts of climate change. Extreme weather events such as droughts, floods, and storms are increasing and threatening agricultural yields:

- **Drought and Water Scarcity:** According to IPCC reports (2023), water availability in many parts of Africa, South Asia, and Latin America is expected to decrease by up to 25% by 2050, which could significantly impact agricultural production.
- **Loss of Arable Land:** Soil degradation, desertification, and rising temperatures are causing increasingly less land to be suitable for food cultivation. Regions such as the Sahel are particularly affected, where the advancing climate change is already leading to massive crop losses (FAO, 2022).

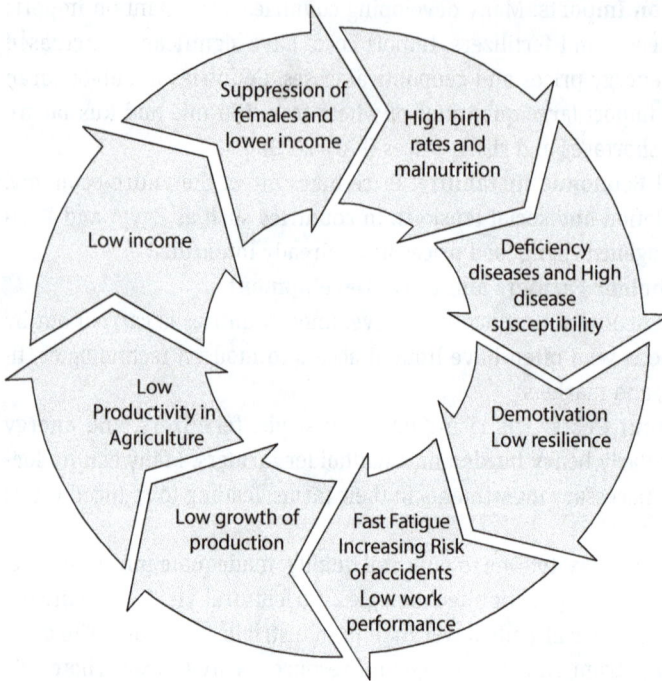

Figure 2.7: Relationship of various factors between low income and low productivity.
Figure illustrates the complex interrelationship between low income and low productivity, highlighting how these factors reinforce each other in a cyclical and self-perpetuating manner. The diagram depicts both the direct and indirect consequences of low income, which can limit access to critical resources such as quality inputs, technology, education, and healthcare, thereby reducing labor efficiency and overall productivity. In turn, low productivity results in limited earnings and economic opportunities, perpetuating poverty and constraining investments in improvements. This creates a vicious cycle where the lack of financial means and low output mutually exacerbate one another, making it difficult for individuals, households, or communities to break free from persistent economic hardship. The figure underscores the need for integrated interventions that simultaneously address income constraints and productivity barriers to foster sustainable development and economic resilience.

Consequences for the Agricultural Industry

The global agricultural industry also faces profound challenges. Increasing uncertainty in raw material supply, regulatory adjustments, and the demand for more sustainable production methods are compelling companies to develop new strategies.

Raw Material Scarcity and Rising Production Costs

The agricultural industry is heavily reliant on the availability of inputs such as fertilizers, pesticides, and energy, which makes it particularly vulnerable to price fluctuations and geopolitical tensions. The aforementioned fertilizer shortage represents a significant challenge: phosphorus, potassium, and nitrogen are essential nutrients for agricultural production, and their global reserves are limited. The production costs of these fertilizers are rising due to increasing energy prices, which places countries like Morocco, China, and Russia – who control a large portion of the world's phosphate reserves – in a geopolitically dominant position (World Fertilizer Report, 2023). This concentration of raw material reserves creates dependencies that can destabilize the global market. Furthermore, rising energy prices and transportation costs have direct impacts on agricultural production and logistics. Countries with long supply chains and a high proportion of imported food are particularly affected.

Adaptation to Sustainable Production Methods

The growing demand for more environmentally friendly farming methods, along with political measures aimed at reducing environmental impacts, has profound consequences for the agricultural industry. In response to these demands, companies must increasingly invest in sustainable technologies. Consumers and governments are calling for resource-efficient and sustainable food production, putting pressure on companies to reduce the use of pesticides and fertilizers. At the same time, digitalization and precision agriculture are gaining increasing importance in order to enhance production efficiency and minimize resource consumption. Technologies such as drones, AI-assisted crop monitoring, and water-saving irrigation systems enable more precise and resource-efficient farming practices.

Market Concentration and Structural Changes

The agricultural industry is experiencing increasing market concentration as large multinational corporations acquire or push out smaller businesses. A prominent example is the oligopolization of the agrochemical sector, where a few companies such as Bayer, Syngenta, and Corteva dominate the global market for seeds and pesticides. This market power exerts price pressure and creates dependencies for farmers, who have limited alternatives to the products offered by these corporations. Additionally, there is a growing trend of vertical integration in food production, with large food corporations increasingly investing in their own agricultural operations to secure their supply chains and reduce costs.

Future Perspectives

To mitigate the negative impacts on developing countries and the agricultural industry as a whole, various approaches are being discussed:

- **Promotion of Sustainable Agriculture**: Expanding regenerative agriculture and integrating agroforestry systems could contribute in the long term to maintaining soil fertility and reducing dependence on synthetic fertilizers.
- **Investments in Infrastructure and Technology**: Expanding irrigation systems, storage technologies, and agricultural research is considered crucial to stabilize agricultural productivity and optimize resource use.

– **International Cooperation and Trade Agreements**: Stabilizing global supply chains and lifting export restrictions on food are essential measures to ensure long-term global food security.
– **Promotion of Food Sovereignty**: Diversifying farming systems and strengthening regional markets could help reduce dependence on global markets and increase resilience to geopolitical and climatic risks.

2.5.5 Consequences for Global Food Production

The global food production system is increasingly caught in the tension between limited resources, rising demand, and geopolitical uncertainties. Competition for arable land, water resources, and strategically important inputs such as fertilizers and seeds has intensified significantly over the past few decades. Scientific studies indicate that these distribution conflicts have social, economic, and ecological consequences at both national and international levels, manifesting in trade disputes, land grabbing, and increasing social unrest (FAO, 2023; IPCC, 2022) – see also Figure 2.8.

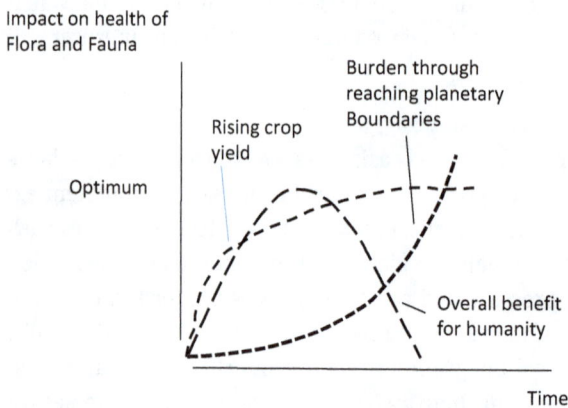

Figure 2.8: Impact on the health of the overall system from flora and fauna.
While growing crop yields initially provide significant benefits to humanity, reaching and surpassing the planetary boundaries leads to a flattening of yields and, beyond a certain point, results in a decline in the overall benefits of agriculture instead of further increases.

The current challenges facing the agricultural sector, such as climate change, resource scarcity, geopolitical tensions, and increasing production costs, have profound implications for global food production. These factors not only affect the quantity and quality of agricultural outputs but also influence the stability and sustainability of food systems and thereby the total benefit for human society around the world. Below are some of the key consequences for global food production:

1. **Decreased Agricultural Productivity**: Climate change, water scarcity, soil degradation, and rising input costs (such as fertilizers and energy) are likely to reduce agricultural yields. Areas already vulnerable to extreme weather conditions, such as droughts and floods, will experience even more significant production losses.
2. **Supply Chain Vulnerabilities**: Geopolitical conflicts have highlighted the fragility of global food supply chains. Disruptions in key exporting countries can lead to price volatility and food shortages, affecting both importing and producing nations. The reliance on a few global suppliers for critical inputs, such as fertilizers, also makes the agricultural sector highly vulnerable to global market fluctuations.
3. **Increased Food Prices**: Rising production costs, including those associated with labour, energy, and raw materials, are expected to lead to higher food prices. In addition, decreased availability of key ingredients due to environmental and geopolitical factors will put upward pressure on prices. Higher food prices have a direct impact on food security, particularly in developing countries, where a large portion of household income is spent on food.
4. **Increased Land Use Change**: As agricultural productivity decreases in certain areas, there may be increased pressure to convert natural ecosystems, such as forests and wetlands, into agricultural land. This could lead to deforestation, loss of biodiversity, and further degradation of ecosystems, ultimately exacerbating the challenges facing global food production.
5. **Regional Disparities in Food Security**: The global disparities in food production capabilities will widen. While developed countries may have the resources to invest in technology and infrastructure to adapt to the challenges, many developing nations will struggle to keep pace. This could exacerbate the already significant inequalities in food access and increase the number of people facing hunger and malnutrition, particularly in low-income countries.

The Increase in Distribution Conflicts over Agricultural Resources

The increase in distribution conflicts over agricultural resources is attributed to a variety of factors, both natural and anthropogenic in origin. Key causes include population growth, climate change, shifting dietary patterns, geopolitical tensions, and the ongoing industrialization of agriculture.

Population Growth and Rising Demand for Food

With the growing global population, which is projected by the UN to reach nearly 10 billion people by 2050, the demand for agricultural land and inputs continues to increase. This is particularly evident in emerging economies such as India, China, and Nigeria, where rising incomes are associated with a shift in dietary patterns, leading to higher demand for protein-rich foods such as meat and dairy products (FAO, 2023). Another issue arises from the competition for land between food and feed production, as approximately 77% of global agricultural land is used for feed production or as pastureland for livestock, while only a small fraction is directly available for growing plant-based foods (Foley et al., 2011). Simultaneously, the expansion of industrial agriculture, driven by the increasing demand for agricultural raw materials, leads to the intensification of farming, exacerbating the resource competition between large-scale industrial agricultural operations and smallholder farming systems.

Climate Change and Ecological Constraints

Climate change exacerbates existing distribution conflicts over agricultural resources, as rising temperatures, altered precipitation patterns, and extreme weather events alter the availability of arable land. Desertification and soil degradation particularly affect regions such as the Sahel in Africa or parts of India, where climatic changes and overuse lead to the loss of valuable agricultural land. It is estimated that approximately 33% of global soils are already degraded (FAO, 2021). Additionally, water scarcity is becoming an increasing conflict factor: agricultural use accounts for approximately 70% of global freshwater resources. In regions experiencing growing water shortages, such as the Middle East or parts of South Asia, the unequal distribution of water resources leads to conflicts between countries and within societies (IPCC, 2022).

Geopolitical Tensions and Trade Conflicts

Control over agricultural resources is increasingly gaining importance on the geopolitical stage. In recent years, several countries have imposed export restrictions on key agricultural commodities such as grains, oilseeds, and fertilizers to secure their own supply. Resource nationalism and export controls have particularly affected countries like Russia, China, and Indonesia, which have implemented measures to limit the export of fertilizers and agricultural products. This has led to global price increases and uncertainties in supply chains (World Bank, 2023). Furthermore, agriculture is increasingly being used as a tool of power: major powers such as China are securing access to fertile agricultural land in Africa and Latin America through long-term investments, creating tensions with the affected countries (GRAIN, 2022).

Forms of Distribution Conflicts over Agricultural Resources

The growing competition for agricultural resources manifests in various forms of conflict, ranging from economic power struggles and state-driven land grabs to social unrest. One of the most prominent forms is "land grabbing"" and foreign investment in agricultural land. "Land grabbing" refers to the large-scale acquisition of agricultural land by states or multinational corporations, often at the expense of local communities. Between 2000 and 2020, an estimated 40 million hectares of agricultural land in developing countries were acquired by foreign investors, particularly in sub-Saharan Africa, Southeast Asia, and Latin America, according to the Land Matrix Initiative.

These land grabs often lead to the displacement of small farmers, exacerbation of so-cial inequalities, and destabilization of rural areas (Borras et al., 2011). Another source of distribution conflict is water resources. The competition for water is one of the main causes of conflict in agriculture. Rivers such as the Nile, Indus, and Mekong are essential water sources for the agriculture of multiple states, often leading to political tensions. In countries like Chile and South Africa, the privatization and control of water resources have resulted in agricultural corporations disproportionately benefit-ing from water supplies, while small farmers suffer from water scarcity (Shiva, 2002).

Another phenomenon contributing to distribution conflicts is speculation on agri-cultural commodities and the resulting price increases. Financial markets are playing an increasingly influential role in global food production, as institutional investors pour capital into farmland, agricultural commodities, and food corporations. The trading of food commodities on international stock exchanges can lead to price vola-tility, disproportionately affecting low-income countries and vulnerable populations (Clapp & Helleiner, 2012). At the same time, a small number of multinational corpora-tions control significant portions of global seed, fertilizer, and pesticide production, creating a high level of dependency for farmers on a few dominant market players.

Approaches to Reducing Distribution Conflicts

To mitigate the negative impacts of distributional conflicts over agricultural resources, a combination of political, economic, and technological measures is necessary. Promot-ing sustainable agriculture through the expansion of agroecological farming methods can contribute to a more efficient use of natural resources and increase agricultural resilience (FAO, 2021). Furthermore, stricter regulation of land investments could help prevent land grabbing at the expense of local communities (Borras et al., 2011). Interna-tional agreements and stronger national legal frameworks are required to ensure a more equitable global distribution of agricultural resources (GRAIN, 2022). Improved water management systems – such as drip irrigation and rainwater harvesting – can reduce competition over water and promote more sustainable water use (Shiva, 2002). Finally, increasing transparency in agricultural markets and more stringent regulation of speculative financial activities involving agricultural commodities could help prevent extreme price volatility and limit the influence of financial markets on agricultural pro-duction (Clapp & Helleiner, 2012).

2.5.6 Transformation of Agriculture – Lengthy and Complex

As indicated in the previous chapters and in Figure 2.9, yields are affected by several factors. In addition to that about a third of all food becomes waste prior to reaching their customer.

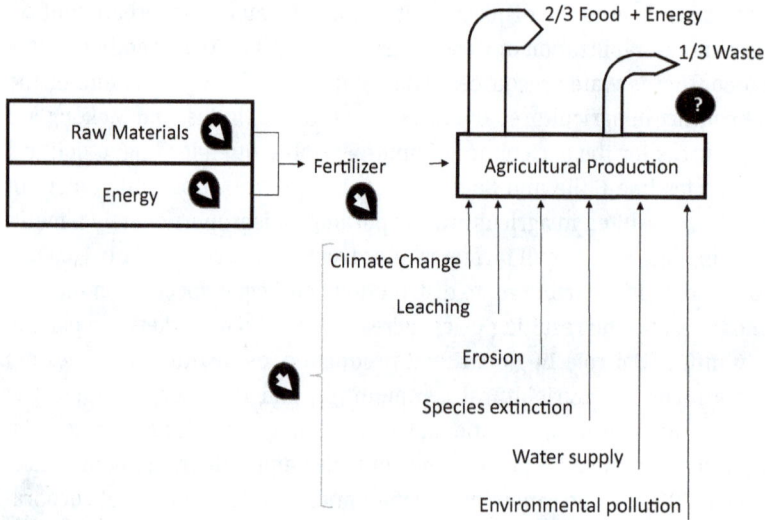

Figure 2.9: Factors affecting yields in agriculture.
Figure illustrating the multifaceted factors that influence agricultural yields, highlighting the complexity of crop production systems. Agricultural yields are affected by a combination of biophysical, environmental factors including soil quality, water availability, climate conditions, pest and disease pressures, farming practices, and technological inputs. The diagram emphasizes how these variables interact to reduce overall productivity on farms. Additionally, the figure draws attention to the significant challenge of food loss and waste, noting that approximately one-third of all food produced is lost or wasted before it reaches the end consumer. This loss occurs at multiple stages including harvesting, storage, transportation, and processing, further exacerbating the pressure on agricultural systems to meet global food demand sustainably. Understanding these factors is crucial for developing strategies aimed at improving yield efficiency, minimizing waste, and enhancing food security in the face of growing population and environmental challenges.

Addressing temperature change, soil acidification, erosion, and climatic effects is a lengthy and complex process that requires far more than just short-term adaptation strategies. The duration of the transformation and the timing of when it must begin depend on a variety of factors, including the specific region, the intensity of climate change, the existing agricultural infrastructure, and the political and social frameworks (Lipper et al., 2014; Howden et al., 2007).

Duration of Transformation

A fundamental transformation of agriculture in response to the aforementioned climatic challenges can take years or even decades. Particularly, the shift towards more sustainable and resilient agricultural practices not only requires technological innovations but also profound changes in education, political and economic systems as well as societal structures (Thornton et al., 2014).

Long-Term Adaptations (10–30 Years)

Soil Improvement and Humus Formation: Building and stabilizing humus in the soil, such as through agroforestry or the use of organic fertilizers, is a long-term process. Humus takes years to decades to form in sufficient quantities and sustainably improve soil fertility (Lal, 2015).

Crop Rotation and Variety Selection: Transforming cultivation methods to introduce drought-resistant varieties or deep-rooted plants must occur over several growing cycles. New varieties must be adapted to regional climatic conditions, which includes research and breeding projects that may take several years to decades (Fischer et al., 2010).

Medium-Term Adaptations (5–15 Years)

Soil Management: The introduction of conservation tillage practices, such as mulch seeding or direct seeding, can show quicker results in terms of reducing erosion and protecting the soil. However, widespread sustainable implementation will require several years (Pimentel et al., 2011).

Water Management and Irrigation Technologies: The shift to more efficient irrigation systems can occur in a relatively short time frame (5–10 years) when the necessary resources and technologies are available. However, large-scale implementation of drip irrigation or water-saving techniques requires extensive infrastructure development and corresponding investments (Rockström et al., 2009).

When Should the Transformation Begin?

To minimize the negative effects of climate change, such as rising temperatures, soil acidification, and erosion, the transformation of agriculture must begin immediately to ensure the transition to a more sustainable and resilient agricultural system. This is especially crucial when considering climate change projections for the coming decades (IPCC, 2021):

1. **Early Start:** An early start to the transformation is ideal, beginning at least 10–20 years before significant climatic changes are projected to start. Through proactive adaptation strategies, farmers can better prepare for the expected conditions and build resilient agricultural systems (Howden et al., 2007).
2. **Political and Economic Factors:** The transformation also requires political support and investments, which may take longer to materialize. Providing subsidies, expanding research, and creating incentives for sustainable agricultural practices must be coordinated and implemented, which can also take years (Lipper et al., 2014).
3. **Research and Innovation:** Building knowledge and developing new technologies to adapt agriculture to changing climatic conditions is an ongoing process that must take place over decades. Climate scientists and agronomists must continuously develop and refine new methods (Thornton et al., 2014).

The transformation of agriculture in response to climate change must be long-term, with initial measures ideally being initiated now. The development and implementation of sustainable agricultural practices and infrastructures require at least 10–30 years, with the speed of adaptation varying depending on the region and available resources. The earlier the implementation begins, the better the negative impacts of climate change can be mitigated, and the long-term resilience of agricultural systems can be ensured (Rockström et al., 2009).

There are indications that through improved farming methods, optimized fertilization strategies, and more resilient crop varieties, agricultural yields can be stabilized despite challenging conditions (Ray et al., 2013). The use of digital agriculture and precision irrigation may allow for more efficient resource utilization (Zhao et al., 2018). Several sources suggest that the integration of circular economy principles and nutrient recycling, such as phosphorus recovery from wastewater, is seen as a long-term approach to reduce dependence on synthetic fertilizers (Cordell et al., 2009; Tilman et al., 2011). Furthermore, a more diversified food system, extending beyond the dominant staples like wheat, rice, and maize, could increase resilience to climatic and geopolitical shocks (Foley et al., 2011; Pretty, 2008).

However, the recent wars made clear that stable global trade relationships, avoiding export restrictions are far from reality and the ever tighter budgets and investment in aged economic concepts prevent from investing in agricultural research despite they are critical to ensuring long-term food security (Headey & Fan, 2010). Additionally, and to help prevent social unrest policy measures such as price stabilization and subsidies for particularly affected households may be supportive (Trostle, 2010) – but are hardly lesser likely to be applied in times of tighter budgets.

Insert

The above chapters describe why human society is dependent on a continuous supply of mineral, fossil, and biogenic resources. Furthermore, it is shown that a portion of the fossil and mineral resources is nearly exhausted. It is possible to conceptualize the human, animal, and plant populations of Earth as a socio-economic system, with the provision of resources to humans serving as the foundation of its contribution to this system – its social metabolism. As raw materials become scarcer, it follows that the dynamics of the social metabolism will change. To understand how this change occurs, the Hubbert model for describing resource extraction characteristics and the Adaptive Cycles model have been introduced. The combination of both models allows for the interpretation that a collapse in energy and resource supply could lead to the collapse of the current form of the socio-economic system. The available data on resource supply from fossil, mineral, and agricultural sources suggest that the supply of certain resources is approaching its end, which, if the model assumptions provide accurate predictions, significantly increases the risk of a societal collapse.

The discussion of raw materials is incomplete without considering the development of agriculture. However, it becomes apparent that the challenges to agricultural production – caused by climate change, geopolitical crises, resource scarcity, and increased operating costs coupled with a growing population – will also lead to supply shortages in food. Particularly, the availability and price of inputs such as fertilizers, water, and energy have direct impacts on crop yields and food prices. These developments show that long-term consequences for global food security are inevitable.

On one hand, it is therefore tempting to attempt to compensate for the increasingly evident shortages of fossil and mineral resources by using biogenic raw materials. On the other hand, an analysis of the supply situation for agricultural, biogenic products reveals that, even in this category, the supply situation is far from relaxed, and shortages are either already foreseeable or have already occurred.

Nevertheless, it is worthwhile to take stock of whether biogenic raw materials can serve as substitutes for mineral and fossil resources. This question will be addressed in the following chapter. However, even if fossil resources could contribute to some relief, the risk that too much will be consumed, leading to a collapse of the social metabolism, is high. Therefore, it is logical to

1. massively reduce consumption to mitigate risk and
2. seek alternatives in the supply of mineral and fossil resources.

The first path, consumption reduction, seems obvious and is ultimately unquestionable. Therefore, the second path will be pursued further.

The fundamental question is:

What biogenic raw materials can the Earth produce with the help of the sun, and which of these raw materials are, according to current knowledge, likely or possibly available – or, conversely, not available?

Given that the Earth's surface is finite and this area represents the habitat for the human, animal, and plant populations of the Earth, it is clear that the goal cannot be to discover further growth potentials but rather to assess how the potential of biogenic raw materials should be evaluated from today's perspective.

Chapter 3
Biogenic Raw Materials

3.1 Overview

The recognition that fossil and mineral resources are not only finite but, when har-vested, refined and used, also pose significant environmental challenges, has driven re-search and industry increasingly toward a more sustainable and resource-efficient econ-omy (OECD, 2011). Since the early 2000s, the development of biofuels, bio-based plastics, and biomass as alternatives to fossil resources has made remarkable progress (IEA, 2020). These advancements have largely been driven by the urgent need to reduce de-pendence on fossil fuels and to identify more sustainable and environmentally friendly alternatives. This urgency has been further intensified by the mounting pressure of the climate crisis, rising energy prices, and global resource scarcity (UNEP, 2019).

Research on biogenic resources is promising and, through the development of new technologies and efficient biotechnological processes, could reduce industrial de-pendence on fossil raw materials. In particular, bio-based plastics, lignocellulose-derived materials, and the biotechnological production of chemicals hold the potential to provide more sustainable alternatives in the future (Isikgor & Becer, 2015). How-ever, in order to become competitive with fossil-based materials, these technologies must be further optimized to improve efficiency and reduce costs. Overall, the transi-tion to biogenic resources demonstrates significant potential to reduce land use de-mand, though challenges remain concerning agricultural utilization and production efficiency (Cherubini, 2010). In addition, one-to-one replacements are often impossible which leads to the need to rethink module and product-concepts.

History – and economic research – demonstrates that the availability and utiliza-tion of natural resources are closely linked to economic development and technologi-cal progress (Tilton, 2003). While earlier societies often relied on locally available re-sources, the modern economy is characterized by a globally interconnected network of raw material markets. However, the growing scarcity of certain resources necessi-tates the adoption of sustainable resource management strategies, including recycling, efficiency improvements, and the development of alternative materials. The future stability of the global economy will largely depend on the successful implementation of these measures and the extent to which a sustainable and resilient supply of raw materials can be ensured (UNEP, 2011).

Although the development of biogenic alternatives to fossil and mineral resources is steadily advancing, it still faces significant challenges. Technological innovation, political frameworks, and market conditions must continue to evolve in order to support the transition toward a more sustainable and resource-efficient economy. In light of the growing urgency surrounding climate change and resource scarcity, biogenic resources are expected to play an increasingly important role in the global economic landscape.

https://doi.org/10.1515/9783112218747-003

3.2 Classifications

The targeted development of biogenic resources as substitutes for fossil and mineral raw materials has a long-standing tradition. Particularly since the 1970s, when the oil crisis and increasing resource scarcity began to disrupt global markets, efforts to develop alternative, renewable energy sources and raw materials have intensified significantly (Mittlefehldt, 2018). This crisis prompted a fundamental shift in the energy policies of many industrialized nations, which began seeking new sources to reduce their dependence on fossil fuels. It was during this period that the first serious scientific and industrial initiatives aimed at developing biogenic resources were launched (Mittlefehldt, 2018).

These early scientific and industrial initiatives laid the foundation for the current development of biogenic resources and contributed to establishing the concept of biofuels and bio-based plastics as genuine alternatives to fossil resources (see Figure 3.1). While the 1970s still presented numerous technical and economic obstacles, subsequent research efforts initiated during that time paved the way for the more intensive developments in subsequent decades, which are now reflected in modern biotechnology and sustainable materials (Cherubini, 2010). The 1970s can therefore be considered a turning point in the history of alternative raw materials, as they laid the groundwork for later advancements in biotechnology and renewable energy (Mittlefehldt, 2018).

In the 2000s, the focus on sustainability and climate protection intensified. As a result, the development of materials increasingly considered criteria such as the origin of raw materials (biogenic or fossil-based) and their biodegradability. Figure 3.2 illustrates this structure and will later be refined with classifications of plastic types. The two-dimensional nature of the diagram also suggests that some biogenic substances are chemically and physically indistinguishable from their fossil-based counterparts, while others represent entirely different materials. With regard to biodegradability, it is important to note that some forms of biological degradation – such as composting – do not occur under typical environmental conditions and require industrial composting facilities.

Biodegradability poses a particular challenge because certain plastics cannot be easily distinguished from one another, yet they require different disposal and recycling pathways. Combined recycling is not always feasible, which necessitates knowledgeable waste separation – something that often overwhelms consumers, both in terms of understanding and willingness to engage.

3.2.1 Bio-based, Non-biodegradable Plastics

Bio-based, non-biodegradable plastics include materials produced from renewable resources that lack the ability to biodegrade. They offer a sustainable alternative to fos-

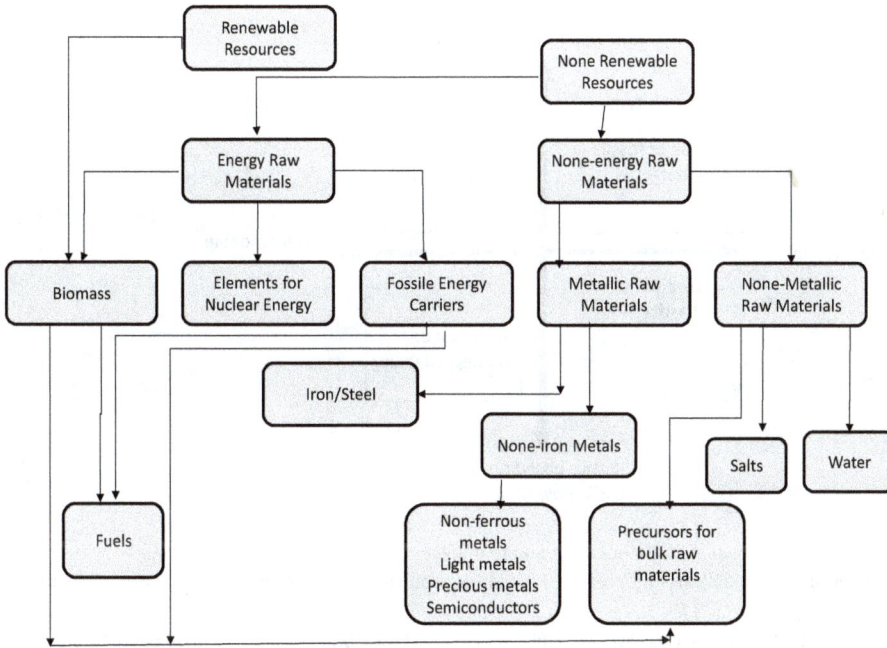

Figure 3.1: Classification of renewable and non-renewable resources and their derivative products (adapted from Angerer et al. (2016)).
Detailed schematic representing the classification of natural resources into renewable and non-renewable types, along with the spectrum of derivative products originating from each category. Renewable resources include biomass from agriculture and forestry, solar and wind energy, and water, all characterized by their capacity for natural regeneration. Non-renewable resources, such as coal, oil, natural gas, and mineral ores, are depicted as finite stocks that contribute to the production of conventional fuels, plastics, metals, and other industrial materials. The figure outlines how these inputs feed into manufacturing chains to produce bio-based products, fossil-based goods, and mineral-derived materials, thereby illustrating the environmental and economic implications of resource utilization patterns. This classification aids in understanding the trade-offs between resource availability, sustainability, and product life cycle impacts. It is worthwhile to be noted that at the time this classification scheme has been developed the option to use biomaterials as source of raw materials beyond serving as energy supplier was not widely accepted. Hence, in the figure links between renewable resources and iron/steel and non-iron metals are missing.

sil-based raw materials, but their disposal and recycling require the same systems as conventional plastics. Two prominent examples of this group are bio-polyethylene (Bio-PE) and bio-polyethylene terephthalate (Bio-PET):
– **Bio-polyethylene (Bio-PE)** is derived from sugarcane and has the same properties as conventional polyethylene (PE), produced from petroleum. Since Bio-PE shares the same chemical structure as its petrochemical counterpart, it can be used in the same applications, such as packaging, containers, and plastic films. By using sugarcane as the raw material, the dependence on fossil resources is reduced, as sugarcane is a renewable resource. However, the plastic itself remains

Plastics made from
renewable raw materials

Bioplastics   Bioplastics

No use of fossil
raw materials   Optimum

None   Biodegradable
Biodegradable

Traditional
Plastics

Advantage:
Biodegradability

Bioplastics

Plastics based on
fossile Raw Materials

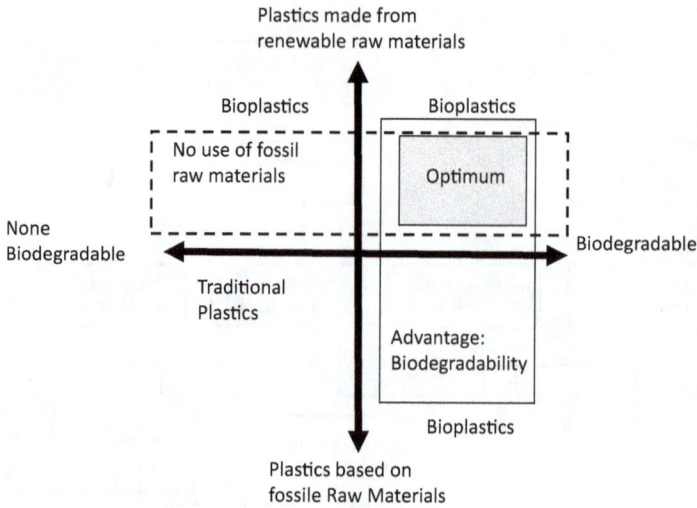

Figure 3.2: Classification of plastics according to origin and biodegradability.
Categorization of bioplastics with respect to feedstock source and biodegradability (H. Endres, A. Siebert-Raths, 2009).

Figure depicting the classification of plastics based on their origin – whether fossil-based or bio-based – and their biodegradability properties. The diagram categorizes plastics into distinct groups, illustrating that bio-based plastics can be divided into two main categories, each with distinct properties and areas of application: bio-based, non-biodegradable plastics and bio-based, biodegradable plastics. Both groups offer more environmentally friendly alternatives to conventional plastics derived from fossil resources such as petroleum or natural gas, thereby contributing to the reduction of the ecological footprint (Spierling et al., 2018). This classification framework helps clarify the complex landscape of plastic materials, providing insights into their environmental impact, potential for recycling or degradation, and relevance in addressing plastic pollution challenges through sustainable material design.

non-biodegradable, which means it relies on existing recycling systems to ensure sustainable disposal (Kiss et al., 2019).

– **Bio-polyethylene terephthalate (Bio-PET)** is a variant of traditional PET, partially made from biogenic raw materials such as corn or sugarcane. It is primarily used for the production of beverage bottles, such as the well-known Coca-Cola PlantBottle™. Bio-PET offers the advantage of relying largely on renewable resources, while still being fully recyclable. This characteristic enables a circular economy, in which Bio-PET can be reused through established recycling processes, further enhancing its environmental compatibility (Spierling et al., 2018).

3.2.2 Bio-based, Biodegradable Plastics

The second main category of bio-based plastics includes materials that are not only made from renewable resources but are also biodegradable. These plastics break down under natural conditions, making their disposal significantly more environmentally friendly, as they do not persist as plastic waste in the environment. Two widely used bio-based biodegradable plastics are polylactic acid (PLA) and polyhydroxyalkanoates (PHA):

–   **PLA** is a bio-based plastic produced from corn starch or sugarcane through a fermentation process. PLA is biodegradable and is commonly used in the packaging industry, textiles, and even 3D printing materials. Due to its degradable nature and the ability to be produced from renewable resources, PLA is considered particularly environmentally friendly. However, its compost ability only occurs under specific conditions, which implies that in some applications, it still relies on existing disposal systems (Pérez et al., 2020).

–   **PHAs** are a group of polyesters produced fermentative by certain microorganisms from plant-based raw materials. PHA is biodegradable and is increasingly used in single-use packaging, medical applications, and even disposable tableware due to its versatility. PHA has the advantage of being not only biodegradable but also capable of decomposing in a variety of environments, including marine settings, making it a particularly sustainable solution in the fight against plastic waste (Koller et al., 2017).

Bio-based plastics offer a wide range of alternatives to conventional plastics derived from fossil resources. Bio-based, non-biodegradable plastics such as Bio-PE and Bio-PET help replace fossil raw materials while meeting the same functional requirements and being compatible with existing recycling systems. On the other hand, bio-based, biodegradable plastics like PLA and PHA provide an environmentally friendly solution, composed not only of renewable resources but also offering more sustainable disposal through their ability to biologically decompose. Both groups contribute to reducing the CO_2 footprint of the plastic industry and provide means for a more sustainable future.

Biogenic raw materials can be broadly classified according to the criteria of . . .

–   Biogenic raw materials and bio-based feedstocks (Chapter 4)
–   Substitution of mineral raw materials (Chapter 4)
–   Energy carriers (Chapter 5)

This distinction is accounted for in the following (see Figure 3.3), with the note that some substances can be classified into both categories.

The distinction between raw materials and feedstocks is illustrated in the following representation:

When selecting and developing bio-based materials, the specific requirements and performance parameters of each application are carefully considered. For instance, in the production of bio-based plastics, special attention is given to factors such as formability, stability, water resistance, and mechanical strength to achieve performance comparable to petroleum-based plastics.

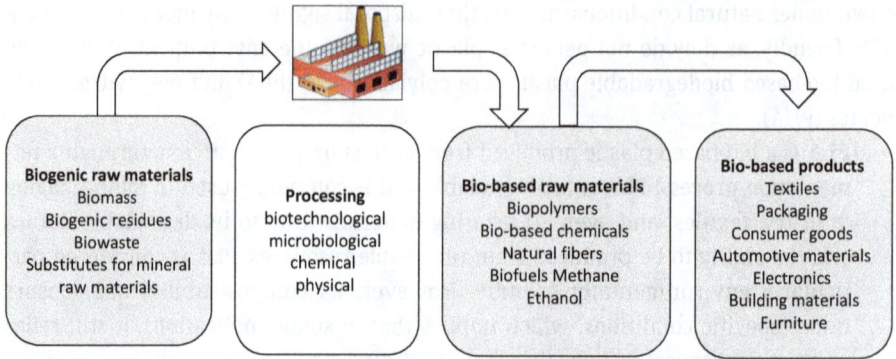

Biogenic raw materials	Processing	Bio-based raw materials	Bio-based products
Biomass	biotechnological	Biopolymers	Textiles
Biogenic residues	microbiological	Bio-based chemicals	Packaging
Biowaste	chemical	Natural fibers	Consumer goods
Substitutes for mineral	physical	Biofuels Methane	Automotive materials
raw materials		Ethanol	Electronics
			Building materials
			Furniture

Figure 3.3: Distinction between biogenic raw materials and feedstocks and their pathway into production. Figure illustrates the distinction between biogenic raw materials and feedstocks to raw materials alongside their respective pathways into the production process of bio-based products. The diagram sketches how raw materials – such as biomass derived directly from plants, animals, or organic waste – to processed or intermediate materials used as inputs in manufacturing bio-based products. The flowchart traces the transformation steps, starting from the sourcing of renewable biogenic resources, often through various stages of processing and refinement, ultimately leading to the creation of bio-based products. The pathway visualization highlights the sequential processes by which these raw inputs are converted, through biochemical or thermochemical methods, into final bio-based products. This conceptual map serves to clarify terminology and production flows.

An example of this is polylactic acid (PLA), which is derived from corn starch or whey and is currently used primarily for plastic packaging such as yogurt cups. The usability of pure PLA is limited due to its low softening point (around 60 °C), but it can be improved through blends (Fraunhofer Institute for Molecular Biology and Applied Ecology, n.d.). A similar situation occurs in the development of bio-based construction materials or fibres, where strength, thermal insulation, and moisture regulation are considered key properties for use in the construction or textile industries.

The EU proposes a classification of biogenic materials based on their origin, as indicated in Figure 3.4.

Another method for classifying biogenic raw materials is their production process, as shown in Figure 3.5.

One could also classify biogenic resources from a functional point of view – in the sense of properties. Often in product design certain properties of materials are used as a starting point for development while pricing traditional materials serves as additional reference (see Table 3.1).

3.3 Examples for Biogenic Substances and Raw Materials

Research on biogenic raw materials as substitutes for fossil and mineral resources has gained significant momentum in recent decades. These approaches aim to reduce

	Biomass containing oil and fat	Sugar and starch- containing biomass	(Ligno)cellulosic biomass	Mixed resource intermediate product
Biogenic Main Products	Oilseeds · Damaged plants	CerealsSugar beet		Cyanobacteria Algae
	Palm oil	Catch crops and plants from heavily degraded areas		
Biogenic Sideproducts	Animal fats		Straw · Cellulose-containing non-food material	Partly other material containing ligno-cellulose
Biogenic waste (and remains)	Used Oils			Liquid manure, sewage sludge · Municipal wastewater
	Biogenic industrial waste		Biogenic municipal waste	

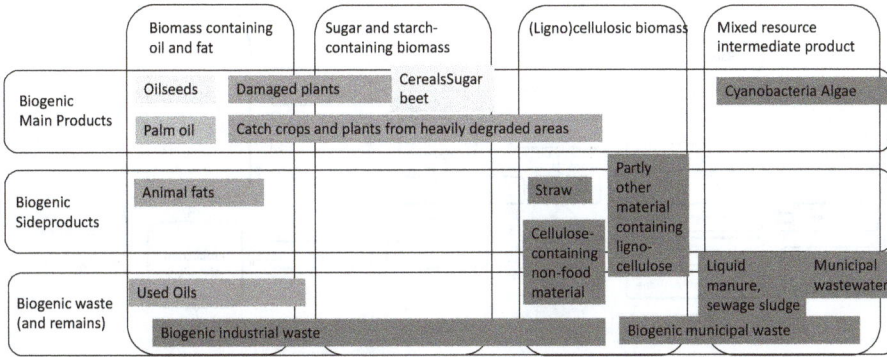

Legend: Bio-based energy sources According to Directive (EU) 2018/2001 and Deliberated Directive (EU) 2024/1405

Food and fodder plants · (EU) 2018/2001 IXA · (EU) 2018/2001 IXB

Figure 3.4: Classification of biogenic raw materials (columns) and bio-based products according to EU Directives 2018/2001 and 2024/1405 (as per Naumann et al., 2025).
Classification diagram showing the organization of biogenic raw materials (represented in columns) and their corresponding bio-based product categories, following the guidelines and criteria set forth by the European Union's Directives 2018/2001 and 2024/1405. The figure details the different types of renewable biological inputs, such as agricultural biomass, forestry resources, and biodegradable waste, and links them to product classifications based on bio-based content and sustainability performance standards. This visual summary facilitates a clearer understanding of how the EU's regulatory framework categorizes and governs bio-based products, emphasizing the importance of source material origin and compliance with updated legislative definitions aimed at promoting environmental sustainability and reducing fossil-based resource dependency.

dependence on finite, environmentally harmful raw materials in various application areas:
–   Biogenic raw materials and bio-based feedstocks,
–   Substitution of mineral raw materials, and
–   Energy carriers.

and to develop more sustainable alternatives (Tritscher et al., 2018). In particular, research focuses on bio-based plastics, that is, polymers either derived from renewable resources or produced through biotechnological processes.

Examples of this include:

PLA and PHA: PLA is produced from lactic acid, which is fermented from plant sugars (usually derived from corn starch or sugarcane) (Arias et al., 2020). PLA is biodegradable and is already used in the packaging industry as well as in medicine (e.g. for resorbable surgical sutures) (Koller et al., 2017).

Another promising biogenic polymer is PHA, which is produced by microorganisms through fermentation (Lee et al., 2021). PHA is characterized by high bio-

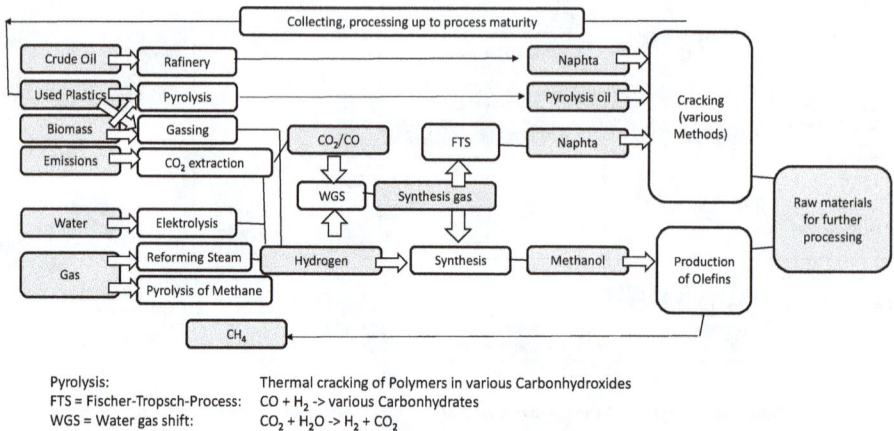

Pyrolysis: Thermal cracking of Polymers in various Carbonhydroxides
FTS = Fischer-Tropsch-Process: CO + H_2 -> various Carbonhydrates
WGS = Water gas shift: CO_2 + H_2O -> H_2 + CO_2

Figure 3.5: Classification of various hydrocarbons.
Comprehensive classification of various hydrocarbons derived from fossil raw materials, organized according to their origin, primary processing methods, and pathways toward high-value and further chemical processing. The figure categorizes hydrocarbons into major groups and traces their production from crude oil, used plastics, Biomass, emissions and gas through refining and petrochemical conversion processes – including distillation, cracking, reforming, and synthesis. It highlights the transformation of base hydrocarbons into intermediate and high-value chemical products such as ethylene, propylene, benzene, toluene, and xylene (BTX), which serve as foundational building blocks for plastics, solvents, synthetic fibers, and other industrial materials. The diagram also outlines secondary refining stages and specialty chemical production routes, emphasizing the central role of hydrocarbons in the fossil-based chemical value chain. This classification underscores the complexity and integration of fossil fuel processing systems.

degradability and can be used in a variety of applications, from packaging to medical devices.

Lignocellulose as a raw material: Lignocellulose is the main component of plant cell walls and consists of cellulose, hemicellulose, and lignin. It is the primary component of the biomass of many plants and a potentially vast source for the production of biofuels, biochemicals, and biomaterials (Himmel et al., 2007). Lignocellulose is particularly interesting because it is found in large quantities in agricultural waste (e.g. straw and wood chips) and even in so-called non-food crops like *Miscanthus* or nettles (Börjesson & Mattiasson, 2014).

The challenge that research is trying to address is to efficiently convert lignocellulose into valuable products, as its structure is difficult to degrade, and it requires complex processes during conversion (Sanchez et al., 2017). Research is primarily focused on finding ways to convert lignocellulose into sugars and other chemicals without the use of fossil energy carriers and with minimal environmental impact, which can then be used for bioplastic production or biochemicals (Liao et al., 2016). Lignocellulose-oriented bio refining processes aim to process the entire biomass in a cascading utili-

Table 3.1: Properties of traditional materials often used as a starting point for the use of biogenic raw materials.

Property	Detail	Examples
Mechanical properties	**Tensile strength**	Materials like hemp or flax have high tensile strength and can be used in composite materials (e.g. hempcrete and hemp composites) to replace materials such as steel or aluminium in lightweight structural components or vehicle constructions.
	Compressive strength	Materials such as concrete or wood are often replaced by biogenic building materials. Hempcrete, for example, offers low compressive strength for lightweight building applications but provides high thermal insulation and CO_2 storage.
	Bending stiffness	Biogenic fibres such as flax and hemp are increasingly used as reinforcement fibres for plastic composites or lightweight components to achieve bending stiffness similar to fibreglass or carbon fibre.
Thermal properties	**Heat resistance**	In the electronics industry or building materials, biogenic plastics or biogenic composite materials could provide heat resistance or temperature stability similar to PET or polycarbonate. Biogenic plastics like PLA (polylactic acid) offer moderate heat resistance, though research aims to improve this.
	Thermal conductivity	Biogenic insulation materials (e.g. hempcrete and flax insulation) offer low thermal conductivity, replacing mineral insulation materials like Styrofoam or mineral wool. These materials also provide advantages in CO_2 storage and sustainability.
Chemical properties	**Corrosion resistance**	Biogenic materials like hemp or flax are corrosion-resistant in specific applications, making them suitable as replacements for metals in the construction and automotive industries. These materials are chemically inert and are less affected by moisture or chemicals than some mineral or fossil materials.
	Reactivity	In the construction industry, replacing cement with biogenic raw materials like hemp and flax can reduce reactivity in alkaline environments. Cement reacts with water during hydration, releasing CO_2. Biogenic alternatives like hempcrete have CO_2 emission-free benefits and are pH-neutral.

Table 3.1 (continued)

Property	Detail	Examples
Optical and aesthetic properties	Colour and transparency	Biogenic plastics like PLA can exhibit similar optical properties to conventional plastics (e.g. PET) through colouring and processing. These properties are especially important in the packaging industry, where optical appeal is a key criterion.
	Texture and surface	Biogenic textiles like hemp or flax offer high breathability, durability, and a natural surface texture, making them suitable for the fashion industry. These textiles can serve as eco-friendly alternatives to cotton or synthetic fibres.
Electrical and magnetic properties	Electrical conductivity	In some applications, such as batteries or super capacitors, biogenic materials like algae or sugarcane products could serve as electrolytes or electronic components, replacing materials like lithium or cobalt. Biogenic materials like graphene from biomass or biogenic polymer capacitors could offer enhanced conductivity.
	Magnetic properties	In the media industry or for the production of magnetic materials, biogenic alloys with enhanced magnetic stability may have the potential to replace minerals like neodymium. However, biogenic materials in these fields are still in the development phase.
Physical properties	Density	Many biogenic materials like hemp and flax have low density, making them attractive lightweight materials for the automotive and aerospace industries. These materials offer similar strength properties to metals but at lower weight, leading to reduced fuel consumption and CO_2 emissions.
	Strength	Biogenic plastics and fibres (e.g. flax and hemp) achieve similar strength values to fossil-based plastics in many cases. They are already being used in the automotive and packaging industries as alternatives to polypropylene and polyethylene.
Functional properties	Biodegradability	One of the main advantages of biogenic materials like PLA or PHA is their biodegradability. Unlike fossil plastics, which take hundreds of years to decompose, biogenic plastics break down within months to years without causing long-term environmental harm.
	CO_2 sequestration	Biogenic materials like hempcrete or wood can act as carbon sinks, thus contributing to a positive climate balance. These materials absorb CO_2 during their growth, which helps make building projects more climate-friendly.

Table 3.1 (continued)

Property	Detail	Examples
Sustainability and resource conservation	Water consumption	Biogenic materials like hemp or flax have significantly lower water requirements compared to conventional crops, enhancing the sustainability of their cultivation and conserving water resources.
	Energy consumption	The production of biogenic plastics and textiles often requires less energy than the production of fossil plastics or metals. For example, PLA production consumes less energy compared to conventional PET production.

The development of biogenic raw materials as substitutes for mineral or fossil-based materials requires careful consideration of the material properties, the technical feasibility of manufacturing processes, and an assessment of ecological impacts to provide long-term sustainable solutions. It is not a single quality that is crucial, but their complexity and combinatorial possibilities.

zation approach, allowing multiple value-added products (such as sugars, bioethanol, bioplastics, and other chemicals) to be derived from the same raw material (Awasthi et al., 2020).

3.3.1 Renewable Resources for Energy Generation and Transport

The use of bioenergy from biogenic raw materials represents one of the most important approaches to replacing fossil fuels. Among the most significant sources of bioenergy are biomass (e.g. wood and agricultural waste) and algae. Bioethanol and biogas are produced from organic materials through fermentation or digestion and can be used as direct fuels or in other energy-intensive sectors such as transportation. Algae, in particular, offer enormous potential due to their high growth rate and low land usage.

Research in this field focuses on improving crop yields and optimizing conversion processes to enhance the efficiency of bioenergy production. The development of dual-purpose biomass utilization methods, which not only generate energy but also high-value chemicals and bioplastics, is also being increasingly promoted. In the long term, bioenergy could play a crucial role in the global energy mix, as it may provide a climate-neutral and sustainable alternative to fossil fuels.

Current research focusses on (Sánchez et al., 2020)
– Improvement of production efficiency,
– Reduction of production costs, and
– Expansion of the application scope of the produced materials.

3.3.2 Biotechnological Processes

Biotechnology plays a key role in the development of biogenic raw materials. Microorganisms such as bacteria, fungi, and algae are increasingly being used to produce biochemical substances that are traditionally derived from fossil resources (López-Gallego et al., 2012). A notable example is the production of bioethanol and biodiesel through microbial fermentation, where sugars or vegetable oils are converted by microorganisms into process able and valuable fuels (Zhang et al., 2020). The production of biochemicals such as acetone, butanol, and lactic acid via microbial fermentation is also being intensively researched (Yang et al., 2019).

Another emerging field is synthetic biology, where microorganisms are genetically modified to produce highly specialized products that are currently made from fossil resources (Shi et al., 2022). The biological production of polymers (e.g. PHA or PLA) and pharmaceutical active ingredients offers potential for a more sustainable and environmentally friendly production process. Research is focused on increasing the productivity and efficiency of microorganisms, optimizing resource utilization, and reducing the costs of bioproduction (Vargas-Tah et al., 2021).

3.3.3 Waste Management and Upcycling

One research approach focuses on utilizing waste streams as a source of raw materials. This includes the recycling and upcycling of materials traditionally considered as waste, transforming them into new, high-quality products. An interesting example is the use of wood waste or straw as feedstock for the production of biogenic plastics, bioenergy, or chemical intermediates. Wood waste, like algae, is a potentially valuable raw material for the production of biomaterials and biogenic plastics due to its high cellulose and lignin content.

Another key area of research is the utilization of waste streams specifically from the food industry. These wastes, which often contain organic substances such as sugars, starch, or vegetable oils, can be converted into biochemicals, biodiesel, or bioplastics. The upcycling technology aims to use these materials as efficiently as possible while simultaneously reducing the volume of waste.

The bendability of material properties of biogenic plastics are also a focus of research to enhance their performance, making them suitable for a wide range of industrial applications (Cheng et al., 2018).

In order to effectively implement alternative materials and production methods in industrial practice, two additional factors, alongside technological feasibility and environmental aspects, must be considered: scalability and economic efficiency. These two aspects play a crucial role in how sustainable alternatives can gain widespread adoption and compete long term with conventional raw materials.

Another important research topic concerns the challenges in scalability and costs. While many biogenic raw materials show promise in pilot projects or small-scale production, questions arise regarding production costs, raw material availability, and the long-term sustainability of production processes when it comes to broader market acceptance.

Furthermore, research is being conducted on how land use and fertilizer requirements can be optimized to maximize resource utilization in the production of biogenic raw materials and avoid conflicts with food production. Sustainable and efficient use of biotic resources is urgently needed to reduce the environmental impact of production and meet the growing demand for biomass amid limited availability (Environmental Protection Agency, 2021).

3.4 Scalability, Availability, and Economic Viability

A key research topic in the field of biogenic materials is the scaling of production, as current production methods are often significantly more expensive than the production of petroleum-based plastics (Liu et al., 2019), which may make their use economically unfeasible at present.

3.4.1 Scalability of Alternative Raw Materials and Technologies

Scalability refers to the ability to transfer new technologies or materials, initially developed on a small or experimental scale, to industrial production levels. This presents companies and researchers with various challenges that need to be overcome in order to enable efficient and cost-effective mass production. Several factors must be taken into account, including the availability of raw materials, infrastructure, and technological requirements.

Availability and Raw Material Base

A key aspect of scalability is the availability of the required raw materials. Fossil resources such as petroleum are globally abundant and can be relatively easily and consistently utilized for various industrial purposes. In contrast, many sustainable alternatives are still available on a much smaller scale, and their production has not yet reached the necessary economic profitability.

An example of this situation is hydrogen technology. The use of green hydrogen as an energy carrier is considered one of the most promising alternatives to fossil fuels. However, scaling up electrolyser capacities – the process that splits water into hydrogen and oxygen – requires a significant increase in production capacities. Currently, the production of green hydrogen is still expensive and limited to relatively

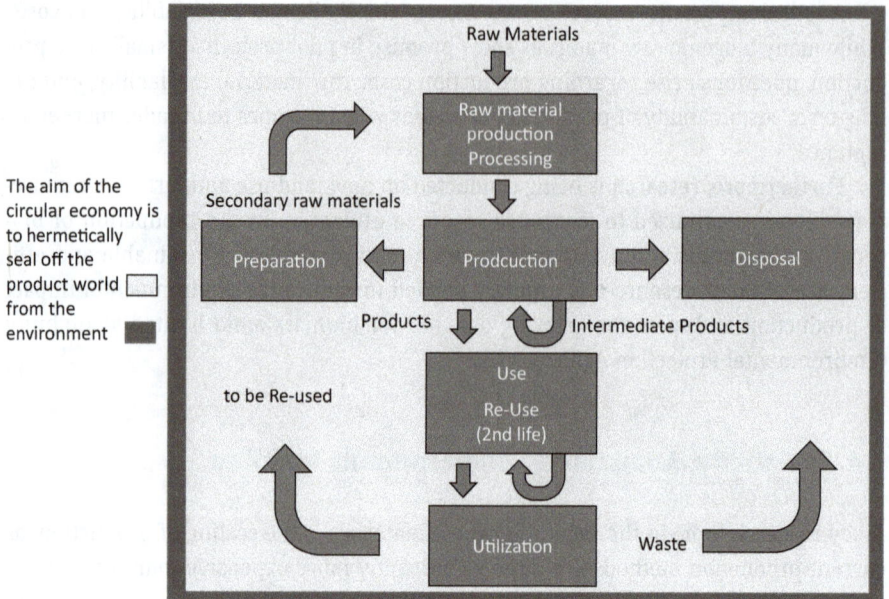

Figure 3.6: Resource use in the life cycle, exemplified by material flows.
Illustration of resource use across the life cycle of a product, exemplified through representative material flows from raw material extraction to end-of-life processing. The figure maps the sequential stages of a typical product life cycle – including resource extraction, material processing, manufacturing, use and re-use, to disposal or recycling (via preparation to Re-use the now secondary raw materials). If applied to a specific material and by tracing material flows, the figure provides insight into the cumulative resource intensity of products and systems, identifying critical points for efficiency improvements, substitution, or circular interventions. The depiction reinforces the importance of life cycle thinking in environmental assessment and sustainable product design, where minimizing resource use, maximizing recyclability, and extending product longevity are key strategies to reduce ecological footprints and promote circular economy principles.

small quantities. To make the technology economically competitive, production must be expanded to a much larger scale, which requires substantial investments in infrastructure and research. According to the International Renewable Energy Agency (IRENA), green hydrogen production could become more cost-effective through scaling and technological improvements, potentially competing with fossil alternatives by 2030.

Recycling

Recycled materials also present a challenge for scalability. Although recycling of plastics and metals is considered an important component of the circular economy, the recovery and sorting of industrial and consumer waste is still not efficient enough to achieve near-complete substitution of primary raw materials (see also Figure: resource use in the life cycle, exemplified by material flows). For example, plastic waste must be sorted into different types, which is a significant challenge due to the variety of plastics and often insufficient recycling infrastructure. Similarly, there are obstacles in metal recycling related to the purity and quality of the recycled materials, making it difficult to use them as direct raw material alternatives.

The proportion of recycled raw materials varies significantly between different countries and regions. In the European Union (EU), an average of 41% of plastic packaging waste was recycled in 2022 (Eurostat, 2024a). Slovakia achieved a recycling rate of 60%, followed by Belgium at 54%, and Germany and Slovenia at 51% each. In contrast, the recycling rate in Malta was just 16%, in Denmark 23%, and in France and Austria, it was 25% each (Eurostat, 2024a).

In 2021, the EU generated an average of 34.6 kg of plastic packaging waste per person, of which 13.0 kg was recycled (Eurostat, 2022). Between 2010 and 2020, the amount of plastic packaging waste per capita increased by 23%, while the recycled volume rose by 32% during the same period. Despite these advancements, the non-recycled amount increased by 3.4 kg per person (Eurostat, 2022).

These differences can be attributed to various factors, including the industries generating the waste, as well as the efficiency of the respective waste management systems. Countries with high recycling rates often implement strict recycling regulations and effective collection systems.

It is important to note that recycling rates have declined in some countries in recent years. For example, the recycling rate in the EU decreased from 41% in 2019 to 38% in 2020, which can be attributed to stricter regulations regarding the reporting of recycling (Eurostat, 2022).

These declines in recycling rates in certain countries can be attributed to a variety of factors. Some of the key reasons include:

Changes in consumer waste behaviour: In many countries, consumer behaviour has shifted. The increasing use of single-use plastics and packaging materials that are more difficult to recycle (such as composite materials) complicates recycling efforts. Often, products or packaging are designed in ways that make it difficult to separate them into their individual components, thus hindering recycling.

Problems with waste collection and sorting: One of the greatest challenges for recycling is the inefficient collection and sorting of waste. In some countries, there are issues with the infrastructure for waste separation, making it more difficult to correctly sort recyclable materials. As a result, many materials end up in general waste instead of being recycled.

Changes in international waste trade: Since China significantly restricted the import of many recyclable waste materials (especially plastics) in 2018, many countries have faced difficulties exporting their recyclable waste. As a consequence, recycling costs have risen, and some waste has been sent to landfills or incinerated instead of being processed.

Lack of market demands and incentives: The demand for recycled materials can fluctuate significantly. When the price of recycled raw materials is higher than that of primary raw materials, companies have fewer incentives to invest in recycling technologies or use recycled materials. In some cases, this can lead to reduced effort in the recycling process.

Lack of innovation and technology: In many instances, recycling technologies have not advanced quickly enough to keep pace with increasingly complex waste streams. The development of new technologies to process even difficult-to-recycle materials, such as mixed plastics, is crucial. In the absence of such innovative solutions, recycling rates can decline.

Political and regulatory changes: Some countries have adjusted their recycling policies and regulations or have enforced them less stringently, which can lead to a decrease in recycling rates. Additionally, a lack of consistent or effective implementation of recycling policies may result in countries failing to meet their recycling targets.

Economic crises or pandemics: Economic crises or global challenges, such as the COVID-19 pandemic, have impacted many industries, including the recycling sector. During the pandemic, for example, there was an increased demand for single-use packaging, coupled with a slowdown in recycling processes due to restrictions or disruptions in the supply chain. This led to a decline in recycling activities.

All of these factors contribute to fluctuations and declines in recycling rates, highlighting the need to invest in sustainable waste management, innovative recycling technologies, and stricter regulations.

To ensure the scalability of sustainable alternatives in industry, the availability of raw materials and technological infrastructure are key factors. These challenges can be overcome through technological innovations, investments in infrastructure, and the creation of economies of scale to achieve a sustainable and resource-efficient industry:

Technological Scalability

A significant barrier to the industrial application of alternative materials is technological scalability. This involves adapting processes that work effectively on a small scale for large-scale production. The transition to mass production often requires substantial technical adjustments and incurs high costs, which can affect widespread market availability and economic competitiveness.

A particularly clear example of this issue is the circular economy in the chemical industry, especially the chemical recycling of plastics (see also Figure 3.6). Although chemical recycling is considered a promising technology that enables plastics to be returned to their original materials and recycled, its large-scale implementation has been limited so far. The processes of chemical recycling are often highly energy-intensive and require complex plant technologies that are not yet sufficiently available. This leads to high operating costs and complicates economic implementation on a larger scale. Currently, few facilities exist that can efficiently implement these processes in practice (Rahimi & García, 2017).

Another example of challenges in technological scalability can be found in the field of bio-based construction materials. Materials such as hempcrete or geopolymer cement offer a sustainable alternative to conventional types of concrete and cement, as they cause lower CO_2 emissions and consist of renewable raw materials or industrial by-products. Although these bio-based construction materials have already been successfully used in smaller markets or specialized construction projects, their widespread use is still highly limited. This is mainly due to the lack of infrastructure required for large-scale industrial production. The manufacturing of these alternatives requires specific production processes and resources that are often not available on a large scale, which complicates their cost-effective and efficient implementation in the construction industry (Pacheco-Torgal et al., 2013).

Implementation

In the ideal implementation process of biogenic substances and materials, the first step involves the provision of a material – either from an existing portfolio or as a newly designed compound. Subsequently, its compatibility with established processes is evaluated. These processes span the entire life cycle, including production, utilization, disposal, recycling, reuse, and biological degradation. A critical aspect of this assessment involves examining both the individual biodegradability of the material and its degradability in the context of other co-utilized substances. Biodegradability testing for plastics, in particular, often employs standardized protocols such as the OECD Test Guideline 301 B. This method is designed to determine whether a chemical substance can be classified as "readily biodegradable," which is a key environmental indicator (OECD, 1992). Under this test, the substance is incubated for up to 28 days in a closed system containing microorganisms derived from activated sludge – typically sourced from wastewater treatment facilities – along with a nutrient solution. The degradation process is monitored indirectly by measuring the amount of carbon dioxide (CO_2) produced, under the assumption that greater CO_2 evolution correlates with higher biodegradability. To meet the criterion for ready biodegradability, at least 60% of the theoretical maximum amount of CO_2 must be released within the 28-day period, with specific requirements regarding the degradation curve within the first ten days after reaching 10% mineralization. A positive result in the OECD 301 B test can facilitate compliance with regulatory frameworks in various jurisdictions, including the EU's REACH regulation (ECHA, 2023; OECD, 1992).

3.4.2 Economic Viability

Economic viability is a crucial factor for the widespread adoption of alternative materials and technologies. Even if these alternatives are technologically feasible and environmentally sustainable, they must also be economically competitive with conventional materials and processes. Several factors contribute to these higher costs, including the complexity of production processes, the need for specialized infrastructure, and the relatively smaller scale of current production compared to traditional materials.

To make alternative materials economically viable, it is necessary to achieve economies of scale, reduce production costs, and improve efficiency in the manufacturing processes. Additionally, government incentives, subsidies, or regulatory frameworks can play a key role in making sustainable alternatives more competitive. However, in the long term, the economic success of alternative materials depends on their ability to deliver comparable or superior performance at lower or comparable costs, as well as on their capacity to meet market demand while minimizing environmental impact.

An important aspect of the cost structure of alternative raw materials is that their production costs are often higher than those of traditional fossil or mineral materials. This cost disparity arises from several factors:

– **Higher raw material and production costs:** Biogenic materials, such as bio-based plastics or sustainable construction materials often require more elaborate and complex manufacturing processes compared to conventional products. These production processes are frequently more energy-intensive or demand specialized technologies, which leads to higher production costs. For example, the production of bio-polyethylene (Bio-PE) or PLA necessitates the use of renewable raw materials such as sugarcane or corn, which are more expensive compared to fossil feedstocks like petroleum or natural gas (Clark et al., 2017).

– **Lack of economies of scale:** Another significant factor is that alternative raw materials are often produced in small quantities. In contrast to fossil materials, whose production have been highly streamlined over decades of industrial processes and established infrastructure, many alternative materials are still in the development or pilot phase. This means they are not yet produced on a large scale and thus cannot benefit from the economies of scale that arise with mass production. Economies of scale lead to cost reductions per unit produced in established industries, as fixed costs are spread across more units with increasing production. However, this cost advantage has not yet been realized for alternative raw materials, which are produced in small quantities.

– **Infrastructure costs:** Introducing new raw materials and materials into industry often requires adjustments to existing production facilities. Many traditional production sites are designed for processing fossil materials and need to be upgraded to work with biogenic or recycled raw materials. These infrastructure costs can be very high, as new machinery, production processes, and possibly new logistics structures may be required. An example of this is the shift from conventional to sustainable production processes in industries such as construction or chemicals, where existing plants must be adapted to process bioplastics or bio-based construction materials (Friedrich & Almajid, 2013).

The economic efficiency of alternative raw materials is heavily dependent on production costs. These higher costs, which result from more expensive raw materials, the absence of economies of scale, and necessary investments in infrastructure, pose a significant barrier to making these materials competitively viable against traditional raw materials in the long term. Without targeted measures to reduce production costs or government subsidies that support the transition to the use of sustainable materials, it is challenging for companies to integrate these materials on a large scale and at low cost into their production processes. The long-term marketability of alternative raw materials is largely dependent on their price development. While fossil raw materials such as oil and gas often experience significant price fluctuations, biogenic materials are often influenced by agricultural yields and the related fluctuations in supply and price. This dependency can affect the price stability of alternative raw materials and, thus, determine their competitiveness in global markets. The marketability and price development of alternative raw materials are influenced by various interconnected factors:
- price fluctuations of fossil raw materials,
- political framework conditions, and
- technological developments.

- **Price fluctuations of conventional raw materials:**
  One of the largest influencing factors on the price development of bioplastics and other biogenic materials is the price fluctuation of fossil raw materials, especially oil. When oil prices fall, this can significantly impact the competitiveness of bioplastics, as the production costs of bio-based materials are often higher than those of fossil alternatives. For example, when oil prices decline, the production costs of bio-polyethylene (Bio-PE) or PLA could become less competitive compared to petroleum-based plastics, which may reduce market demand (Spierling et al., 2018). Such price fluctuations make long-term planning and investments in alternative raw materials more difficult for companies, as pricing is strongly dependent on global markets and geopolitical factors (Awasthi et al., 2020).
- **Political framework conditions:**
  Political decisions and frameworks play a crucial role in promoting the marketability of alternative raw materials. Governments can improve the economic attractiveness of biogenic materials through various measures. Subsidies and government funding programs for sustainable production methods or the use of bioplastics can offset the initial higher production costs and create incentives for

companies to invest in these technologies. Furthermore, carbon-pricing systems and stricter recycling quotas for plastic waste can further stimulate the market for biogenic materials by making fossil raw materials less attractive and increasing demand for more environmentally friendly alternatives (Aresta et al., 2019). Political decisions thus have a direct impact on price development and can help strengthen the competitive advantage of biogenic raw materials, especially when environmental protection and sustainability are promoted (Pereira et al., 2020).

– **Technological developments:**
Another key factor for the price development and marketability of alternative raw materials is technological progress. New production methods and improvements in process technology can lead to a reduction in the production costs of biogenic materials in the long run. Through innovative manufacturing technologies and optimization of production processes, it may be possible to increase production efficiency and reduce the costs of bioplastics or other biogenic materials (Huang et al., 2021). Technological breakthroughs in biotechnology and the circular economy, such as more efficient recycling processes or the development of bioplastics from new raw material sources, could also contribute to cost reduction and thus improve marketability. Advances in production could offset the price advantage of fossil raw materials and accelerate the transition to more sustainable alternatives (Cherubini et al., 2016).

A stable political support, technological innovations, and the ability to compete with fossil fuels on global energy markets are crucial for the success of substituting mineral and fossil raw materials with sustainable alternatives.

3.4.3 Measures to Gain Attractivity for Biogenic Raw Materials

To make alternative raw materials economically competitive, various measures can be implemented to ensure not only the technological feasibility but also the economic efficiency of these materials. The economic scaling of alternative materials is a complex challenge influenced by numerous factors. One of the key measures to promote successful scaling is targeted investment in research and development (R&D). Through investments in R&D, significant advancements can often be made that not only reduce production costs but also open up new application areas for alternative materials. This is a crucial step in strengthening the competitiveness of biogenic or recycled materials on the global market and ensuring their long-term establishment (Arista et al., 2019; Worrell & Reuter, 2020).

Promotion of R&D

Investments in R&D of new technologies and processes are crucial to overcoming the challenges of economic scaling. Various research fields play a key role in this regard:

- **Improvement of recycling processes:**
  A key approach to reducing costs and increasing the efficiency of recycling processes is the further development of sorting discipline (to raise the purity of the source) and technologies and chemical recycling (often accepting the price of higher concentration of chemicals used). Advances in these areas can lead to higher recovery rates by making the separation and recycling of materials more efficient. Chemical recycling, in which plastics are broken down into their basic building blocks and then processed into new plastics, offers significant potential in this regard. By advancing these technologies, not only can costs be reduced, but the material's resource retention could also be significantly increased. This would, in turn, reduce the demand for primary raw materials and strengthen the circular economy (Jehanno et al., 2022). More intensive research in this area could lead to recycling techniques that can be efficiently and cost-effectively applied on a large scale, greatly improving the competitiveness of recycled materials compared to fossil raw materials (Garlotta, 2018).

- **Optimization of biogenic materials:**
  Another area that often can be significantly advanced through research is the optimization of biogenic materials production, such as bioplastics. The use of new biotechnological processes, as well as the breeding of plant species that provide particularly high yields for bioplastic production, could play a crucial role in reducing production costs. In particular, methods for producing polymers from renewable raw materials such as corn or sugarcane could be made more efficient through genetic modifications and biotechnological innovations, further reducing manufacturing costs (Rujnić-Sokele & Pilipović, 2017; Singh et al., 2021). Such advancements would not only help expand the production capacity for bioplastics but also increase the availability and reliability of the raw materials required for their production (Martin et al., 2020).

It can safely be stated that promoting R&Din various areas, from recycling technology to the optimization of biogenic raw materials, plays a key role in the economic scaling of alternative materials. Targeted investments in these research fields can both reduce production costs and open up new and more efficient applications for sustainable materials. These advancements are crucial in establishing alternative raw materials not only as an environmentally friendly option but also as an economically competitive alternative.

Political Measures to Promote Alternative Raw Materials

The economic scaling of alternative raw materials is heavily dependent on the regulatory framework shaped by political decisions. The right political support can significantly contribute to making sustainable materials not only technically feasible but also economically competitive. Targeted political measures can create incentives that increase the demand for alternative raw materials while expanding the production capacities and infrastructure required for their manufacture. Various political instruments are particularly well-suited for this purpose, including carbon pricing, subsidies for sustainable production methods, as well as recycling quotas and circular economy strategies (Georgiou & Manousakis, 2019; Fu et al., 2020).

Carbon Pricing: An important political lever for promoting sustainable materials is carbon pricing. By imposing higher taxes on fossil raw materials or implementing a targeted CO_2-tax on emissions from CO_2-intensive materials, the costs of fossil raw materials could significantly increase; thereby improving the competitiveness of bio-based or recycled materials. This measure would raise the production costs of fossil materials, thus reducing the price advantage that conventional materials currently have over sustainable alternatives. At the same time, it would incentivize businesses and industries to transition more toward sustainable materials to save costs and benefit from tax incentives. Carbon pricing is therefore a crucial measure for promoting a more environmentally friendly and sustainable production process, strengthening the market for eco-friendly raw materials in the long term, and reducing dependence on fossil fuels (Huang et al., 2021; Stede et al., 2021).

Subsidies for Sustainable Production Methods: Another important political tool for promoting alternative raw materials is the provision of start-up assistance for sustainable production methods. Financial incentives in the form of grants or tax benefits can encourage companies to invest in innovative and environmentally friendly technologies that enable scalable production of sustainable raw materials. These financial support measures could, especially in the early stages of developing and establishing new technologies, ease market entry and offset the initial investments in modern production facilities and infrastructure. Such subsidies would be particularly crucial in areas such as bioplastics, renewable energy, and the circular economy, to enhance competitiveness against conventional, fossil-based materials.

Recycling Quotas and Circular Economy Strategies: Another important political tool for promoting alternative raw materials is the implementation of recycling quotas and circular economy strategies. The introduction of stricter recycling requirements could significantly increase the availability of secondary raw materials by obligating companies to use more recycled materials. This could be achieved by setting recycling quotas, which would encourage companies to either source more recycled materials or develop their own recycling processes. At the same time, promoting circular economy strategies that extend the life cycle of materials and maximize their reuse could reduce the need for primary raw materials and minimize waste generation. Political measures that promote recycling and the creation of a closed-loop material cycle therefore have the potential to not only make alternative raw materials economically competitive but also position them as an integral part of a country's or company's modern sustainability strategy.

In summary, political measures such as CO_2 pricing, subsidies for sustainable production methods and recycling quotas can play a crucial role in promoting the economic scaling of alternative raw materials. If applied well these measures can reduce the production costs of alternative materials and enhance their competitiveness against conventional raw materials. Political support thus provides the necessary foundation to establish sustainable materials not only from an ecological perspective but also from an economic standpoint in the global market over the long term.

Chapter 4
Biogenic Raw Materials and Replacement of Mineral Resources

4.1 Overview

The growing scarcity of fossil and mineral raw materials, as well as the need for a sustainable economy, has led to an increased focus on researching and testing alternatives based on bio-based resources in recent decades. Particularly in the fields of plastics and fuels, there are promising bio-based solutions that can help reduce the consumption of fossil raw materials and lower CO_2 emissions. As described in the introduction, the use of bio-based raw materials in various industrial sectors has been outlined below (see Table 4.1).

Table 4.1: Contribution of bio-based raw materials to resource supply in various industries.

Industry sector	Share of biogenic ingredients worldwide	Typical biogenic raw materials
Pulp and paper industry	~95–100%	Wood, cellulose, and waste paper
Textile industry	~30–40%	Cotton, linen, hemp, and viscose
Construction industry	~10–15%	Wood, cork, straw, and natural insulation materials
Chemical industry	~10–15%	Starch, sugar, vegetable oils, cellulose, and glycerine
Cosmetics and personal care	~20–30%	Vegetable oils, fats, extracts, and essential oils
Plastic industry	~2–5%	Bioplastics made from starch, PLA, PHA, and cellulose
Energy (biomass and biogas)	~15–20% (higher regionally)	Wood, energy crops, manure, and organic waste
Pharmaceutical industry	~5–10%	Plant-based active ingredients, fermentation products

In light of the increasing scarcity of mineral and fossil resources, along with the associated ecological and economic challenges, the use of bio-based raw materials as an alternative ingredient is becoming increasingly significant. Bio-based raw materials, in particular, offer promising substitution potentials in fields such as materials science, energy supply, and the chemical industry (Zhao et al., 2021). However, there are

https://doi.org/10.1515/9783112218747-004

also technological, ecological, and economic limitations that make the complete replacement of conventional raw materials challenging (Yang & Liu, 2021).

4.1.1 Bio-based Plastics as an Alternative to Fossil-Based Plastics

Bio-based raw materials have been the subject of research projects since at least the 1970s. The focus has been on a wide range of materials that could potentially replace fossil-based plastics. The key areas of focus include:

- **Polylactic Acid (PLA)**: PLA is a biodegradable plastic derived from sugarcane, corn, or other plant-based raw materials. It possesses properties similar to conventional plastics but is produced from renewable resources and is fully biodegradable. PLA is primarily used in packaging, textiles, and medical applications (Ruthven et al., 2020).
- **Polyhydroxyalkanoates (PHAs)**: PHA is a family of polymers produced by microorganisms from organic waste. These plastics are biodegradable and find applications in the packaging industry, as well as in agriculture and medical fields. Research focuses on optimizing the production processes and modelling the material properties of different types of PHA (López et al., 2021).
- **Polybutylene Adipate Terephthalate (PBAT)**: PBAT is another biodegradable plastic used as an alternative to non-biodegradable plastics like PET. PBAT is often combined with PLA to improve the flexibility and process ability of bioplastics. Research is focused on improving mechanical properties and developing environmentally friendly production processes (Li et al., 2021).
- **Biodegradable Polyamides (Bio-PA)**: Bio-polyamides are made from renewable raw materials such as plant oils or sugars. They offer an alternative to traditional petrochemical polyamides, which are used in textiles, automotive parts, and electronics. Research on bio-PA aims to enhance their strength, durability, and resistance (Schwager et al., 2020).
- **Starch-Based Plastics**: Plastics made from natural starches, such as potato or corn starch, are biodegradable and can be used in packaging, disposable cutlery, and other applications. However, these plastics are often less stable than their petrochemical counterparts, and research focuses on improving their strength and resistance to moisture and temperature (Papageorgiou et al., 2021).
- **Cellulose-Based Plastics**: Cellulose, the main component of plant cell walls, is a promising material for bioplastics. It can be used in various forms, including cellulose acetate and other modifications, for applications in packaging, films, and textiles. Research is exploring how cellulose can be used more effectively and cost-efficiently for plastic production (Hafner et al., 2021).
- **Lignin-Based Plastics**: Lignin, a complex organic polymer found in plant cell walls, is increasingly being researched as a raw material for bioplastics. Lignin has the potential to serve as a sustainable alternative to petrochemical plastics, as

it is abundant, inexpensive, and chemically resistant. The challenge lies in making lignin accessible in a form suitable for industrial applications (Dourado et al., 2020).

– **Biogenic Elastomers**: Biogenic elastomers, made from renewable raw materials such as plant oils or types of rubber, are a promising alternative to synthetic rubbers. They are used in the tyre industry, seals, and medical devices. Research focuses on improving the elasticity, durability, and process ability of these materials (Möller et al., 2020).

Although the relative market share of these materials in the plastic market is in the low single-digit percentage range (see Table 4.2), the mentioned materials have already established themselves.

Table 4.2: Proportional consumption of bio-based raw materials (European Bioplastics, 2020).

Material	Weight percentage
PLA	26
Starch	20
PBAT	13
PE	12
PHA	5
Other biodegradable materials	5
Other non-biodegradable materials	20

Particularly, the limited recycling infrastructure for bio-based and biodegradable polymers currently represents significant hurdles (Singh et al., 2019). For fossil resources, nature took over the task to transfer the originally biogenic material to become a fossil resource. The price for speeding up this process is paid in terms of production cost. Due to this, costs are among the greatest challenges with bio-based plastics (Zhao et al., 2020).

4.2 Market Size and Market Development Estimates

The market for bioplastics has gained significant importance worldwide in recent years. These plastics, made from renewable resources such as plants and microorganisms, are considered an environmentally friendly alternative to fossil-based plastics, whose production is associated with high CO_2 emissions and a heavy dependence on non-renewable resources (European Bioplastics, 2020). The growing demand for sustainable products and increasing consumer awareness of environmental issues have significantly influenced the development of this market (MarketsandMarkets, 2020). In this context, it is crucial to examine both the current market size and the long-term market developments for bioplastics to provide a well-informed assessment of the prospects and challenges of this market.

The market size for bioplastics has grown considerably in recent years and is expected to continue experiencing growth. In 2020, the global bioplastics market was estimated at approximately 7.5 million tonnes, with an expected annual growth rate of 20% until 2026 (Grand View Research, 2020). This reflects increasing demand on both the production and consumer side. The main driver behind this growth is the shift of companies towards more environmentally friendly alternatives with less harmful environmental impacts (Grand View Research, 2020). Bioplastics are increasingly being used in industries such as packaging, automotive, construction, and healthcare (European Bioplastics, 2024). According to current forecasts, the global production capacity of bioplastics will rise from around 2.47 million tonnes in 2024 to 5.73 million tonnes by 2029, representing a doubling within 5 years (PlastXnow, 2023).

The share of bioplastics in the global plastics market is approximately 2%, with an estimated growth to around 3% by 2028 (Newsroom, 2023) equalling an annual growth rate similar to that of conventional plastics. In absolute terms, it is expected that global production of bioplastics will rise from 2.1 million tonnes in 2018 to 2.6 million tonnes by 2023. In comparison to the annual production capacity of conventional plastics, which amounts to 335 million tonnes, the share of bioplastics appears to be just a drop in the ocean. Reasons for the relatively slow increase in the production capacity of bioplastics include low oil prices, slow political support, and limited market access.

Nevertheless, demand for advanced bioplastics, applications, and products is increasing, driving continuous market growth. The largest share of the bioplastics market currently consists of drop-in bioplastics – bio-PET, bio-PE, and bio-PA, which account for 48% of the bioplastics market alone. It is expected that bio-PE will continue to expand its production capacity in the future. Great expectations are also placed on a new bioplastic – polyethylene furanoate (PEF). PEF is 100% bio-based and shows better barrier and thermal properties than bio-PET. Particularly in the packaging sector, this bioplastic could gain traction and partially replace bio-PET. PEF is expected to enter the market in 2023.

The greatest innovation potential and market growth in the future are expected particularly in novel bioplastics such as PHA and PLA. Especially due to their functional properties, these materials open up new application fields beyond the scope of conventional plastics. The production capacity of PHA is expected to quadruple in the next 5 years. One of the reasons for this development is the independence of these bioplastics from the price pressure of low oil prices, as there are no conventional counterparts. However, their success also largely depends on how quickly these bioplastics can be established in the market. Obviously, support from both politics and society could accelerate this development.

4.2.1 Industry Segments as Consumers of Biogenic Plastics

The growing demand for more environmentally friendly alternatives is being driven by increasing government regulations and consumer trends toward more sustainable products, which will become evident across various markets. At present, these effects are likely not attributable to a shortage of fossil raw materials. The most important industry segments using bioplastics are the

- **Packaging Industry**: This segment is the largest consumer of bioplastics, accounting for approximately 45% of the global production capacity of biogenic plastics in 2024 (PlastXnow, 2023; Grand View Research, 2020). Forecasts indicate that by 2029, the share of PLA will rise from 37.1% to 42.3%, while the share of PHA will increase from 4.1% to 17% (PlastXnow, 2023; European Bioplastics, 2020).
- **Textile Industry**: Bioplastics such as PLA and polytrimethylene terephthalate (bio-PTT) are used in the textile industry. By 2028, it is expected that the share of bio-PTT will decrease in favour of PLA fibres (Bayern Innovativ, 2021).
- **Consumer Goods**: Bioplastics are used in products such as toys, household items, and electronic housings, making up about 13% of production in 2024 (PlastXnow, 2023).
- **Fibres**: Bioplastics account for 20% of the textile and nonwoven sectors (PlastXnow, 2023).
- **Agriculture and Horticulture**: Bioplastics are used in mulch films and plant protection products, accounting for approximately 5% of production in 2024 (PlastXnow, 2023).

The largest market share is held by biodegradable plastics such as PLA and PHA, which are used for both packaging and applications in agriculture and medical technology (Rujnić-Sokele & Pilipović, 2017).

Ongoing research and development in the field of bio-based plastics are expected to further improve both the quality and cost structure of these materials. Advances in polymer chemistry – such as the development of novel, high-performance bioplastics and improvements in biodegradability – will likely facilitate broader acceptance of bioplastics in market segments that have so far remained underdeveloped (Garlapati, 2021). In this context, the global market for bio-based plastics is anticipated to grow significantly in the coming years. A report by MarketsandMarkets projects that the bioplastics market will increase from approximately USD 11.9 billion in 2022 to USD 22.7 billion by 2030. Although this represents an annual growth rate of around 8.7%, it remains relatively modest in light of the still small overall market share (MarketsandMarkets, 2022).

This growth will primarily be driven by the increasing demand for sustainable packaging solutions and the expanding use of bioplastics in other sectors, such as textiles and consumer goods (Garlapati, 2021).

The development of bio-based raw materials not only involves the search for alternatives to mineral or fossil resources but also requires identifying limitations and opportunities for optimization. Both research and industry are exploring ways to improve the mechanical properties, process ability, and environmental impact of bio-based materials. This includes the modification of biogenic resources through biotechnological processes, as well as the use of composite materials in which bio-based substances are combined with other renewable raw materials or additives to achieve desired properties. For instance, polymer blends made from natural and synthetic polymers can enhance material performance while remaining biodegradable (Kunststoff Magazin, n.d.).

4.2.2 Regional Distribution

In 2023 and according to the Institute for Bioplastics and Bio composites at the Hannover University of Applied Sciences, approximately 50.9% of the bioplastic production capacities were located in Asia, 20.1% in North America, and 16.9% in Europe (Newsroom, 2023). Europe holds particular significance due to its high awareness of sustainability and environmental issues. Additionally, production capacities are increasing in countries like the United States and China, which are large markets for bioplastics.

From a geographical perspective, Asia is expected to remain the largest market for bioplastics, maintaining a market share of approximately 50% by 2028. China and India, in particular, are key growth markets due to increasing demand for sustainable materials and the implementation of environmentally friendly regulations by their governments (Zhao et al., 2020). In the EU, the demand for bioplastics is primarily driven by regulatory frameworks and packaging regulations that mandate a reduction in the consumption of non-recyclable plastics. According to an estimate by MarketsandMarkets, the bioplastics market in North America will grow at an annual rate of approximately 17% from 2020 to 2025 (MarketsandMarkets, 2020). Strong growth is anticipated in North America, especially in the packaging sector, where many companies and governments have already begun adopting more sustainable solutions. The European Union is expected to take a leading role, as stricter regulations on plastic waste and greater support for a circular economy are likely to accelerate the adoption of bioplastics (European Bioplastics, 2020).

4.2.3 Long-Term Market Development

The long-term development of the bioplastics market depends on several key factors, including technological advancements, political and regulatory frameworks, economic viability, and the availability of raw materials. The market size is expected to continue growing in the coming decades due to the following trends:

- **Technological Innovations:**
  The advancement of bioplastics, particularly biodegradable plastics such as PLA and PHA, is critical to ensuring their long-term competitiveness. Currently, bioplastics are still produced in relatively small quantities, and with respect to their use in some applications, their properties are not yet fully comparable to those of conventional plastics (Zhao et al., 2020). However, ongoing research and development in areas such as biotechnological processes, material modification, and recycling technologies are expected to significantly improve the performance of bioplastics. These advancements could reduce production costs and facilitate the adoption of bioplastics across a broader range of market segments (Garlapati, 2021).

- **Legislative and Political Support:**
  Legislation within the EU and globally will play a decisive role, as the demand for bioplastics is often driven by regulatory measures such as packaging directives, carbon pricing, and bans on conventional plastics (European Bioplastics, 2020). In Europe and North America, strict regulations aimed at reducing single-use plastics and promoting sustainable alternatives are already being implemented. These measures are expected to further boost the demand for bioplastics in the coming years. Political initiatives promoting sustainable materials could not only expand the market for bioplastics but also accelerate investment and production capacity in this sector (Kümmerer & Zeise, 2019).

- **Cost Reduction Through Economies of Scale:**
  A key factor for the long-term market development of bio based plastics will be the ability to reduce production costs through economies of scale. While the current production costs for bioplastics are generally higher than those for fossil-based plastics, increased production volumes and improved manufacturing processes could lead to significant cost reductions over time (Shen et al., 2010). Furthermore, the market for recycled bioplastics could expand through innovative recycling technologies, increasing the availability of secondary raw materials and reducing overall costs. The growing adoption of bioplastics across various industries may thus contribute to the sustainable establishment of the bioplastics market.

- **Availability of Raw Materials:**
  Another critical factor in the long-term development of the market is the availability of raw materials. Bioplastics, such as PLA and PHA, are derived from plant-based feedstocks like corn, sugarcane, or other sustainable agricultural products. As global demand for agricultural raw materials rises due to the increased use of biofuels and bioplastics, it will be essential to address land use, water consumption, and resource availability through innovative agricultural practices and efficient production processes (Bayer et al., 2021). Research into genetic optimization and the efficient utilization of agricultural residues could offer new potential in this area.

The long-term development of the market for bio-based plastics will be significantly influenced by technological innovations, political support, shifts in consumer demand, increasingly stringent environmental regulations, and improvements in production capacities. Key forecasts and trends shaping the market's evolution include:

4.2.4 Challenges and Uncertainties

Despite the significant growth potential of bioplastics, several notable challenges remain:

- The production costs of bioplastics are often still higher than those of conventional plastics, which can impair their competitiveness (Shen et al., 2010).
- The availability of feedstocks, such as renewable raw materials for bioplastics, is a limiting factor, as demand for these resources is also increasing in other sectors (e.g., food production and the textile industry) (Bayer et al., 2021).
- The limited compatibility with established value chains, particularly recycling infrastructure, poses an additional obstacle (Garlapati, 2021).

It should be noted that it is very likely that it will take time for comparatively new materials to enter the established ecosphere of materials with known properties and long-term behaviour. However, the outlook for the bioplastics market is positive, with substantial growth potential in the coming years. This development is supported by technological advancements, increasing demand for sustainable solutions, and political initiatives and regulations. Bioplastics are expected to play an increasingly important role in the global plastics industry, particularly in the areas of packaging, textiles, and consumer goods.

4.3 Structural Challenges

As previously mentioned, the replacement of a single property should not be interpreted as an indication of the comprehensive substitution of one raw material by another. Rather, there are a variety of technical properties that must be considered when substituting mineral or fossil raw materials with biogenic materials in specific products. These include mechanical properties such as strength and tensile strength, thermal and chemical properties, optical and aesthetic characteristics, as well as functional properties like biodegradability or CO_2 sequestration. The challenge lies in developing biogenic materials that offer corresponding or entirely new properties to a sufficient degree while reducing resource consumption. Research plays a crucial role in this, as each biogenic material source offers different properties and advantages depending on the application.

4.3.1 Recycling

Despite increasing environmental awareness and the willingness of many consumers to separate waste, the effective recycling of plastic waste remains a central challenge for the circular economy (see Figure 4.1 below). For metallic raw materials, methods are more advanced than for plastics: the use of secondary raw materials reduces the need to extract new natural resources. Recycled aluminium or steel are examples of materials that can be reused through recycling processes. Recycling these materials not only protects the environment but also reduces the energy and resource expenditures associated with the extraction and processing of new raw materials. This lowers the demand for primary raw materials and enhances resource efficiency, contributing to a more sustainable production process (Schiavoni et al., 2016).

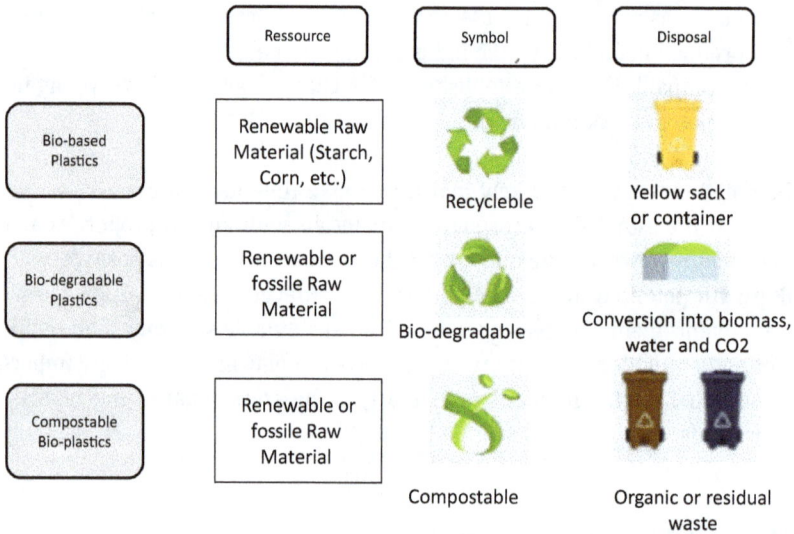

Ressource	Symbol	Disposal
Bio-based Plastics — Renewable Raw Material (Starch, Corn, etc.)	Recycleble	Yellow sack or container
Bio-degradable Plastics — Renewable or fossile Raw Material	Bio-degradable	Conversion into biomass, water and CO2
Compostable Bio-plastics — Renewable or fossile Raw Material	Compostable	Organic or residual waste

Figure 4.1: Separation of biogenic waste materials.
Schematic representation of the separation and categorization of biogenic waste materials, illustrating the processes and pathways through which organic waste streams are identified, collected, and prepared for subsequent valorization. The figure distinguishes between various sources of biogenic waste – including food waste from households and the hospitality sector, agricultural residues, forestry by-products, green waste from landscaping, and organic industrial waste – and outlines the steps required for their effective separation from non-organic or contaminant materials. Key stages such as source separation, mechanical sorting, and pre-treatment are highlighted as essential for maintaining the quality and usability of biogenic inputs in downstream applications. The separated materials can then be directed into appropriate recovery routes such as composting, anaerobic digestion for biogas production, or conversion into bio-based materials and fuels. This figure underscores the critical role of waste segregation infrastructure and public participation in enabling the circular use of biogenic waste, reducing landfill dependency, minimizing greenhouse gas emissions, and contributing to a more sustainable and resource-efficient bioeconomy.

While bioplastics such as PLA are theoretically biodegradable, they cannot always be processed according to established methods in existing recycling facilities. These methods are based on long-term production optimization and are not easily adaptable. This results in market resistance, as the technology is not yet widely established, and start-up costs make changes difficult. Moreover, there are recycling issues when bioplastics are mixed with fossil-based plastics, as they not only have different physical properties (such as melting points) but also distinct chemical properties, which complicate or make recycling in established process steps impossible. In practice, bioplastics are also not always biodegradable and often require specialized industrial composting facilities for proper disposal, which adds additional infrastructure and costs.

Especially in the case of so-called mixed plastics, efficient and economically viable separation into pure fractions has so far only been limitedly possible. These structural barriers hinder the closure of material loops and highlight the need for innovative solutions both at the design level and in recycling technologies. With the widespread use of biogenic raw materials, the lack of recyclability of mixtures of certain fossil-based and biogenic plastics occasionally poses an additional problem:

4.3.2 Chemical Industry and Bioplastics Recycling

The chemical industry plays a crucial role in recycling plastic waste by breaking it down into its original chemical components, thus reducing the need for new raw materials. Technologies such as pyrolysis and hydrolysis enable the efficient recycling of plastic waste and the conversion of it back into usable chemical feedstocks. These processes help close the loop of plastics by decomposing waste into its monomeric components, which can then be used again to manufacture new products. Pyrolysis and hydrolysis are particularly promising for recycling plastics that were previously difficult to process due to their complex structure (Jehanno et al., 2021). These technologies enable a much more effective utilization of plastic waste, contributing to reducing landfill burdens and minimizing the use of new fossil raw materials.

Many bioplastics, although biodegradable, do not fit into the existing recycling streams designed for conventional plastics. This means they cannot be processed in traditional recycling plants set up for plastics such as polyethylene (PE) or polypropylene (PP). Instead, these bioplastics require specialized composting facilities or must be treated under controlled conditions. However, these specialized facilities are still not sufficiently developed in many regions, leading to bioplastics being improperly disposed of, either remaining in the environment or being lost in existing waste streams (Häußler et al., 2021; Walker & Rothman, 2020). To solve the recycling issue, larger investments in infrastructure would be necessary to recycle bioplastics efficiently and sustainably. Furthermore, research needs to progress further to develop biogenic materials that are more compatible with conventional recycling processes.

4.3.3 Carbon Cycle Technologies in the Chemical Industry

An exciting area of research in the chemical industry is the so-called carbon cycle technologies. These technologies aim to use CO_2 as a raw material for chemical processes rather than emitting it as a waste product. Carbon cycle technologies refer to a broad set of methods and systems that manage, influence, or mimic the natural carbon cycle – the movement of carbon through the atmosphere, oceans, soil, plants, and animals – to reduce, capture, store, or reuse carbon dioxide (CO_2) and other carbon compounds. They are increasingly important in the fight against climate change because they help balance or reduce excess CO_2 in the atmosphere. Core categories of carbon cycle technologies aim at capturing and storing CO_2, such as

Carbon capture and storage (CCS)
Direct air capture (DAC)
Enhanced weathering
Soil carbon sequestration
Ocean-based carbon removal

Or using it as raw material or as a resource, for example, in

Carbon capture and utilization (CCU)
Bioenergy with carbon capture and storage (BECCS)

It should be noted that carbon capture technologies are designed to be mounted to a strong CO_2 emitter in order to capture the CO_2 generated immediately after it was produced. While technically working, taking CO_2 out of the air at places, randomly chosen, as not been seriously considered due to cost reasons. Today, most of these technologies are regarded as too expensive to be seriously considered.

The conversion of CO_2 into useful chemical products (carbon capture and utilization (CCU)), such as plastics, fuels, or chemicals, offers the potential to close the carbon cycle and utilize CO_2 as a valuable resource. While these technologies are still in the development phase, they could play a key role in the future in reducing CO_2 emissions and producing products from renewable sources (Aresta et al., 2019). By using CO_2 as a feedstock, the chemical industry could not only reduce its dependence on fossil fuels but also make a significant contribution to reducing global greenhouse gas emissions.

The substitution of fossil raw materials in the chemical industry offers numerous opportunities to reduce resource consumption and minimize environmental impacts. Biogenic plastics and chemicals, sustainable solvents, and innovative recycling technologies are promising approaches to reduce the CO_2 footprint of the chemical industry. The development of new synthesis pathways and the integration of carbon cycle technologies further open up potential for a sustainable future. The chemical industry must continue to invest in research and development to optimize these technologies and enable their broader application. Only through the successful implementation of these alternative approaches can the industry contribute to reducing global environmental burdens and take a sustainable path into the future.

Initial results from life cycle assessments indicate that bioplastics can save fossil raw materials and release less CO_2. However, due to their origin, they contribute more to eutrophication (the addition of nutrients to an ecosystem) and soil acidification than plastics originating from fossil resources. Another factor that could significantly improve the sustainability of bioplastics in the future is waste disposal. Currently, there is a lack of a suitable waste management system for bioplastics, with effective recycling infrastructures, particularly for biodegradable bioplastics.

In principle, the recycling of bioplastics is possible. Due to their similar properties, drop-in bioplastics such as bio-PE and bio-PET can be integrated into the recycling streams of conventional plastics. For biodegradable bioplastics like PLA and PHA, entirely new recovery pathways may be developed. These can, in theory, also be subjected to composting and anaerobic digestion due to their properties.

However, in practice, the recovery and recycling of bioplastics look different. Due to existing uncertainties among plant operators and insufficient sorting technologies, bioplastics are mostly sorted out and incinerated before recycling or recovery. Biodegradable bioplastics, in particular, are often regarded as contaminants. The reasons for this include their impact on the quality of stable plastic recyclates, which can affect the marketability of recycled products. Furthermore, the entry of micro- and macro-plastics into the environment is possible, as the biological degradation of biodegradable bioplastics is often insufficient under the process conditions and residence time in industrial plants, preventing complete degradation. Still, even if bioplastics are sorted out and incinerated before recovery, this is more beneficial than conventional plastics, as significantly less CO_2 is released. Specifically, only as much carbon is released in the form of CO_2 as was previously bound in the biogenic material – provided that the carbon in the bioplastic comes entirely from renewable raw materials (Bertling et al., 2018).

Bioplastics will not be the solution to the large plastic waste gyres in the oceans. However, they can represent a sustainable alternative to conventional plastics if their biodegradability is further improved and if a suitable waste and disposal infrastructure is in place. Until then, incinerating bioplastics is still better than producing no bioplastics at all. Second-generation bioplastics, plastics made from waste and residual materials, could also alleviate concerns about land use and competition with food and feed production.

4.3.4 Strategies for Substitution

The substitution of mineral and fossil raw materials with bio-based raw materials is a critical step towards a more sustainable economy. In addition to the use of bio-based plastics discussed here, there are other substitution strategies based on a variety of innovative ideas that aim to reduce both the ecological footprint and dependence on finite natural resources. The key strategies include:

Use of New Material Combinations and Composites: Advances in materials science enable the creation of innovative materials that offer both comparable and improved properties compared to traditional materials. A well-known example of this is the use of carbon fibres, which are increasingly replacing metals in applications such as aerospace and the automotive industry. Carbon fibres are characterized by their high strength combined with low weight, leading to a significant reduction in vehicle weight and, consequently, CO_2 emissions during use (Friedrich & Almajid, 2013). These new material combinations not only provide more powerful properties but also offer a more sustainable alternative to conventional, heavier, and resource-intensive materials.

Technological Process Adjustments in Production: An example of such an adjustment is the use of hydrogen instead of coke in steel production. This technology enables a significant reduction in CO_2 emissions, as hydrogen is used as a reducing agent in iron production instead of carbon, which releases CO_2 when burned in traditional processes. Such innovative procedures can revolutionize production processes in energy-intensive industries like steel and cement manufacturing, significantly reducing the reliance on fossil fuels (Aresta et al., 2019). However, it needs to be noted that the replacement of the previously used material goes along with significant changes in composition of materials used in steel production (i.e., carbon encapsulated in steel may react with the hydrogen causing an unwanted and potentially disabling loss of strength of the steel).

In practice, it is often necessary to combine all of these strategies. However, it must not be forgotten that the established production strategies have been successful for many decades, and any intervention in these strategies can have fundamental and far-reaching consequences, potentially even leading to changes in entire products.

4.4 Examples – Substitution of Fossil Raw Materials

The substitution of mineral and fossil raw materials with biogenic alternatives in plastic production, construction, and the chemical industry already involves established or promising approaches for utilizing biogenic alternatives. Research in these areas aims to develop biogenic plastics with improved properties in terms of strength, flexibility, and durability, enabling them to compete with current petrochemical plastics. Particularly important is the optimization of production processes to make them more cost-effective while simultaneously minimizing the ecological footprint, ensuring that biogenic plastics can be established as a sustainable alternative to fossil-based plastics in the long term (Shen et al., 2020).

4.4.1 Alternative to Petrochemical Polymers

The global consumption of plastics derived from fossil sources has not only led to a dramatic increase in plastic waste but has also left behind a substantial CO_2 footprint. These plastics persist for long periods as they do not degrade easily. They remain in the environment for extended durations, resulting in a continuous burden on ecosystems and living organisms. The increasing pollution of micro plastics, which arises from the breakdown of larger plastic pieces into tiny, invisible particles, represents one of the greatest ecological challenges of our time. Micro plastics enter food chains and potentially have harmful effects on animals, plants, and humans (Chamas et al., 2020).

Biogenic plastics represent a promising, sustainable alternative to traditional petro chemically produced plastics, which are currently manufactured in large quantities worldwide. They are characterized by being derived from renewable raw materials such as plant oils, starch, or sugars, or by possessing the ability to be biologically degraded – or even combining both properties. This fundamentally distinguishes them from conventional plastics, which are based on fossil resources like petroleum and natural gas, causing significant environmental issues, particularly in terms of micro plastic pollution in the oceans and the associated ecological damage (Emadian et al., 2022). These characteristics make biogenic plastics a promising solution for mitigating the harmful environmental impacts of conventional plastic production. They not only provide a more environmentally friendly option for packaging, single-use products, and textiles but also for durable products used across various industries.

Biogenic plastics offer a sustainable alternative to traditional plastics. They are also a step towards a circular economy that relies on renewable resources to reduce plastic waste and CO_2 emissions significantly. However, it is important to note that current circular economies are not yet prepared for biogenic plastics. In extreme cases, products made from biogenic raw materials can disrupt the processes of the established circular economy (Walker & Rothman, 2020) – for example, as bio-based plastics in waste plastic streams make the recycling of the mixture impossible.

4.4.2 Use in Construction

Substituting mineral and fossil raw materials in the construction industry is a crucial step toward sustainability. Wood, biogenic-building materials, recycled concrete, geopolymers, biogenic insulation materials, and recycled plastics offer numerous alternatives to conventional building materials, which are still predominantly made from non-renewable resources. These substitutions help reduce CO_2 emissions while promoting sustainable resource use and waste management. However, it is important to continue researching these innovative materials, optimizing their production processes, and expanding the infrastructure for recycling and utilizing waste materials to enable broader application. The construction industry is challenged with the large-scale integration of alternative materials to make meaningful contributions to sustainability.

Mineral-based construction materials like cement, concrete, and steel are indispensable for their strength, stability, and durability. However, the production of these materials has significant drawbacks due to high CO_2 emissions and natural resource consumption. The cement industry accounts for about 8% of global CO_2 emissions, more than the entire international aviation industry (Klimakiller Zement, n.d.; KEI, n.d.). The search for environmentally friendly alternatives has highlighted the lower environmental impact and resource conservation potential of bio-based raw materials.

Promising alternatives are wood and wood-based materials. Wood has gained new significance in the construction industry through modern building methods, such as Cross Laminated Timber (CLT). CLT is a modular, lightweight, and robust material for multi-story buildings and larger construction projects, significantly reducing CO_2 emissions (Frangi et al., 2019). Wood's advantage lies in its ability to sequester carbon and its regenerative properties. Particularly in multi-story buildings and larger construction projects, wood has gained significant importance in recent years as a sustainable building material (Geng & Chase, 2019).

Recycled concrete is a promising building alternative that utilizes construction waste and old concrete structures as aggregates. This method decreases the demand for materials like sand and gravel, which are crucial for concrete production. Additionally, using recycled concrete reduces CO_2 emissions and conserves energy, as less material needs to be extracted and transported. Recycled concrete has demonstrated effectiveness in resource efficiency while minimizing the ecological footprint of construction (Pacheco-Torgal et al., 2013).

Another innovative material is bio-concrete, which develops unique properties through microorganisms or biogenic additives. Bio-concrete can self-heal when cracks form by activating microorganisms that produce calcium carbonate to close the cracks. This self-healing property reduces the need for repairs, lowers material consumption, and extends the lifespan of buildings. Such technological advancements could significantly decrease material requirements and minimize the construction industry's environmental impact (Jonkers, 2010; Baustoffwissen, 2023).

Natural fibre-reinforced building materials represent a sustainable option. Materials like hemp concrete and straw bale construction serve as alternatives to conventional insulation materials and non-load-bearing components. These materials are ecologically beneficial, more cost-effective, and energy-efficient to produce. Hemp and straw help minimize the environmental impact by minimizing the use of fossil resources (Walker et al., 2014).

Hemp and flax require less nitrogen fertilizer than cotton. NPK fertilizer used for hemp cultivation is approximately 50–70 kg of nitrogen per hectare versus 100 kg for cotton. Reducing fertilizer use contributes to a lower environmental impact and minimizes eutrophication.

The land area needed for producing hemp concrete is relatively small, as approximately 1 hectare produces 10 tonnes of hemp. Therefore, less agricultural land is needed, resulting in a reduced environmental impact.

Geopolymers as a Cement Replacement

Geopolymers are a class of inorganic polymers formed by the chemical reaction of aluminosilicate materials (such as fly ash or metakaolin) with alkaline activators (like sodium hydroxide or potassium silicate). Geopolymers can serve as an alternative to conventional cement. They are produced from industrial by-products such as fly ash or blast furnace slag, with the potential to replace cement while significantly reducing CO_2 emissions. Geopolymers are a category of synthetic inorganic materials formed through the polymerization of aluminosilicate substances in the presence of an alkaline solution. They are considered "green materials" because, compared to conventional cement, they exhibit lower CO_2 emissions during production and are potentially more environmentally friendly. While traditional cement manufacturing is a major contributor to global CO_2 emissions, geopolymers produce substantially fewer emissions. They also provide an opportunity to utilize waste products from other industries, further improving their environmental impact. Geopolymers can be used not only in concrete production but also in brick manufacturing and various other building materials, thereby reducing reliance on traditional CO_2-intensive materials (Davidovits, 2015).

Despite the many advantages of these biogenic building materials, there are practical and organizational challenges that need to be addressed to facilitate their widespread adoption.

A critical aspect is adapting building regulations, which often still focus on traditional, mineral-based materials. A broader acceptance and use of biogenic building materials requires changes to normative standards and a thorough evaluation of their long-term stability and safety. Additionally, ensuring the use of sustainable forestry and agricultural practices is crucial. Otherwise, negative impacts on land use could arise, such as overuse of land for industrial cultivation, leading to changes that could harm the environment. Therefore, it is important that the use of biogenic building materials is accompanied by responsible resource management and sustainable planning.

Substitution of Insulation Materials and Plastics

In the field of insulation materials, numerous sustainable alternatives to synthetic, petroleum-based options exist. Hemp, sheep wool, and cellulose are excellent biogenic insulation materials that provide outstanding thermal and acoustic insulation while being made from renewable raw materials. These materials are biodegradable, contributing to a lower environmental impact. Compared to traditional petroleum-based insulation materials, biogenic options have a significantly lower CO_2 footprint and do not emit harmful substances. For instance, hemp is a fast-growing raw material that can be cultivated without pesticides or chemical fertilizers, making it particularly environmentally friendly (Asdrubali et al., 2015). Sheep wool is noted for its ability to absorb and neutralize pollutants like formaldehyde from indoor air, enhancing the indoor climate (Mühlethaler & Haas, 2006). These natural insulation materials not only provide a sustainable alternative but also offer better indoor climate and health protection.

Biogenic building materials, such as wood, flax, and hemp, are increasingly being explored as sustainable alternatives to concrete and steel in the construction industry. The land requirements for these biogenic materials depend on the necessary building capacity. Estimates suggest that the production of bio-concrete could reduce agricultural land demand to about 100 to 300 hectares per year to replace a specific construction volume. This represents a significant saving compared to the land area required for producing mineral construction materials (Li et al., 2020).

Overall, biogenic building materials offer a promising opportunity to reduce CO_2 emissions in the construction industry and significantly lower its ecological footprint. Despite these advantages, the increased use of biogenic building materials requires adjustments to building regulations and the assurance of sustainable forestry and agricultural practices to prevent negative land-use changes. The widespread adoption of biogenic building materials not only demands technological innovations but also legal and economic adjustments, alongside a heightened focus on sustainability in agriculture and forestry. Recycled plastics are increasingly utilized in construction projects, including PVC pipes, window frames, and insulation materials. Using recycled plastic materials in the construction industry helps minimize waste while simultaneously lowering the demand for primary plastics derived from fossil resources. By recycling plastics, valuable resources can be reused, which reduces the need to produce new plastics. Furthermore, the adoption of recycled plastics contributes to decreased CO_2 emissions, as producing recycled plastic generally requires less energy than manufacturing new plastic from petroleum (Stichnothe & Azapagic, 2013). This development fosters the circular economy in construction and further lessens the ecological footprint of construction projects.

4.4.3 Use in the Chemical Industry

The chemical industry has traditionally relied heavily on fossil raw materials, particularly oil and natural gas, which serve as feedstocks for various chemical products. In light of global efforts toward sustainability and climate protection, replacing these fossil resources is becoming increasingly urgent. The substitution of fossil raw materials can be achieved in various ways, including using bio-based feedstocks, recycling processes, or new synthesis routes. These alternatives offer ecological benefits and new opportunities for the chemical industry to optimize resource consumption and reduce environmental impacts.

Conventional plastics are primarily produced from fossil raw materials such as oil and natural gas, leading to environmental burdens and the depletion of non-renewable resources. Bio-based plastics are a promising alternative, produced from renewable raw materials such as starch, cellulose, or lignin. These plastics help reduce the use of fossil resources while providing a more sustainable solution for plastic production (Singh et al., 2019).

An example of bio-based plastics is PLA, which is derived from corn starch or sugarcane. PLA is biodegradable and is increasingly used in the packaging industry, textiles, and medical products (Pérez et al., 2020). Another example is PHA, a group of biopolymers produced by microorganisms from plant-based raw materials. PHAs offer a promising alternative to petroleum-based plastics and can be utilized in various applications, such as packaging and medical devices (Koller et al., 2017).

Bio-PE is a bio-based plastic derived from sugarcane. It possesses the same properties as conventional polyethylene but has a significantly lower CO_2 footprint, making it a more environmentally friendly option (Kiss et al., 2019).

Bio-based Catalysts from Microbes or Enzymes
In the case of bio-based catalysts and fuel cells, the potential to minimize agricultural land requirements is particularly pronounced. Using plant-based enzymes or microorganisms as substitutes for platinum or cobalt could result in a significantly lower land area requirement, as these catalysts are based on microbes or algae, requiring minimal land space (Li et al., 2020). However, there are still insufficient concrete estimates, as these technologies are still in the early stages of research (Yang & Liu, 2021):

Land Use Requirement: Microbes and enzymes offer promising bio-based alternatives to catalysts such as platinum or cobalt in the chemical and energy industries. These biocatalysts can be produced under controlled conditions in bioreactors and do not require large agricultural areas (Li et al., 2020).

Fertilizer Requirement: The fertilizer requirements for producing microbes or enzymes are generally minimal, as many microbes can grow in fermentation processes based on waste products or residual nutrients. As a result, the amounts of fertilizer needed are low, further minimizing the ecological footprint of production (Wang et al., 2020).

Bio-based Chemicals as an Alternative to Petrochemical Substances
In the chemical industry, bio-based platform chemicals offer new opportunities to replace fossil resources such as oil and natural gas, significantly reducing dependence on non-renewable resources. These bio-based alternatives not only contribute to the sustainability of the industry but also provide the opportunity to reduce the environmental impacts of chemical processes. Unlike traditional petrochemical processes that

rely on fossil fuels, bio-based platform chemicals utilize renewable feedstocks, offering a more environmentally friendly production potential.

A significant example of bio-based platform chemicals is lignin-based aromatics. Lignin, a naturally occurring polymer found in plant cell walls, serves as a sustainable substitute for petrochemical aromatics widely used in the plastics and pharmaceutical industries. Aromatics are essential building blocks for the production of a variety of products, ranging from plastics to pharmaceuticals. Lignin, which is produced in large quantities as a by-product of the paper and pulp industry, can be converted into aromatic compounds through specialized processes, exhibiting the same properties as petrochemical aromatics. These bio-based lignin-derived aromatics offer a more environmentally friendly alternative and contribute to the valorization of a previously underutilized waste product (Rinaldi et al., 2016).

Another example is bio-based surfactants derived from plant oils or sugars. Surfactants, commonly used in detergents and cleaning products, are typically made from petroleum-based sources. Using plant oils or sugars as feedstocks for surfactants represents a sustainable alternative that can significantly reduce the CO_2 footprint of these products. These bio-based surfactants exhibit similar or even superior cleaning properties compared to their petrochemical counterparts. They offer the advantage of being derived from renewable sources and are biodegradable, thus enhancing their environmental compatibility (Campos et al., 2019).

Finally, fermentatively produced organic acids represent another essential element of bio-based platform chemicals. Organic acids such as lactic acid and succinic acid are increasingly made through biotechnological processes, where microorganisms ferment plant-based raw materials. These biotechnologically produced acids serve as feedstocks for plastics, solvents, and other chemical compounds. Lactic acid, for example, is used in the production of PLA, a bio-based plastic. In contrast, succinic acid is utilized in plastics, food additives, and pharmaceutical products. The fermentative production of these organic acids is not only more energy-efficient but also reduces reliance on fossil fuels, as it is based on renewable raw materials (Werpy & Petersen, 2004).

In summary, bio-based platform chemicals offer the chemical industry a sustainable alternative to traditional petrochemical feedstocks. Using lignin-based aromatics, bio-based surfactants, and fermentatively produced organic acids can not only help reduce environmental burdens but also contribute to creating a circular and resource-efficient production model. The chemical industry plays a key role in reducing global CO_2 emissions and promoting a more sustainable economy by replacing fossil raw materials with renewable, bio-based materials. The challenges with bio-based chemicals primarily lie in the often still inadequately established production processes and scalability.

Individual Bio-based Plastics and Chemicals

One of the most significant substitution opportunities in the chemical industry lies in using bio-based plastics and chemicals derived from renewable feedstocks, offering a more sustainable alternative to petroleum-based products.

Bio-polyethylene (bio-PE) and bio-polypropylene (bio-PP) are two examples of bio-based plastics that are increasingly used in industry. These polymers are primarily derived from plant-based feedstocks such as sugarcane or corn and replace traditional fossil-based polymers in a variety of applications, including packaging, containers, and fibres. Bio-PE and bio-PP offer the advantage of possessing the same physical properties as their petrochemical counterparts but with a lower CO_2 footprint (Shen et al., 2009). By using these bio-based plastics, companies can significantly reduce the environmental impact of their production processes.

Another significant example is **polylactic acid (PLA)** and **polyhydroxyalkanoates (PHA)**, which are also made from renewable feedstocks such as sugarcane, corn, or plant oils. These plastics are biodegradable and provide an environmentally friendly alternative to conventional plastics, particularly in packaging and medical technology. PLA is increasingly used in single-use packaging; while PHA is utilized in medical applications such as resorb able implants or surgical sutures (Künkel et al., 2016). Using these bio-based plastics helps address the plastic waste problem, as they degrade under natural conditions, thus providing a sustainable solution for the environment.

Besides plastics, sustainable solvents are becoming an increasingly important alternative to petrochemical solvents. The chemical industry uses a variety of solvents that are essential in many production processes. Using water-based or bio-based solvents, derived from plant-based feedstocks, reduce the negative environmental impacts caused by volatile organic compounds (VOCs) and helps lower CO_2 emissions. These bio-based solvents can be used in many chemical processes that traditionally require petrochemical solvents, thus contributing to sustainable production (Clark et al., 2012).

4.4.4 Use in Automotive Industry

In light of increasing demands for sustainability and efficiency, the automotive industry faces the challenge of reducing both the weight of its vehicles and the environmental impacts associated with their production and use. An important strategy to address these challenges is the increased use of alternative materials that ensure the performance and safety of vehicles and reduce their environmental footprint. Innovative materials, particularly in bodywork and interior, play a crucial role in this regard.

Alternative Materials for Bodywork and Interior

A critical approach to weight reduction in the automotive industry is the use of fibre-reinforced plastics, primarily carbon fibre-reinforced plastics (CFRP) and glass fibre-reinforced plastics (GFRP). These materials offer excellent weight reduction while maintaining high strength and stiffness, making them ideal candidates for use in automotive production. CFRP and GFRP have increasingly replaced traditional metal parts in the automotive industry in such areas as bodywork, bumpers, or structural components. The use of these composites leads to a significant reduction in vehicle weight, which lowers fuel consumption and, in turn, reduces CO_2 emissions over the entire lifespan of a vehicle. This is especially important as the automotive industry focuses on developing lighter vehicles to meet the demands for fuel efficiency and emission-free mobility (Friedrich & Almajid, 2013). However, it should be acknowledged that the same automotive industry, in terms of product policy, tends to prioritize higher-margin, less environmentally friendly SUVs over smaller cars, which may be more cost-effective from the customer's perspective.

In addition to fibre-reinforced plastics, natural fibre-reinforced plastics (NFRP) are increasingly used as lightweight materials. These plastics consist of plant fibres such as flax, hemp, or jute, which are incorporated into a polymer to improve the material's mechanical properties. Natural fibres offer a more environmentally friendly alternative to synthetic fibre composites, as they are based on renewable resources and have a lower CO_2 footprint. In the automotive industry, NFRP are primarily used in interior components such as door panels, dashboards, or trunk covers. The advantages of these materials include their high availability, low weight, and positive ecological footprint. Furthermore, they contribute to creating more sustainable production processes, as they are biodegradable and reduce dependence on fossil resources (Pickering et al., 2016).

The increased use of these bio-based and fibre-reinforced materials also represents a response to the growing demand for sustainable automotive components. This is especially crucial in the future of mobility, which relies on electric vehicles and emission-free technologies. By implementing these alternative materials in vehicle production, automotive manufacturers can make a valuable contribution to reducing environmental impacts while simultaneously enhancing the efficiency and performance of their products.

4.4.5 Use in the Packaging Industry

The increased use of bio-based feedstocks in the packaging industry is increasingly seen as a key strategy for reducing reliance on fossil resources and combating pollution. Various studies have shown that bio-based polymers, such as PLA and PHA, have significant potential to replace conventional petroleum-based plastics on a large scale (Gandini & Lacerda, 2015; Niaounakis, 2013). Both materials are notable for being made from renewable feedstocks, such as corn starch, sugarcane, or other sugary plants, and are biodegradable under certain conditions.

As a result of these technological advancements, large companies are increasingly turning to bio-based plastics to make their packaging systems more sustainable. Prominent examples include corporations like Coca-Cola, Danone, and Nestlé, which have launched initiatives to integrate bio-based materials into their product packaging. PLA, for instance, is primarily used for beverage bottles, food packaging, and single-use cups due to its good mechanical properties and transparency. PHA, on the other hand, is preferentially used in single-use packaging, biodegradable films, and plastic bags, as it can be broken down more rapidly in natural environmental conditions, such as in the sea or soil (Niaounakis, 2013).

These developments are part of a broader trend towards more sustainable packaging solutions, with the aim of minimizing the negative environmental impacts of conventional plastics – particularly their contribution to global plastic pollution. The shift to bio-based plastics offers companies the opportunity to significantly reduce their CO_2 footprint, decrease the consumption of fossil resources, and simultaneously respond to changing consumer preferences and regulatory requirements (Rujnić-Sokele & Pilipović, 2017).

However, despite these promising approaches, challenges remain concerning the cost-effectiveness, availability of raw materials, and the technical properties of bio-based plastics, which have so far limited their broader market penetration.

4.4.6 Use in Batteries

Bio-based batteries, also known as *bio-batteries* or *biological batteries*, represent a promising technology that utilizes the principles of conventional batteries but employs biological materials as electrodes or electrolytes. This technology uses biodegradable or renewable raw materials to provide a more sustainable energy source that is less environmentally harmful compared to conventional batteries (Cheng et al., 2021). There are various types of bio-based batteries, but the basic mechanism is largely the same: they store and release energy through chemical reactions that rely on biological materials (Zhao & Liu, 2023).

Functioning of Bio-based Batteries

1. **Electrode Materials:** In conventional batteries, metals such as lithium or cobalt are typically used for the electrodes. In bio-based batteries, however, materials such as sugars, cellulose, proteins, or enzymes are often employed. These bio-based materials (e.g., enzymes like glucose oxidase) react with the electrolytes to generate an electrochemical reaction that releases electrical energy (Müller et al., 2021).

2. **Electrolyte:** The electrolyte in a biological battery is often a liquid or gel-like medium that allows ion flow, similar to that in conventional batteries. Organic substances or biological solutions, such as saltwater or organic acids, can be used to enhance conductivity and enable ionic currents between the electrodes (Xie et al., 2020).

3. **Battery Types:**
   – **Microbial Fuel Cells (MFCs):** This type of bio-battery uses microorganisms to produce electrons. Certain bacteria or algae absorb organic substances and release electrons, which then flow through an external circuit. MFCs are known for their ability to directly convert biomass into electrical energy (Wang et al., 2022).
   – **Bio-polymer Batteries:** These batteries use natural polymers (e.g., cellulose) to form the battery structure. In some cases, a biodegradable plastic is used as the electrolyte (Zhang et al., 2021).
   – **Enzymatic Fuel Cells:** These cells use enzymes to accelerate chemical reactions that release electrons. One example is the use of glucose in enzymatic fuel cells, where glucose is broken down by the enzyme glucose oxidase into electrons and protons (Li et al., 2020).

4. **Reaction Mechanism:**
   Bio-based batteries operate like other batteries by converting chemical energy into electrical energy. When the battery is discharging, electrons flow from the negative terminal (anode) to the positive terminal (cathode) through an external circuit, thereby providing energy. Oxidation reactions occur at the anode, where, for example, sugar or other organic substances donate electrons. At the cathode, the electrons react with oxygen or other molecules, enabling the flow of current (Liu et al., 2020).

In the field of bio-based batteries, such as sodium-ion or magnesium-ion batteries, research shows that the use of bio-based materials for electrolytes requires significantly less land area compared to the land demand for lithium-ion batteries (Cheng et al., 2021). In the commercial production of these bio-based materials, the land requirements for growing raw materials like sugarcane or rapeseed must also be taken into account (Bhowmik et al., 2023). Land consumption for replacing lithium with sodium can be relatively low due to the reduced mining needs and less intensive agricultural use, particularly when bio-based materials are reused in a circular approach through recycling (Wang et al., 2022). For example, replacing cobalt, which is used in batteries, with bio-based materials would likely require several million hectares of agricultural land, as cobalt is widely used in lithium-ion batteries (Zhao & Liu, 2023). In contrast, replacing aluminium with bio-based lightweight materials such as plant fibres could require less agricultural land, as these materials do not rely on intensive cultivation (He et al., 2023).

Sodium-ion batteries with bio-based materials, such as sugarcane and algae, as electrolytes and electrode materials could offer an alternative to lithium-ion batteries (Wang et al., 2021). The land use for algae production is minimal, as algae grow in water tanks or marine systems without occupying agricultural land (Müller et al., 2022). Algae cultivation for battery production could take place on non-agriculturally used land or in saline water, thus minimizing land requirements (Liu et al., 2022).

Fertilizer Requirements for Algae: The fertilizer demand for algae is relatively low, as algae grow in a water-based environment and can obtain nutrients from seawater or wastewater (Müller et al., 2022). In contrast, sugarcane requires a significant amount of nitrogen and phosphate fertilizers for high productivity, as it is an intensive crop (Bhowmik et al., 2023).

Bio-based batteries offer an environmentally friendly alternative to conventional batteries, which often rely on non-biodegradable materials and environmentally harmful chemicals. A major advantage of this technology is its sustainability. Many of the materials used in bio-based batteries, such as sugars, cellulose, or organic enzymes, are derived from renewable resources. This makes these batteries potentially renewable and helps reduce dependence on fossil and non-renewable resources. Moreover, bio-based batteries, unlike traditional batteries, are often biodegradable, meaning they cause less environmental pollution after disposal. This contrasts sharply with conventional batteries, which contain heavy metals and can cause severe environmental problems if improperly disposed of (Yang et al., 2021).

Despite their promising advantages, the development of bio-based batteries is still in its early stages and faces fundamental challenges.

One of the biggest disadvantages is their currently lower performance and efficiency compared to conventional batteries. The energy density of many bio-based batteries is not yet sufficient to be used on a large scale for applications such as electric vehicles or portable devices (Zhao et al., 2021). Additionally, scalability presents a significant hurdle. The production of bio-based batteries at an industrial scale is not yet economically viable due to the lack of cost-effective production methods (Wang & Li, 2020). Another issue is the shorter lifespan of many bio-based batteries compared to conventional batteries, which currently limits their application to short-term usage scenarios (Sun et al., 2020). It is unclear how the practical applications of this technology will evolve in the coming years.

Although bio-based batteries offer great potential for the future of sustainable energy use, they are still in the early stages of development. Advances in materials science, enzyme technologies, and scaling production processes will be crucial for determining how successfully bio-based batteries can be deployed as an alternative to fossil and conventional chemical batteries (Chen et al., 2021). Given the increasing importance of sustainable technologies, however, research in this field is likely to gain more significance in the coming years (Zhao et al., 2021).

4.4.7 Further Examples for the Use of Biogenic Plastics

The application of bio-based plastics has gained significant importance in recent years across various industries, as an increasing number of companies are turning to sustainable and environmentally friendly materials. Bio-based plastics offer a promising alternative to conventional plastics made from fossil resources and are already being used in numerous areas to reduce environmental impacts. Below are some notable practical examples of the use of bio-based plastics.

Biogenic Textiles

Biogenic textiles, based on materials such as hemp, flax, and bamboo, also offer significant potential for reducing land use compared to conventional cotton-based textiles. For example, hemp grows faster and requires less water than cotton. Estimates suggest that replacing cotton with hemp globally would require approximately 7 million hectares of land, whereas cotton production currently occupies around 35 million hectares. This land savings illustrates the potential of biogenic textiles to significantly reduce the demand for agricultural land (Barrett et al., 2021) associated with textiles. Additionally, biogenic building materials such as wood, flax, and hemp are increasingly being explored as sustainable alternatives to concrete and steel in the construction industry. The land requirement for biogenic building materials depends on the required building capacity. Estimates indicate that the production of bio-concrete could reduce the agricultural land demand to about 100 to 300 hectares per year to replace a specific volume of construction (Li et al., 2020).

3D-Print

Another application area of biogenic plastics is 3D printing, which is increasingly establishing itself as a key technology in modern manufacturing. PLA is the most commonly used material in this field, as it does not rely on fossil raw materials and degrades relatively quickly under the right conditions (Tanzi et al., 2019). Due to its good process ability and the fact that it is available in various forms, such as filaments for 3D printers, PLA has found widespread use in additive manufacturing. It is employed for the production of prototypes, models, tools, and even functional parts. The use of PLA in 3D printing not only provides a more sustainable alternative to traditional petroleum-based materials but also contributes to reducing waste and environmental impacts (Gantenbein et al., 2018).

Medical Applications

Biogenic plastics also hold promising applications in the medical industry, as they offer a more environmentally friendly option for medical devices and implants. PLA and PHA are increasingly used in the production of resorbable implants, surgical sutures, and drug packaging (Avérous & Pollet, 2012). Particularly, PLA is significant in the medical field due to its biocompatibility and ability to degrade in the body. Resorbable PLA implants can be completely broken down after tissue healing, thereby eliminating the need for a second surgery to remove the implant. Additionally, PHA, which is produced by microorganisms, has proven useful in medical research, especially in wound healing and the manufacturing of medical products such as dressings and implantable devices that degrade within the body (Zhao et al., 2020; Chen & Wu, 2005). These biogenic materials offer a substantial improvement over conventional plastics used in the medical field, as they are not only derived from renewable resources but are also safe for the human body and cause less environmental impact.

The increasing use of biogenic plastics across various industries demonstrates that they represent a promising approach for reducing plastic waste and promoting a more sustainable economy. In the packaging industry, they enable the production of biodegradable products that reduce the carbon footprint. In 3D printing, they offer an environmentally friendly alternative to conventional materials, while in the medical industry; they provide innovative and biocompatible solutions for medical applications. These practical examples highlight the potential of biogenic plastics to deliver both ecological and economic benefits, fostering a more sustainable future.

4.5 Examples – Substitution of Mineral Raw Materials Contained in the Critical Raw Materials List of the U.S. Geological Survey

The distinction made here between the substitution of fossil raw materials and mineral raw materials is not absolute. For example, when hemp replaces iron in concrete as a building material, it is just as much a substitution of mineral raw materials as the use of certain biomaterials as catalysts.

As outlined in Chapter 1, the supply of not only fossil but also mineral raw materials is increasingly affected by the scarcity of these materials. Therefore, it seemed appropriate to draw a distinction here between the replacement of fossil raw materials and the replacement of mineral raw materials from the Critical Raw Materials List of the U.S. Geological Survey. The metals and ores listed there are raw materials that serve as the basis or ingredients for production in many industries. The appendix provides examples of mineral raw materials and sector-specific approaches for their biogenic substitutes. The aim of this extensive list was to compile a range of examples for biogenic alternatives or substitutes and to demonstrate that biogenic raw materials could support or replace the use of mineral raw materials in various areas. The 2022 Critical Minerals List of the U.S. Geological Survey was used as a reference for this list (U.S. Geological Survey, 2022). It must be emphasized that some of the examples listed are purely research activities and not yet viable industrial-scale solutions. However, this limitation should not obscure the fact that this area of research is still relatively young, and the body of experience is correspondingly limited.

The table presented in the Annex shows that there are a variety of biogenic materials with the potential to replace mineral raw materials in specific applications or reduce their use in certain industries. These biogenic alternatives could not only reduce pressure on the mineral markets but also contribute to a more sustainable, environmentally friendly, and cost-efficient future. Research in this area is promising and may lead to changes in raw material usage in the coming years. The development of biogenic materials could lead to far-reaching changes in industrial production processes while significantly reducing the ecological footprint of many products.

Nevertheless, the development and use of these materials are still in their infancy. Furthermore, competition for land use with food production and the need to avoid energy consumption for fertilizer production require caution and the necessity of resource conservation.

Chapter 5
Biogenic Raw Materials and Replacement of Fossil Energy

5.1 Overview

Biomass and biogenic fuels, which are produced from biomass such as plants, algae, and waste, offer the potential to replace fossil energy carriers while promoting a CO_2-low or CO_2-neutral energy supply (Demirbas, 2009). The market for biogenic fuels includes a variety of products, including biofuels, biogas, biomass, and green hydrogen technologies, and plays a crucial role in the global transition strategy to renewable energy sources (Grand View Research, 2020). These products can be structured as displayed in Figure 5.1 with biogenic raw materials shown in the "biomass" column and biogenic feedstocks in the "fuel" column.

Biomass refers to organic materials derived from plant and animal sources that can be used as fuels. Biomass is one of the oldest sources of energy and is utilized in various forms:

– **Wood and wood waste:**
Wood has been used as a fuel for centuries and remains a significant source for biomass for energy production. Wood pellets, wood chips, and wood briquettes are common forms of biomass used for electricity and heat generation (Bauen et al., 2018).

– **Agricultural waste:**
This includes plant residues such as straw, crop residues, and unused agricultural land. This biomass can be utilized for energy production through combustion or conversion into biogas (Gheorghe et al., 2014).

– **Algae:**
Algae are an emerging source for biomass from which biofuels such as biodiesel and biogas can be produced. They have the potential to replace fossil resources due to their rapid growth cycles and high lipid content (Li et al., 2018).

– **Animal waste:**
This includes organic waste from animal husbandry, such as manure, which can be used for biogas production (Krämer et al., 2020).

Biomass is primarily converted into biological fuels, such as wood pellets, biogas, biomethane, or through thermochemical processes like pyrolysis and gasification (Zhang et al., 2020).

Biogenic fuels, derived from biogenic raw materials such as plants or waste, offer a promising alternative to fossil energy sources. They contribute to the reduction of CO_2 emissions by decreasing dependence on non-sustainable resources such as coal,

https://doi.org/10.1515/9783112218747-005

Biomass	Process	Intermediate Product	Process	Fuel

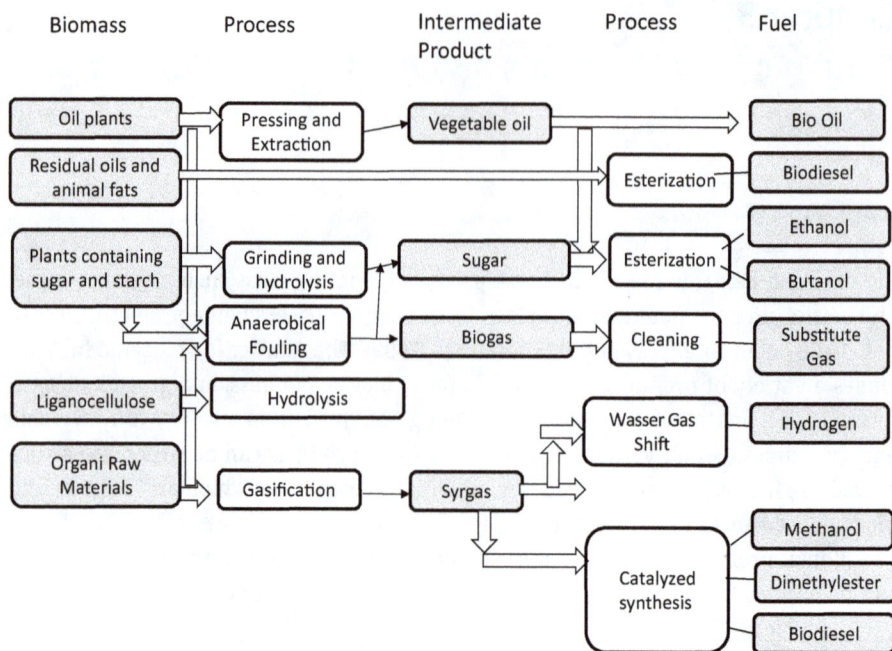

Figure 5.1: Pathways of biogenic raw materials (in the "biomass" column) to biogenic feedstocks (in the "fuel" column).

Flow diagram illustrating the transformation pathways of various biogenic raw materials – grouped under the "Biomass" column – into biogenic feedstocks used in the production of bio-based materials, chemicals, and energy carriers. The figure maps out the conversion routes for different types of biomass, including lignocellulosic residues (e.g. wood and straw), carbohydrate-rich crops (e.g. corn, sugarcane), oil-bearing plants (e.g. rapeseed and soy), and organic waste streams (e.g. food and green waste). These raw materials are processed through a range of thermochemical, biochemical, and mechanical processes – such as fermentation, transesterification, anaerobic digestion, pyrolysis, and hydrolysis – to yield diverse biogenic feedstocks like bioethanol, biodiesel, biogas, syngas, biopolymers, and platform chemicals. The figure emphasizes the versatility and complexity of biomass valorisation, demonstrating how renewable biological resources can replace fossil-based inputs in industrial and energy systems. It also underscores the importance of optimizing conversion efficiency, selecting sustainable feedstocks, and integrating cascading use principles to maximize environmental and economic benefits within the emerging bioeconomy.

oil, and natural gas. The most well-known biogenic fuels (also known as biofuels) include biodiesel, bioethanol, as well as biomethane and biogas.

Algae as a Raw Material for Biofuels

Algae Oil as an Energy Source: The idea of using algae as a raw material for biofuels gained momentum in the 1970s. Algae, as oil plants, offer a promising source for biodiesel due to their high lipid content and rapid growth. An early initiative in this direction was the Aquatic Species Program, launched by the U.S. Department of

Energy (DOE). The goal of this program was to research and commercialize biodiesel production from algae.

Wood Gas as an Alternative Fuel

Wood gas was an important energy source in Europe during the 1940s but was largely replaced by fossil fuels after the World War II. During the oil crisis of the 1970s, initiatives to revive wood gas as an environmentally friendly alternative were launched in several European countries (especially in Scandinavia) as well as in the United States. Interest in using wood and other biomass for energy generation was renewed in the 1970s. Research projects focusing on the efficient use of biomass for energy production were expanded, which subsequently led to a greater use of wood pellets and other biomass fuels in the 1980s and 1990s.

Biofuels – The Development of Ethanol and Biodiesel

Ethanol Production from Corn (USA): In the 1970s, the U.S. government began promoting ethanol as a biofuel. Initially, ethanol was primarily produced from corn to reduce dependence on petroleum. A notable example of an early initiative was the Energy Policy and Conservation Act (1975) of the U.S. government, which encouraged the cultivation of crops for biofuel production.

Biodiesel from Vegetable Oils (Europe): In Europe, researchers and companies began exploring the possibility of converting vegetable oils, particularly rapeseed oil, into biodiesel. These research initiatives eventually led to the commercial development of biodiesel as a sustainable alternative to fossil diesel.

Biofuels can be classified into different types (generations):

- **First Generation (1G):** These biofuels are derived from food crops such as corn (bioethanol) or rapeseed (biodiesel). They are primarily used in the automotive industry to partially replace fossil fuels.
- **Second Generation (2G):** These biofuels are based on lignocellulose materials such as straw, wood, or waste wood, which are not used for food production. The production of these fuels is more technically demanding but potentially offers a more sustainable solution, as they are based on waste products rather than food crops.
- **Third Generation (3G):** These biofuels are derived from algae. Algae have the potential to produce large amounts of oil in a relatively short period, making them a promising source for biofuels. However, this type of biofuel is still not widely used.
- **Fourth Generation (4G):** These biofuels utilize advanced technologies for CO_2 capture and utilization to reduce emissions during the production process and make the biofuel life cycle carbon-neutral.

5.1.1 Biofuels: From Research to Commercialization

Research and development in the field of biofuels experienced rapid acceleration since the 2000s, as biofuels emerged as a promising way to reduce dependence on fossil energy sources (Demirbas, 2007). As early as the late 1990s, the production of biofuels such as bioethanol and biodiesel from plant materials like corn, sugarcane, or rapeseed was already established. The years following 2000 were characterized by the search for more innovative and sustainable biofuel alternatives:

- An important milestone was the increased research in the area of **second-generation biofuels.** This generation of biofuels, derived from non-edible plants such as *Jatropha*, algae, or straw, was intended to address the ethical and ecological concerns associated with using food crops for fuel production (Meyer, 2017). In particular, algae biodiesel research gained momentum during this period, as algae have a high lipid concentration and grow relatively quickly (Chisti, 2007). Projects such as the Aquatic Species Program of the DOE, which was active until 1996, laid the foundation for many of the current activities in this area (DOE, 1998). In the 2000s, both industry and academia made significant efforts to further develop algae as a sustainable source of biofuels. However, commercial realization has been only partially achieved due to high production costs and technological challenges (Knothe and Krahl, 2010).
- The use of **cellulose ethanol**, produced from plant cellulose rather than sugar, was another significant research focus in the 2000s. Cellulose is one of the most abundant biological substances on Earth, and converting cellulose into biofuels could significantly increase the sustainability of biofuel production, as it relies on waste products rather than food crops (Ragauskas et al., 2014). Advances in enzyme technology and the development of efficient fermentation processes have made cellulose-based biofuel production increasingly profitable (Yang et al., 2009).

5.1.2 Biofuels and Their Applications

Biofuels are particularly important for the transportation sector, as well as for industrial applications and energy production. Biogas is a gaseous fuel primarily produced through the anaerobic digestion of organic materials such as agricultural waste, organic household waste, or sewage sludge (López et al., 2015). The main components of biogas are methane (CH_4) and carbon dioxide (CO_2), with methane being used as the fuel. Biogas can be utilized for electricity and heat production or as fuel for vehicles. It can also be upgraded to produce biomethane, which can either be injected into the natural gas grid or used as fuel for vehicles (Lund, 2002). Biomethane represents a valuable renewable alternative to fossil natural gas and offers the potential to reduce dependence on fossil fuels (Bertoldi et al., 2014).

5.1.3 Biomethanol and Biochar

Biomethanol is another bio-based energy carrier that can be derived from biomass. It is produced through the gasification of biomass into synthesis gases, which are then converted into methanol (Zhu et al., 2014). Biomethanol has a wide range of applications, such as in fuel cells or as a feedstock for the production of chemical products. It is also increasingly being explored as an alternative fuel for vehicles (Vargas et al., 2014).

Biochar is a carbon-rich solid produced through the pyrolysis of biomass in the absence of oxygen (Lehmann et al., 2006). This form of biomass can serve as a carbon storage medium and is increasingly being used as a sustainable way for agriculture and energy production (Atkinson et al., 2010). Biochar can be applied in various areas, such as a soil enhancer such as in Terra Preta or as fuel (Sohi et al., 2010).

5.1.4 Market Size and Assessments of Market Development for the Use of Biogenic Energy Sources as an Alternative to "Fossil Fuels"

The growing concern about climate change and the harmful environmental impacts of fossil fuels, along with the need for a sustainable energy supply, have intensified interest in biogenic energy.

Table 5.1: Energy use of bioenergy in Germany in 2022 (AGEE-Stat, 2023) – data in GWh.

	2020	2021	2022
Transport			
Biodiesel	30,170	25,072	24,518
Vegetable oil	21	21	21
Bioethanol	8,014	8,412	8,692
Biomethane	884	965	1,061
Gross electricity generation			
Solid biogenic fuels	11,306	11,028	11,187
Liquid biogenic fuels	307	202	175
Biogas	28,757	28,189	28,471
Biomethane	2,914	3,133	2,964
Sewage gas	1,579	1,576	1,575
Landfill gas	247	229	202
Biogenic share of waste	5,820	5,792	5,607
Heat generation			
Solid fuels (households)	66,874	78,559	80,021
Solid fuels (commerce and services)	19,101	22,086	20,671
Solid fuels (industry)	23,279	24,820	23,171

Table 5.1 (continued)

	2020	2021	2022
Solid fuels (CHP/heat plants)	6,296	6,831	6,588
Liquid biogenic fuels	3,217	2,601	2,455
Biogas	13,603	13,393	13,611
Biomethane	4,023	4,751	4,761
Sewage gas	2,378	2,368	2,412
Landfill gas	95	85	83
Biogenic share of waste	15,060	15,650	15,073

Sources as alternatives to fossil fuels (IEA, 2020). An analysis of market size and an assessment of long-term market development are therefore essential for understanding the prospects and challenges of this sector.

5.1.5 Market Size of Biogenic Energy Sources

As given in Table 5.1, the market for biogenic energy sources is currently experiencing significant growth and is expected to continue expanding in the coming years. In 2020, the global biofuels market was valued at approximately USD 142 billion and is projected to grow at a compound annual growth rate (CAGR) of 5.6% through 2030 (Grand View Research, 2020). This growth forecast reflects the global trend driven by increasing demand for sustainable and renewable energy sources. Biofuels – particularly bioethanol and biodiesel – constitute a substantial share of the market and are primarily used in the transportation and industrial sectors (IEA, 2020). According to estimates by the International Energy Agency (IEA), biofuels are expected to account for approximately 5% of total global energy production by 2030, with further increases in production anticipated in the years thereafter (IEA, 2020).

In addition to biofuels, the markets for biogas and biomethane are also experiencing significant growth. In 2019, approximately 19.5 million tonnes of biogas were produced globally, and this market is projected to grow at a CAGR of 6.5% through 2027 (Markets and Markets, 2020). The use of biogas, particularly as a substitute for natural gas in industrial applications and electricity generation, is regarded as a key technology for the energy transition (Lund, 2002). Biogas is also increasingly being utilized in combination with combined heat and power plants, which further enhances the efficiency of energy production (Bertoldi et al., 2014). Another significant market for biogenic energy carriers is green hydrogen, which is produced through the electrolysis of water using renewable energy sources. Green hydrogen is considered a key technology for decarbonizing hard-to-abate sectors such as steel and cement production, as well as the transport sector (Hydrogen Council, 2020). The market for green hydrogen is projected to reach USD 180 billion by 2050 (Hydrogen Council, 2020), with many governments around the world supporting projects that promote hydrogen technologies (IEA, 2020).

5.1.6 Long-Term Market Development and Drivers

The long-term market development for biogenic energy carriers is influenced by several key factors. The most important drivers include:

1. **Political and regulatory frameworks**: The implementation of climate goals and the adoption of CO_2 pricing models are key drivers of the market development for biogenic energy carriers. At the international level, binding climate protection agreements are increasingly being adopted, aiming to raise the share of renewable energy in total energy production. The European Union (EU) has set the goal of becoming climate-neutral by 2050, which will favour the introduction and expansion of biogenic energy carriers (European Commission, 2020). Many countries have also enacted regulations aimed at reducing greenhouse gas emissions and promoting renewable energy, which are further stimulating the market for biofuels and other biogenic energy sources (IEA, 2020).

2. **Technological innovations**: Advances in biotechnology, algae research, and recycling technologies for biomass waste are expected to further drive growth in the market for biogenic energy carriers (Miao et al., 2004). New methods for producing biofuels – such as second- and third-generation biofuels that do not compete with food production – could play a key role in the long term (Olsson et al., 2018). These technologies have the potential to significantly reduce production costs over the coming decades, thereby increasing the market share of biofuels in overall energy consumption (Demirbas, 2009).

3. **Economic incentives and investments**: The economic conditions for the production and use of biogenic energy carriers will also be critical for future market development. Subsidies and tax incentives for the production and utilization of biofuels and biomass can help reduce production costs and improve market competitiveness. Many countries have already implemented support programs for biofuels and biogas usage, which can significantly expand production capacities (REN21, 2020).

4. **Availability of feedstocks**: The availability of raw materials for the production of biogenic energy carriers – such as agricultural residues, energy crops, or algae – will be a crucial factor for market development. First-generation biofuels (such as biodiesel and bioethanol derived from food crops) have faced criticism regarding food security and land-use competition (Searchinger et al., 2008), whereas the development of second- and third-generation biofuels, based on non-edible plants or waste materials, is considered more sustainable (Chisti, 2007). Ongoing improvements in feedstock utilization efficiency and the development of residue-based technologies could significantly expand the biofuels market over the long term (Hamelinck et al., 2005).

5.1.7 Challenges for the Market

Despite the significant growth of the market for biogenic energy carriers, several challenges may hinder long-term market development – for example:

1. **Competition with other renewable energy sources**: Biogenic energy carriers compete with other renewable technologies – such as wind, solar, and hydropower – for market share. As these technologies become increasingly cost-effective and efficient, biogenic energy carriers must strengthen their market position, particularly in sectors such as transportation and heavy industry, to remain competitive in the long term (Jacobson et al., 2017). The challenge of defending market share against other renewable technologies calls for continuous technological and economic adaptation (Lund et al., 2010).

2. **Land use and environmental concerns**: The cultivation of energy crops for biofuels can lead to land-use changes, contributing to deforestation and soil degradation. It is therefore essential that the production of biogenic energy carriers follows sustainable practices and maximizes the use of waste- and residue-based feedstocks to minimize environmental impacts (Searchinger et al., 2008; Tilman et al., 2009). The development of second- and third-generation biofuels, which do not compete with food production, offers a potential way to reduce negative environmental effects (Chisti, 2007).

3. **Cost structure**: The production of biogenic energy carriers remains more expensive than that of fossil fuels. As long as the production costs of biofuels and other biogenic energy carriers are not significantly reduced, the market will continue to rely on political incentives for support (Demirbas, 2009). Improvements in efficiency through larger-scale production facilities, optimized technologies, and economies of scale will be necessary to ensure long-term competitiveness (IEA, 2020).

The market for biogenic energy carriers demonstrates impressive growth potential and is expected to play a key role in the global energy strategy over the long term. Technological innovation, policy support, and increasing availability of feedstocks will contribute to reducing production costs and expanding the market share of biogenic energy carriers. While challenges such as competition with other renewable energy sources and environmental concerns remain, biogenic energy carriers offer a significant contribution to CO_2 emissions reduction and the advancement of a sustainable energy future. However, market development will largely depend on regulatory frameworks, investments in research and development, and the efficient use of feedstocks (REN21, 2020).

5.1.8 Potential Applications of Biogenic Energy Carriers

The substitution of fossil fuels with biogenic alternatives spans several areas, each characterized by its distinct origins and applications. Some of the most promising alternatives include biofuels, hydrogen, and battery technologies, each of which can play a specific role in future energy supply:

Biogenic fuels such as biodiesel and ethanol are already widely used in many countries as alternatives to fossil fuels. These fuels are derived from renewable raw materials like vegetable oils or sugarcane, providing a means to replace fossil fuels in the transportation sector. Biodiesel is commonly used as a substitute for diesel, while ethanol is primarily used as an additive in gasoline mixtures. However, the sustainability of these biofuels is often questioned, as their production can compete with food production and may potentially lead to land-use changes or biodiversity loss (Fargione et al., 2008; Searchinger et al., 2008).

Another application of biogenic energy carriers is hydrogen, particularly in conjunction with fuel cell technology. Ultimately, the process of obtaining hydrogen presents a challenge – biogenic energy carriers are burned to generate electricity, and this electricity is then used to power the electrolysis of water. Each of these processes has a limited efficiency, so the end product – hydrogen and oxygen – becomes energetically suboptimal. However, hydrogen has the potential to replace fossil fuels, especially in the areas of mobility and stationary energy supply. In fuel cells, hydrogen reacts with oxygen from the air, producing only water as a by-product, making it a particularly environmentally friendly energy carrier (Sharma and Ghoshal, 2015). This technology could serve as the foundation for emission-free mobility in the transport sector and industry, contributing to the reduction of CO_2 emissions.

The battery technologies mentioned above represent another innovative approach to replacing fossil fuels. Currently, many batteries are based on lithium-ion technology, which is associated with the challenges of lithium extraction and resource scarcity (see also Appendix). As a result, research is increasingly focused on alternatives that are less dependent on critical raw materials like lithium. Sodium- and magnesium-based batteries are promising alternatives, as they are abundantly available and do not cause significant ecological impacts through their extraction (Huang et al., 2021). These batteries could play a vital role in electrical storage and the utilization of renewable energy in the future (Pan et al., 2017).

In addition to these technologies and fuels, there are a variety of bio-based fuels that can substitute fossil fuels. Among the most well-known are biogas and biomethane, which are produced through the fermentation of organic waste. These gases, like fossil fuels, can be used in gas engines or condensing boilers to generate electricity, providing a sustainable alternative to natural gas (Angelidaki et al., 2018; Demirbas, 2009). Biogas, therefore, not only reduces dependence on fossil fuels but also contributes to waste management by converting organic waste into valuable energy. Another significant area of bioenergy is the well-established production of wood pellets and the use of biomass in power plants. Wood pellets are made from renewable raw materials such as wood or agricultural waste and are burned in biomass power plants to generate heat and electricity. This form of energy generation provides a sustainable alternative to coal-fired power plants and is increasingly used in decentralized energy systems, particularly in rural areas or countries with abundant forest resources (Sikkema et al., 2013; Faaij, 2006).

In addition to these established biofuels, vegetable oils and their derivatives serve as alternative fuels. These oils are utilized not only in agriculture but also in aviation and transportation. The use of vegetable oils as biofuels has the potential to reduce greenhouse gas emissions and decrease reliance on fossil fuels (Sims et al., 2008). Especially in countries with significant agricultural production, such as Brazil, vegetable oils present a promising substitute for conventional fuels.

In addition to these technologies and fuels, there is a range of bio-based fuels that can substitute for fossil fuels. Among the most well-known are biogas and biomethane, which are produced through the fermentation of organic waste. These gases, like fossil fuels, can be used in gas engines or condensing boilers to generate electricity, offering a sustainable alternative to natural gas (Angelidaki et al., 2018; Demirbas, 2009). Thus, biogas not only reduces dependence on fossil fuels but also contributes to waste management by converting organic waste into valuable energy.

Another significant area of bioenergy is the established production of wood pellets and the use of biomass in power plants. Wood pellets, derived from renewable raw materials such as wood or agricultural waste, are burned in biomass power plants to generate heat and electricity. This method of energy production provides a sustainable alternative to coal-fired power plants and is increasingly adopted in decentralized energy systems, particularly in rural areas or countries with abundant forest resources (Sikkema et al., 2013; Faaij, 2006).

In addition to these established biofuels, vegetable oils and their derivatives serve as alternative fuels. These oils are used not only in agriculture but also in aviation and transportation. The use of vegetable oils as biofuels has the potential to reduce greenhouse gas emissions and decrease reliance on fossil fuels (Sims et al., 2008). Especially in countries with significant agricultural production, such as Brazil, vegetable oils offer a promising substitute for conventional fuels.

5.1.9 Types and Production of Biofuels from Vegetable Oils

Vegetable oils play a crucial role in the production of biofuels and offer a sustainable alternative to fossil fuels. They serve as the basis for various types of biofuels that can be utilized in mobility and energy supply. These fuels are distinguished not only by their renewable origin but also by their potential to significantly reduce CO_2 emissions compared to conventional fossil fuels (He et al., 2018).

Biodiesel (FAME)

One of the most well-known and widely used forms of biofuels is biodiesel. Biodiesel is produced through transesterification, where vegetable oils such as rapeseed, soybean, or palm oil react with methanol. This process results in the formation of fatty acid methyl esters (FAME), which are referred to as biodiesel. This biofuel can be used either as a pure fuel or as an additive to conventional diesel. Particularly in European countries, biodiesel has found widespread use due to its CO_2-reducing effect. Compared to conventional diesel, biodiesel offers better biocompatibility and reduces emissions of sulphur and soot (Knothe et al., 2015; Hirsinger et al., 2018).

In addition to ecological challenges, there are also technical limitations associated with the use of plant-based fuels, particularly biodiesel. While hydro treated vegetable oil (HVO) can be used in conventional diesel engine systems without significant modifications, biodiesel (FAME) presents several technical challenges for use in vehicles and machinery. Biodiesel often requires engine and fuel system modifications to ensure efficient operation. Especially in older vehicles or machinery, biodiesel use can lead to problems such as clogged fuel lines or corrosive effects. Another issue is the cold sensitivity of biodiesel, as the fuel can gel at low temperatures, impairing engine performance (Atadashi et al., 2010). These technical challenges necessitate additional investments in infrastructure and the development of new technologies to enable the efficient and sustainable use of biodiesel.

In the EU, biodiesel is already regularly blended with conventional diesel to reduce CO_2 emissions and decrease the environmental impact of fossil fuels. A common example of this is the blending of biodiesel with fossil diesel in the so-called B7 fuel, which contains up to 7% biodiesel. This blending is one of the measures implemented by the EU to reduce greenhouse gas emissions and promote renewable energies. By using biodiesel derived from plant-based raw materials such as rapeseed or soybean oil, CO_2 emissions can be significantly reduced compared to pure diesel. This practice has proven to be one of the most effective methods for improving the environmental balance of the transportation sector, particularly in areas where a complete transition to alternative fuels is not yet feasible (European Commission, 2020; Cames et al., 2016).

Hydrotreated Vegetable Oil (HVO)

An advanced form of biofuel is HVO, which is produced through the hydrogenation of vegetable oils or used cooking fats. This process involves converting vegetable oils into a fuel by adding hydrogen, which enhances the saturated bonds in the oil. The result is a fuel with similar properties to mineral diesel but derived from renewable sources. Compared to biodiesel (FAME), HVO offers better combustion efficiency and produces lower soot (black carbon particles) formation and NOx emissions (Kalnes et al., 2009; Ferrari et al., 2018). Due to its performance and compatibility with existing diesel engines, HVO is increasingly used in various sectors of the transportation industry.

In recent years, HVO fuels, produced through the hydrogenation of vegetable oils or used cooking fats, have increasingly been tested as a sustainable alternative to fossil jet fuels. HVO offer significant advantages in the aviation industry, as they have a higher energy density and lower CO_2 emissions compared to conventional kerosene. These properties make HVO a promising candidate for reducing the environmental impact of aviation, one of the most CO_2-intensive sectors. Various test flights and long-term studies have already demonstrated that HVO can be used in aircraft engines without major modifications, significantly simplifying the transition to more sustainable fuels. As a result, the aviation industry has begun to invest more in the research and development of HVO to significantly reduce CO_2 emissions in the coming years (Staples et al., 2018; Vas, 2019).

Vegetable Oil-Based Biofuels

A third type of biofuel directly derived from vegetable oils is vegetable oil fuel. In this case, pure vegetable oil – such as rapeseed oil – is used directly in diesel engines without any conversion processes. This type of biofuel has the advantage of being relatively simple and cost-effective to produce. However, the direct use of vegetable oil in diesel engines presents several technical challenges. Particularly at low temperatures, issues may arise due to the higher viscosity of vegetable oil, which thickens at lower temperatures and can impair engine performance. Furthermore, the ignition quality of vegetable oil is lower compared to conventional diesel, which can lead to incomplete combustion and reduced engine performance (Agarwal, 2007). Due to these drawbacks, the use of vegetable oil as a fuel in diesel engines is less widespread, though it is used in certain niche applications or as a transitional approach in less demanding vehicles.

Not only in the transportation sector but also in agriculture, vegetable oil fuels are increasingly being used. In several countries, agricultural enterprises are utilizing vegetable oils as direct fuel for their tractors and other agricultural machinery. Particularly in remote or rural areas where access to conventional diesel resources is limited, the use of vegetable oil offers an attractive way to increase the energy autonomy of these operations. Farmers can directly fill their machines with rapeseed oil or other vegetable oils after proper preparation. This approach not only helps reduce operating costs but also contributes to the reduction of greenhouse gas emissions. Another important aspect of this practice is the energy independence of agricultural enterprises, as they are no longer reliant on fossil fuels and can produce their own biofuels. This technique is increasingly being adopted in countries with a strong agricultural base, such as Brazil or India, to enhance supply security and sustainability (Prussi et al., 2019; Demirbas, 2018).

Practical examples of the use of vegetable oil-derived fuels in various sectors demonstrate how these sustainable alternatives are already being successfully applied to reduce CO_2 emissions and promote a more sustainable energy supply. Whether as an additive in the fuel sector, as a sustainability approach in the aviation industry, or as a direct fuel in agriculture, the benefits of vegetable oil fuels are diverse. Through ongoing research and development, as well as the growing acceptance of these technologies, vegetable oil fuels could play an even more significant role in the future of the energy transition.

5.2 Biomass as Alternative for Fossil Raw Materials

The use of biomass for energy generation experienced significant growth in the 2000s, as it was considered one of the most viable alternatives to fossil fuels. Biomass utilization encompasses a wide range of materials, including wood, agricultural waste, and industrial by-products, which are used for electricity and heat production (see Table 5.2). This form of energy generation is particularly important in regions with high agricultural activity (McKendry, 2002).

A key area of research in this field is the improvement of biomass gasification and liquefaction efficiency (see Figure 5.1). Companies and research institutions focused on technologies that enable more efficient conversion of biomass into biogas or biomethane. These gases can then be fed into gas grids or used for power generation (Luo et al., 2017). Another growing application area is second-generation biomass utilization, where agricultural and industrial waste is used as raw material for energy production (Sahoo et al., 2018).

5.2.1 Classification of Biogenic Fuels Suitable for Biomass Conversion

Table 5.2: Systematics of biogenic fuels.

Physical state of fuel	Origin	Examples/forms
Solid	So-called energy crops	Annual crops, for example, wheat, rye, barley, and triticale
		Perennial crops, e.g. reed grasses such as *Miscanthus*, or fast-growing woody plants like willows or poplars
	Residual materials	Co-products such as straw and bark
		Small-diameter wood (from thinning) and residual wood from forestry operations
		Residual wood from industrial and commercial wood processing
	Landscape maintenance materials	Pruning from trees and shrubs
		Green waste
Liquid		Alcohols
		Vegetable oils
Gaseous		Biogas
		Pyrolysis gas or low-calorific gas.

The systematics of biofuels that can be processed from biomass encompasses a wide variety of organic materials used for energy production. Biomass is a renewable resource that can come from both plant and animal sources. Here are main categories from a viewpoint of origin and their processing methods:

1. **Plant-Based Biomass**

   Wood: One of the oldest and most widely used forms of biomass. It can be utilized in the form of wood pellets, wood chips, or sawdust.

   Processing: Direct combustion, pyrolysis, gasification, or liquefaction into bio-oil.

   Agricultural Waste: Includes crop residues such as straw, corn stalks, or sugar beet leaves.

   Processing: These wastes can be converted into biogas or biomethane through gasification, fermentation, or direct combustion in biomass power plants.

   Energy Crops: Plants specifically grown for energy production, such as rapeseed, maize, wheat, or algae.

   Processing: Conversion into biofuels such as biodiesel or bioethanol through fermentation or transesterification.

   Algae: Algae is a promising biofuel source because of its rapid growth and high oil content.

   Processing: Algal oil can be converted into biodiesel, and the remaining biomass can be used for biogas production or animal feed.

2. **Animal-Based Biomass**

   Animal Fats and Oils: Waste products from the food industry, such as fat residues from meat processing or used cooking oils.

   Processing: These fats can be converted into biodiesel (HVO or FAME), which can be used in diesel engines.

   Manure: Animal waste can also serve as a biomass source, particularly in biogas production.

   Processing: Manure is fermented in biogas plants to produce methane, which can be used as a renewable energy source.

3. **Industrial Wastes**

   Paper and Cardboard Waste: Residues from the paper industry can be used in biomass power plants.

   Processing: Combustion or gasification for electricity and heat generation.

   Food Waste: Waste from food production or household food waste.

   Processing: Fermentation in biogas plants to produce methane for energy generation.

4. **Secondary Biomass Sources**

   Forest Residues: Includes bark, branches, and roots left over after timber harvesting.

   Processing: These residues can be processed in biomass power plants or through gasification and pyrolysis.

Agricultural By-Products and Food Processing Wastes: Wastes arising from the processing of agricultural products or food items, offering significant potential for energy production.

Processing: Biogas production or biomass combustion in appropriate facilities.

5. **Technologies for Biomass Processing**

Combustion: Direct burning of biomass to generate heat and electricity.

Gasification: Conversion of biomass into a combustible gas that can be used for electricity generation.

Pyrolysis: Conversion of biomass into bio-oil, biochar, and gaseous products through heating in the absence of oxygen.

Fermentation: Conversion of sugars (e.g. from corn or sugarcane) by microorganisms into biogas or bioethanol.

Anaerobic Digestion: Biological breakdown of organic material by microorganisms to produce biogas.

In conclusion, biofuels derived from various forms of biomass play a crucial role in sustainable energy production. They not only serve as an alternative to fossil fuels but also provide a means to recycle waste and by-products, optimizing resource use and contributing to reduced environmental impacts.

A key development has been research into biomass power plants, especially in Europe and North America, where biomass has increasingly been used as a substitute for fossil fuels in existing coal-fired power plants. The focus here is not only on the direct combustion of biomass but also on using biomass to produce fuel pellets, which are utilized in decentralized heating systems or larger energy plants (Mörsdorf et al., 2019).

The years since the 2000s have paved the way for the widespread application of biofuels, biogenic plastics, and biomass, even though full market establishment is still facing challenges. Ongoing research and technological development in these areas have laid the foundation for more sustainable resource utilization, which could contribute to the reduction of greenhouse gas emissions and the conservation of natural resources. In the coming years, overcoming the existing technical and economic barriers will be crucial to further establishing biogenic feedstocks as viable alternatives to fossil and mineral resources.

5.2.2 Market Size and Market Development Estimates for the Use of Biomass as an Alternative to Fossil Fuels

The use of biomass as a renewable energy source has gained increasing importance in recent decades and represents a crucial pillar of global efforts to reduce dependency on fossil fuels. The biomass market as an alternative energy source is shaped not only by technological advancements and the availability of biomass raw materials but also by political, economic, and societal frameworks that influence the acceptance and integration of this technology into existing energy markets.

The biomass market has been growing since the early 2000s and has now become a significant segment within the renewable energy sector (see Figure 5.2 as an example). According to the International Renewable Energy Agency (IRENA, 2020), the global biomass market was estimated to be worth around 50 billion USD in 2020, with an annual growth rate of approximately 6%. The increasing demand for clean energy and global efforts to reduce CO_2 emissions are major contributors to this positive market development.

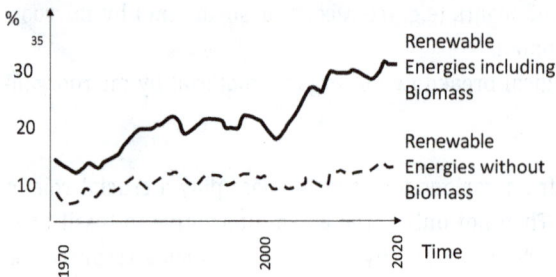

Statistics Austria Energiebilanzen 1970 - 2022

Figure 5.2: Use of bioenergy in Austria.
Overview of the use of bioenergy in Austria, illustrating the contribution of various biomass sources to the country's energy mix across different sectors, including electricity generation, heating, and transportation. The figure highlights the distribution and evolution of bioenergy use over time. Austria has leveraged its resources and strong tradition in sustainable forestry to develop a robust bioenergy sector. The figure also reflects national strategies aimed at reducing greenhouse gas emissions and enhancing energy self-sufficiency by substituting fossil fuels with renewable, locally sourced biomass. Key trends include the continued dominance of solid biomass for heating, the integration of biogas into decentralized energy systems, and the growing role of liquid biofuels in transportation. The depiction underscores bioenergy's central role in Austria's renewable energy policy and its importance in achieving national climate and energy targets, while also pointing to future challenges such as resource competition, air quality concerns, and the need for increased efficiency and innovation in biomass utilization.

The final energy consumption of bioenergy in Austria has increased by 65%, from 137 Petajoules in 2005 to 226 Petajoules in 2021. In 2021, the heating market was the primary sector for bioenergy, accounting for 84% of the market, followed by biofuels with a market share of 9%, and electricity generation from biomass and biogas with 7%. Bioenergy primarily benefits households. An analysis of the biogenic fuels and firewood used in Austria shows that more than 35% of the fuels are used for heating in private households. 21% of bioenergy is used in the manufacturing sector, with 18% alone in the wood and paper industry. 34% of biomass is used in wood power plants or district heating plants for electricity and district heat production, with district heat being used by 46% of private households and 39% of public and private service buildings. Only 1.7% of biomass is directly used for heating in the service sector.

According to a study by Markets and Markets (2024), the biomass market is expected to exceed 80 billion USD by 2025. This development is mainly driven by the increasing integration of biomass into energy production, heating supply, and as an alternative fuel for the transport sector. The growth trend is fuelled by various factors:

Political Support: In many countries, particularly in the EU, the United States of America, and China, political measures have been implemented to promote the use of biomass as a renewable energy source. These measures include moderate subsidies, tax incentives, and renewable energy quotas.

Technological Innovations: Continuous research in areas such as biogas production, bioenergy from algae, biomass pyrolysis, and waste utilization has led to improved and more efficient conversion technologies. These innovations enhance both the profitability and the environmental performance of biomass-based energy processes.

Market Segments for Biomass Energy

Biomass is utilized in various forms, each addressing different markets and applications:

–   **Biomass power plants and district heating:** Biomass is used globally in power plants for electricity and heat generation. In Germany, the share of electricity generation from biomass was just under 10% in 2022 (see Figure 5.3). In Europe, biomass power plants are increasingly being integrated into national power grids. It is estimated that 40% of biomass in Europe is used for electricity generation (European Commission, 2023). In Germany, biomass power plants have become an essential component of the energy transition in recent years. Particularly in the field of district heating, biomass has contributed to CO_2 reduction and increased energy efficiency through the use of wood pellets and agricultural waste (Energy and Management Powernews, 2023).

–   **Biogas and biomethane:** The use of biogas from organic waste for electricity generation is becoming increasingly popular in many countries. According to the Global Market Insights Report, the biogas market is gaining momentum, particularly in Europe, where biogas plays a significant role in achieving climate goals. Biomethane, which is derived from biogas and can serve as a substitute for natural gas, is another growing segment. In countries such as Sweden, Denmark, and Germany, biomethane is already widely used as a fuel and for grid storage (IRENA, 2022).

–   **Bioethanol and biodiesel:** The production of biofuels, particularly bioethanol and biodiesel, has become a significant sub-market of biomass utilization. According to the IEA, approximately 130 billion litres of bioethanol are produced annually worldwide, with the majority derived from corn and sugarcane. These biofuels are particularly important in the transport sector. In the USA, Brazil, and

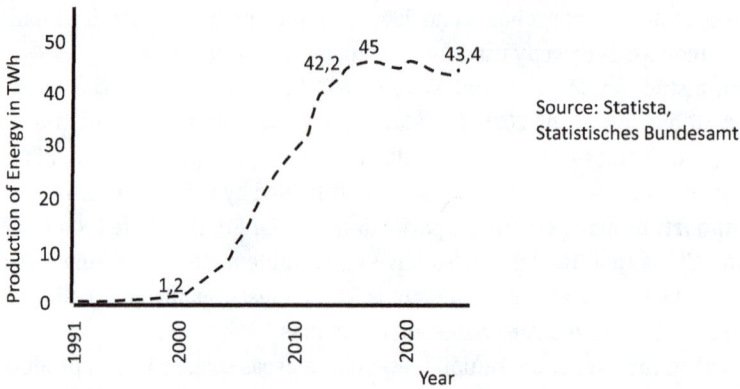

Figure 5.3: Electricity generation from biomass in Germany until 2024.
The statistics indicates the development of gross electricity generation from biomass in Germany from 1991 to 2024. The figure presents annual electricity output derived from various biomass sources – including solid biomass (wood and agricultural residues), biogas, and bio-waste – highlighting trends, growth phases, and periods of stabilization. Key policy milestones, such as the implementation of the Renewable Energy Sources Act (EEG) and subsequent amendments, are reflected in the trajectory of biomass-based electricity production. The data reveal an initial rapid expansion in the early 2000s, followed by a plateauing trend in recent years, due to changing subsidy schemes, increasing sustainability requirements, and competition with other renewable sources. The figure underscores biomass's continued but evolving role in Germany's energy transition ("Energiewende"), balancing dispatchable renewable power generation with environmental considerations, land use constraints, and the broader goal of carbon neutrality. In 2024, the amount of electricity produced from the alternative energy source biomass in Germany is estimated to be around 44 Terawatt-hours (V. Pawlik, Statista, 2024).

the EU, the demand for bioethanol has grown significantly as it is increasingly used as an alternative fuel in vehicles (Wikipedia, 2024).

–   **Industrial and household use:** The market for wood pellets, a form of biomass, has gained significant importance in Europe and North America in recent years. These are used both in private households and in industrial applications for heating and as fuel in biomass power plants. The use of wood pellets as a renewable energy source is also growing in Asia, particularly in China, where large quantities are consumed for heating and energy production purposes.

Market Development Assessments and Challenges

The long-term development of the biomass market, similar to other biogenic raw materials, is influenced by several factors:

–   **Technological Innovations:** Advances in biomass conversion technology to electricity, heat, or fuel are crucial for improving efficiency and reducing costs. In particular, the development of technologies for better utilization of non-edible biomass feedstocks and waste, as well as improving the CO_2 balance of biomass processes, could significantly increase the market potential (Acatech et al., 2019).

- **Political Measures and Regulation:** Governments worldwide are increasingly relying on biomass as part of their strategy to combat climate change and promote renewable energy. Subsidies for biomass power plants and biofuels, as well as CO_2 pricing models that increase the cost of fossil fuels, could further enhance the competitiveness of biomass (BMEL, 2022).
- **Availability of Raw Materials:** A significant barrier to the growth of the biomass market is ensuring a continuous and cost-effective supply of raw materials. Competition for agricultural land between food production and biomass could lead to price fluctuations and availability challenges (Welthungerhilfe, 2020).
- **Sustainability and Environmental Aspects:** The long-term acceptance of biomass as an energy source also depends on the ability to minimize environmental impacts, such as by using waste and residues instead of land used for energy crops. Sustainability certifications and life cycle analyses are playing an increasingly important role in ensuring that biomass production has no negative environmental effects (EU Commission, 2019).

The biomass market as an alternative energy source is growing steadily and offers significant potential for reducing dependence on fossil fuels. Various market segments, from biofuels to biomass power plants, benefit from technological innovations, political frameworks, and the growing demand for sustainable energy. In the long term, biomass is expected to continue playing a key role as an integral part of the global energy transition, with challenges such as raw material availability and sustainability needing to be addressed in order to fully exploit its market potential.

5.2.3 Possible Substitution Areas for Biomass as an Energy Carrier

The use of biomass as a renewable energy source presents an alternative to fossil fuels and offers various substitution opportunities in different areas of energy supply and usage. Biomass, derived from plant, animal, or microbial sources, can be processed into various forms and utilized in numerous industrial and societal applications (IEA, 2020). The substitution of fossil fuels with biomass energy carriers has the potential to contribute to reducing CO_2 emissions while simultaneously decreasing dependence on non-renewable resources (UBA, 2021).

The main substitution areas for biomass as an energy carrier are sketched in some detail below:

Electricity and Heat Production

A significant application of biomass as an energy carrier is in electricity and heat production. Biomass power plants, which burn organic materials such as wood, agricultural waste, or specifically cultivated energy crops like miscanthus or rapeseed, play a crucial role in the energy transition (BMEL, 2022). These power plants can substitute fossil fuels such as coal and natural gas, thereby contributing to the reduction of greenhouse gas emissions:

– **Biomass power plants:** In many countries, particularly in the EU, biomass is used for electricity generation. Biomass power plants rely on the combustion of solid biomass feedstocks, such as wood pellets, wood chips, and other plant-based waste. This technology is particularly well-suited to partially or fully replace existing coal or gas power plants (IEA, 2020). An example of this is the conversion of coal power plants to biomass, as carried out in Denmark or the United Kingdom (UBA, 2021).

– **District heating:** In many regions of Europe, particularly in Scandinavia and the Netherlands, biomass is used not only for electricity generation but also for providing district heating. Biomass serves as a CO_2-neutral energy source for heating households and industrial facilities (BMEL, 2022). In Germany, biomass heating plants also play an important role in decentralized heat supply, especially in rural areas (Federal Environment Agency, 2019).

Substitution in Industry

In industry, biomass offers numerous opportunities to replace fossil fuels in various production processes. Biomass is used not only as an energy source but also as a raw material in the chemical industry (IEA, 2020).

– **Biomass as an Industrial Power Source:** In industrial processes such as the paper and pulp industry, biomass is increasingly being used as a source of heat and electricity. These industries, which have traditionally relied on coal and oil, could contribute to the reduction of their CO_2 emissions through the use of biomass (BMEL, 2022).

– **Biochar and Wood Pellets in Industry:** In the steel industry and other energy-intensive sectors, biomass could be used as an alternative to coal and other fossil fuels. The use of biochar, a solid product derived from the pyrolysis of biomass, offers potential benefits in this regard (UBA, 2021). Wood pellets are also an established energy source for heating systems and industrial operations that depend on a cost-effective and CO_2-neutral approach (IEA, 2020).

Integration into the Circular Economy

The integration of biomass into the circular economy presents a promising opportunity to sustainably substitute fossil raw materials. By increasing the use of biomass from residues, waste, and non-arable agricultural land, biomass can be considered as part of a closed material cycle (BMEL, 2022). This not only reduces dependence on fossil fuels but also contributes to closing material loops and minimizing waste.

Other areas of substitution include:

Transport Fuels: Biofuels such as bioethanol and biodiesel can substitute petroleum-based fuels in the transport sector. Biomass-derived fuels can reduce reliance on gasoline and diesel while providing a renewable option for vehicles, including cars, trucks, and aviation.

Power Storage and Grid Balancing: Biomethane and other biomass-based gases can substitute for natural gas in energy storage and grid balancing applications. Biomethane can be used as a flexible energy source for both electricity generation and heating, enhancing the stability of renewable energy systems.

Biomass as an energy carrier offers numerous opportunities to substitute fossil raw materials in various sectors. From electricity and heat production to the transport sector, industry, and the circular economy, biomass provides a promising alternative to fossil fuels. The widespread use of biomass contributes not only to the reduction of CO_2 emissions but also enables more efficient resource utilization and reduces long-term dependence on non-renewable energy sources (IEA, 2020). Technological innovations and political support will be crucial in further unlocking the potential of biomass as a sustainable energy source.

5.3 The Use of Biomass as a Renewable Energy Source

The use of biomass as a renewable energy source has gained global importance in recent years. It represents a versatile and sustainable alternative to fossil fuels and is increasingly utilized in various sectors to reduce dependence on fossil energy carriers and lower CO_2 emissions. In practice, there are already a variety of successful biomass applications that are used not only in energy generation but also in other sectors such as agriculture and industry. Below are some of the key practical examples.

5.3.1 Biomass for Electricity and Heat Production

The use of biomass for electricity and heat production is one of the most well-known and widely applied uses. Biomass power plants and heating plants are used worldwide to generate energy from organic materials such as wood, agricultural waste, and specific energy crops.

- **Biomass Power Plants in Europe:** In countries such as Denmark and Sweden, biomass is intensively used to replace fossil fuels in electricity and heat generation. A prominent example is the Avedøreværket in Denmark, which operates one of the largest biomass power plants in the world. It uses wood pellets and other biomass sources as fuel and has converted from coal to biomass in order to reduce CO_2 emissions. This transition has helped support Denmark's climate policy and steer the country toward a more sustainable future (Ørsted, 2017).

–   **Biomass Heating Plants in Austria:** In Austria, biomass is widely used in the form of wood pellets and wood chips in heating plants and district heating systems. An example is the biomass power plant in Vienna-Simmering, which, with an electrical capacity of 24 MW, supplies about 12,000 households with district heating and 48,000 households with green electricity. This decentralized energy supply not only reduces the consumption of fossil fuels but also contributes to reducing air pollution and CO_2 emissions (Wien Energie, 2022).

5.3.2 Biofuels in the Transport Sector

Biofuels such as bioethanol, biodiesel, and biogas are significant biomass products that contribute to the reduction of CO_2 emissions in the transport sector. Many countries have already begun producing biofuels on a large scale and integrating them into the transportation sector:

–   **Bioethanol in Brazil:** Brazil is a global leader in the production and use of bioethanol, which is derived from sugarcane. The country has been operating the Proálcool program since the 1970s, which promoted domestic ethanol production and use. Today, Brazil is the world's largest producer of bioethanol and a pioneer in the field of sustainable biofuels. Bioethanol is used not only in vehicles but also as a fuel blend in conventional cars in the form of E85 (85% ethanol and 15% gasoline). This use of bioethanol has helped reduce dependence on fossil fuels and lower CO_2 emissions in the transport sector.

–   **Biodiesel in Europe:** In the EU, biodiesel is used as an additive to conventional diesel to reduce CO_2 emissions in the transport sector. In countries like Germany and France, biodiesel is produced from vegetable oils such as rapeseed or soybean oil and is introduced into the market in large quantities. Biodiesel not only has the potential to replace fossil fuels but also supports the agricultural sectors by providing an additional income source.

–   **Biogas for Public Transport in Sweden:** In Sweden, biogas produced from organic waste, such as food scraps and agricultural waste, is used to fuel buses and trucks. In cities like Gothenburg and Stockholm, public transportation is already largely powered by biogas. This measure has contributed to reducing emissions in urban transport and provides a climate-friendly alternative to conventional fossil fuels.

Biomass in Agriculture

Biomass has significant potential in agriculture, particularly in the form of vegetable oils and biogas. Many agricultural operations worldwide use biomass to increase their energy self-sufficiency and reduce fossil fuel consumption:

- **Vegetable Oil as Fuel in Agricultural Machinery:** In some rural regions, particularly in Germany, vegetable oils such as rapeseed oil are used as fuel for tractors and agricultural machinery. This technology provides farmers with a way to secure their energy supply while reducing dependence on fossil fuels. Some agricultural operations produce their own vegetable oil feedstocks, which strengthens the local economy and reduces the CO_2 footprint (VDI, 2020).
- **Biogas from Agricultural Waste:** Biogas plants are widely used in agriculture, especially in countries like Denmark, where they are employed for electricity and heat production. These plants use manure, crop residues, and other organic waste to generate biogas, which is then used to power the farm from which the materials come or sold to the public electricity grid. The use of biogas offers farmers a sustainable way to recycle waste while also lowering energy costs (Blenker, 2022).

Industrial Applications of Biomass

Biomass is also increasingly used in industrial processes to replace fossil resources and reduce CO_2 emissions. Specifically, biogenic raw materials such as wood, straw, algae, and used cooking oils are being utilized:

- **Biomass in the Cement Industry:** The cement industry is one of the largest CO_2 emitters worldwide. Some companies, including Heidelberg Cement, are increasingly using biomass products such as wood waste and agricultural residues to replace fossil fuels in their production processes. This shift has not only led to a reduction in CO_2 emissions but also promoted the use of waste materials in industrial processes.
- **Biomass in the Chemical Industry:** In the chemical industry, biomass is increasingly used as a raw material for the production of chemicals and plastics. One example is the use of sugarcane to produce bioethanol, which is then used as a feedstock for the production of bioplastics. This replaces the need for petrochemical products and reduces dependence on fossil resources (Wikipedia, 2023, Bioethanol).

Use of Biomass from Waste Recovery

As seen in the examples above the use of biomass in waste recovery represents a practical application that contributes to the substitution of fossil fuels such as

- **Waste-Based Biogas Plants:** In many cities worldwide, including San Francisco, biogas plants are operated that utilize organic waste such as food scraps and garden waste to produce biogas. This technology helps reduce waste volumes, lowers CO_2 emissions, and generates renewable energy (Muetterties, 2015).
- **Wood Waste Recovery:** Wood waste from forestry, furniture production, and the construction industry is increasingly being used in biomass power plants and heating systems for electricity and heat production. This practice contributes to the circular economy by using waste in a value-added manner and replacing fossil fuels (Meinlschmidt, 2013).

The use of biomass in practice demonstrates the versatility and potential of this renewable raw material in various sectors. Biomass is successfully utilized in electricity and heat production, the transport sector, agriculture, and industry to replace fossil fuels and contribute to the reduction of CO_2 emissions. Through technological innovations and political support, biomass is increasingly playing a crucial role in the energy transition and the implementation of sustainable business practices.

Chapter 6
Limits of the Use of Biogenic Raw Materials as Substitutes for Fossil and Mineral Raw Materials

6.1 Overview

The potential to use biogenic raw materials as substitutes for fossil and mineral raw materials is increasingly recognized as part of an approach for reducing dependence on non-renewable resources and mitigating environmental impacts. However, the widespread adoption of biogenic materials, particularly bioplastics, faces several limitations related to technical, economic, ecological, and practical aspects. To begin with the adoption of biogenic raw materials on a practical level is constrained by various practical factors:

- **Infrastructure and Adoption Rates:** The existing infrastructure for the production and processing of fossil-based materials is well-established and may not easily accommodate biogenic alternatives. This can result in a slower rate of adoption, as industries would need to invest in new technologies and adapt supply chains.
- **Consumer Awareness and Acceptance:** There may be resistance to biogenic products due to a lack of awareness or scepticism about their environmental benefits. Consumers may not always be willing to pay a premium for bioplastics, and industries may be hesitant to switch to biogenic materials without sufficient market demand or clear benefits.
- **Limited Availability of Suitable Feedstocks:** Not all regions have access to the necessary feedstocks for biogenic material production. For example, crops like corn and sugarcane may not be viable in all climates, and this can limit the global scalability of biogenic raw materials.

Despite promising progress in replacing fossil and mineral raw materials, significant challenges remain that hinder the widespread adoption of biogenic alternatives. Aside from the above, these challenges primarily concern ecological, economic, and technological factors. The complexity of these issues requires an integrated approach that simultaneously considers environmental sustainability, economic feasibility, and technical viability. A complete substitution of fossil and mineral resources can only be achieved through a combination of technological innovation, political commitment, and economic incentives. Close collaboration between industry, research institutions, and policymakers is essential to develop and implement viable strategies. Only through such cooperation can the sustainable and efficient use of biogenic resources be realized – bringing benefits to the environment, the economy, and society alike.

https://doi.org/10.1515/9783112218747-006

Replacing fossil energy sources with biogenic alternatives is a key goal of the global energy transition. The continued use of fossil fuels such as oil, coal, and natural gas contributes not only to high greenhouse gas emissions but also poses geopolitical and ecological risks. Biogenic energy carriers offer an opportunity to make energy supply more sustainable and environmentally friendly. By utilizing various biomass feedstocks and appropriate technologies, fossil fuels can be replaced across multiple sectors – playing a crucial role in reducing CO_2-emissions and dependency on finite resources.

Biogenic plastics in particular show considerable ecological potential, but several barriers still hinder their widespread adoption. These include high production costs, limited recycling infrastructure, and land-use competition with food production. Successfully establishing biogenic plastics as a sustainable alternative requires comprehensive efforts, such as increased research and development, expansion of recycling systems, and the promotion of sustainable, space-efficient agricultural practices. Only by overcoming these challenges can the full potential of biogenic plastics be realized and a decisive contribution made to the use of plastics and the transformation of the plastics industry. The progressive scarcity of mineral and fossil resources, combined with increasing environmental requirements, has led numerous industrial sectors to seek alternative ways. Substitution strategies play a critical role in ensuring supply security, reducing costs, and promoting ecological sustainability. In particular, the construction, chemical, and automotive industries are already employing or actively developing substitute materials for conventional raw materials.

Although biogenic raw materials offer tremendous potential, challenges often remain hidden when only their advantages are considered. The following section addresses specific aspects from technological, environmental, and market perspectives:

6.2 Technological Limitations

Biogenic raw materials often encounter technological challenges when considered as substitutes for fossil and mineral resources, particularly in industries that require high-performance materials. These issues often occur, however, if one-to-one replacements are sought for – consequently often a wider set of changes in existing products need to be considered. Key technical issues include:

- **Material Properties:** Bioplastics, such as those made from corn, sugarcane, or other biomass sources, sometimes have inferior mechanical properties compared to their fossil-based counterparts. For instance, they may be less durable, less heat-resistant, or more prone to degradation under certain conditions. This restricts their application in industries where strength, longevity, and resistance to environmental factors are critical.
- **Processing Challenges:** The processing of biogenic raw materials may require new or modified machinery, which can increase costs and complexity for indus-

tries accustomed to using fossil-based materials. Additionally, the scalability of biogenic materials is often constrained by their production methods, leading to limitations in meeting the growing global demand.

Biogenic materials often cannot match the quality profile of their fossil or mineral counterparts, particularly in applications involving high temperatures, mechanical stress, or corrosive environments.

For example, biogenic plastics such as polylactic acid (PLA) or polyhydroxyalkanoates (PHA) do not reach the same levels of heat resistance or mechanical strength as conventional plastics like polyethylene or polycarbonate. PLA begins to soften at temperatures as low as 50–60 °C, which limits its applicability in heat-exposed settings (Wikipedia, 2024a). While PHA exhibits high crystallinity, its melting point lies close to the temperature at which thermal degradation occurs, thereby complicating its processing (Wikipedia, 2024b).

Composites made from biogenic materials such as flax or hemp are considered promising alternatives to glass fibre-reinforced plastics in the automotive industry. However, they generally exhibit lower fracture toughness and reduced resistance to moisture.

Hempcrete, a building material made from hemp, offers excellent insulating properties but has a low compressive strength of approximately 0.3 MPa, making it less suitable for load-bearing construction applications (Wikipedia, Hempcrete, 2024).

6.2.1 Scalability

The cultivation and large-scale processing of biogenic raw materials capable of competing with the demand for fossil or mineral resources often present significant logistical and infrastructural challenges. The growth and processing of biogenic feedstocks such as hemp, flax, or sugarcane require specialized knowledge and technology to be efficient and scalable. Currently, biogenic materials are generally available only in limited quantities. Therefore, before they can be adopted for industrial use, their scalability must be ensured – an undertaking that may involve considerable costs and technological barriers.

6.2.2 High Production Effort and Costs

One of the major challenges in the use of biogenic plastics is the high production effort involved. Biogenic plastics are often derived from renewable resources such as corn, sugarcane, or other plant-based materials, which can lead to more complex and energy-intensive manufacturing processes. These processes typically involve fermentation, polymerization, and specific chemical conversions, which not only place technological demands on production facilities but also result in higher production costs.

The production costs of biogenic raw materials are often higher, placing them at a competitive disadvantage compared to well-established fossil-based alternatives. These elevated costs stem not only from current technological limitations but also from the still limited scale of production and the often more demanding processing of biogenic materials. To improve the competitiveness of biogenic resources, economic incentives such as subsidies or tax benefits are necessary to support the transition towards more sustainable resource use. Government funding and investments in research and development could further help to reduce costs and increase market availability (Bröring et al., 2017). Without such incentives, integrating biogenic raw materials into markets on a large scale and at competitive prices remains a significant challenge.

In many cases, the production of biogenic plastics remains more expensive than that of conventional petroleum-based plastics (Shen et al., 2010; Philp et al., 2013). These cost disparities can hinder broader adoption of biogenic plastics in commercial applications, particularly in highly price-sensitive sectors such as packaging or mass-produced goods. To achieve broader market acceptance, it would be necessary not only to reduce production costs but also to phase out subsidies or tax advantages for fossil-based materials in order to facilitate the shift from fossil to biogenic alternatives.

6.2.3 Technological Immaturity

Another significant barrier is the technological maturity of biogenic alternatives and related processes. While certain biogenic materials – such as bioplastics or bio-based chemicals – are already commercially available and used in various industries, many other biogenic substitutes remain in the developmental stage. These materials often require technological breakthroughs to make their production processes more efficient, scalable, and cost-effective.

The technologies involved in producing biogenic materials must be further refined to ensure that the quality and consistency of these products meet the standards of conventional materials. In particular, key innovations are still needed in areas such as biotechnology and fermentation technology in order to expand production capacities and reduce manufacturing costs (Cherubini, 2010).

6.3 Ecological and Environmental Limitations

While biogenic raw materials are often viewed as more sustainable than fossil resources, their large-scale adoption presents several ecological concerns:

- **Land Use and Food Production:** Large-scale cultivation of crops for bioplastics can lead to land-use competition with food production, especially if the crops used for bioplastics are grown on prime agricultural land. This could lead to rising food prices and food insecurity in vulnerable regions.

- **Biodiversity Impact:** Intensive monocultures for biogenic raw materials can also pose a threat to biodiversity, particularly in regions where native ecosystems are replaced with large-scale plantations. The focus on specific high-yield crops could contribute to soil depletion, water-use issues, and other environmental degradation concerns.
- **Carbon Footprint:** While biogenic materials are considered carbon-neutral during their life cycle (due to the carbon dioxide absorbed by plants during growth), the overall carbon footprint can be influenced by factors such as transportation, processing, and agricultural practices. For example, if the energy used in bioplastic production comes from fossil fuels, the carbon savings can be offset.

If biogenic raw materials are used on a larger scale to replace fossil and mineral resources, the demands on land use change significantly. In order to make space for energy crops, land is often diverted from food production, which can potentially affect food supply and prices (Bundestag, 2007; TATuP, 2015).

6.3.1 Land Use Competition

To illustrate the issue, it is helpful to first examine the distribution of land use in more detail – Germany is used as an example:

- **Agricultural Land Use:** Approximately 50.4% of Germany's total land area (around 18 million hectares) is used for agriculture. Of this, roughly 70% is arable land, while about 29% is permanent grassland (Umweltbundesamt, 2025).
- **Forested Areas:** Forests cover around 29.9% of the total land area, which corresponds to approximately 10.7 million hectares (Umweltbundesamt, 2025).
- **Settlement and Transportation Areas:** These combined account for approximately 14.5% of the total land area, with 9.5% used for settlements and 5.1% for transportation infrastructure (Umweltbundesamt, 2025).

These areas are already intensively and optimally utilized, making any restructuring of agricultural land use a complex task with limited flexibility for further optimization. Moreover, Germany is a net importer of agricultural products. Strictly speaking, this means that in addition to the agricultural land used domestically, there is a further area abroad – located in exporting countries – that is required to produce goods for consumption in Germany (Umweltbundesamt, 2025).

Another layer of complexity arises from the fact that, as with all agricultural products, not all cultivated land or local climates are equally well suited for the production of specific biogenic raw materials. The increased use of biogenic resources as substitutes for fossil and mineral materials could have significant implications for global land use. It is therefore essential to develop a balanced and sustainable strategy that takes into account ecological, social, and economic factors in order to minimize negative impacts on food production, biodiversity, and the environment (TATuP, 2015).

The production of bio-based plastics requires large quantities of plant-based feedstocks, which can lead to competition for land with food production. This issue is commonly framed as part of the "food versus fuel" debate, which questions whether agricultural land should be allocated to the production of raw materials for biogenic plastics or for food crops. This land-use competition may result in the diversion of valuable arable land from food production to biomass cultivation. Such shifts could destabilize global food markets and exacerbate ecological issues such as deforestation and soil degradation. The increased cultivation of crops for biogenic raw materials may lead to the clearing of forests to make room for monocultures, with detrimental effects on biodiversity and the CO_2 cycle (Searchinger et al., 2018). These land-use-related challenges must be carefully assessed to ensure that the use of biogenic resources leads to a genuinely positive environmental outcome and does not inadvertently cause harmful side effects.

> For example, crops such as sugarcane, corn, and rapeseed require large areas of land, which can potentially lead to land-use conflicts, especially in regions where agricultural land is already scarce or where food production is in competition. The increased cultivation of energy crops like corn and rapeseed can also have significant impacts on nitrogen balances, leading to environmental issues (Environmental Agency, 2019).
>
> In particular, in regions with limited agricultural land, this can lead to an increase in food prices, as more land is used for the production of raw materials such as corn, sugarcane, or soybeans (Searchinger et al., 2018; Brizga et al., 2020). Consequently, the growing demand for biogenic plastics could jeopardize food security and cause social and economic problems. It would, therefore, be necessary to promote sustainable agricultural practices that meet the demand for biogenic raw materials while ensuring food production.

Thus, land consumption and competition with food production are particularly problematic when biogenic materials such as bioethanol or biogenic plastics are produced in large quantities, as they could divert resources from food crops, potentially leading to shortages and price increases in food. Estimates suggest that global production of biogenic plastics could require around 2–5 million hectares of agricultural land if approximately 1 million tonnes are produced annually. Although this land area is relatively small compared to the areas used for fossil raw material extraction, it is still significant. This is especially true when considering global plastic consumption. The cultivation of raw materials like sugar beets or corn for these plastics could directly compete with food production, leading to land-use conflicts (Zhao & Li, 2019).

6.3.2 Monocultures and Biodiversity

The cultivation of monocultures (e.g. sugarcane, corn, and soybeans) for raw material production for biogenic materials can harm biodiversity and reduce soil quality. There is evidence that intensive monocultures over large areas can lead to soil depletion, erosion, and water pollution through the use of fertilizers and pesticides in the long term (Bund für Umwelt und Naturschutz Deutschland [BUND], 2022). A more sustainable solution would be the cultivation of mixed crops or the use of plant-based waste, but these options are currently and likely in the long term difficult to implement from both a technological and economic perspective.

6.3.3 Water Consumption

Often biogenic raw materials such as hemp and flax often require less water than their mineral competitor, but not all biogenic raw materials offer this advantage. Some biogenic raw materials, such as cotton, sugarcane, and maize hybrids, require large amounts of water, which is especially problematic in regions that are already facing water scarcity (Wissenschaftlicher Beirat der Bundesregierung Globale Umweltveränderungen [WBGU], 2008).

   The increased cultivation of monocultures for raw material production has profound impacts on biodiversity and can significantly affect the stability of ecological systems. Monocultures, which involve the practice of growing a single plant species on large areas over several years, represent one of the greatest threats to biodiversity. This form of agriculture not only reduces the genetic diversity of plant populations but also harms the diversity of associated organisms including insects, birds, soil organisms, and other ecological niches (Tilman, 1997). The significance of this issue justifies a brief digression.

6.3.4 Reduction of Genetic Diversity

The cultivation of monocultures primarily leads to a depletion of genetic diversity within the affected plant populations. When a single plant system dominates large land areas, the genetic variance within the populations decreases. This makes the plants more susceptible to diseases and pests, as a homogeneous plant population can be more easily infected by pathogens or pests. As a result, the entire plant unit is at risk, which not only reduces yields but also jeopardizes the long-term survival of the crops in a changing environment (Altieri, 1999). Furthermore, the low-genetic diversity within monocultures reduces the resilience of the farming system to climatic and ecological fluctuations. The absence of different varieties of a crop, which are adapted to varying environmental conditions, can significantly impair the agricultural system's ability to adapt to climate change (Giller et al., 2015).

6.3.5 Soil Quality and Nutrient Depletion

The continuous cultivation of the same plant species on the same land leads to significant nutrient depletion in the soil, as the same plant species continually absorbs specific nutrients in high concentrations. In monocultures, certain nutrients are often over-exploited, while others remain in the soil and may disrupt the soil microbiome. This one-sided use of soil resources not only reduces soil fertility but also negatively impacts soil biodiversity (Foley et al., 2011). A reduced microbiome leads to a decline in soil organisms such as bacteria, fungi, and soil insects, which are essential for nutrient cycles and soil structure (Bardgett & van der Putten, 2014). The resulting soil compaction and erosion can further degrade the water-holding capacity and structure of the soil, making the cultivation system even more vulnerable to environmental changes. In the long term, this could lead to desertification or a loss of arable land, further threatening the production of biogenic raw materials and food.

6.3.6 Disruption of Food Chains and Habitats

The loss of diversity in agricultural areas has far-reaching impacts on wildlife. Monocultures provide a limited habitat for many animal species, particularly for pollinators such as bees and butterflies, which rely on a diverse plant world (Klein et al., 2008). A single plant habitat does not offer the food diversity necessary to maintain stable food webs. This can lead to a decline in pollinator populations, which can affect agricultural production itself, as many crops depend on insect pollination (Gallai et al., 2009). Furthermore, monocultures result in the loss of stable habitats for wildlife, as the variety of vegetation required by different species is significantly reduced.

Monocultures also lead to a reduction in stable habitats for wildlife, further impacting biodiversity. A reduced habitat for animals not only affects species diversity but also disrupts ecological processes such as plant dispersal (i.e., the process by which plant seeds or plant parts, such as fruits, shoots, or roots, are spread over long distances, particularly for reproduction. This process is a crucial component of plant species' ecology and contributes to biodiversity and ecological dynamics. It enables plants to migrate to new areas, expand their populations, and thrive in different environments) and soil aeration, which are essential for healthy ecosystems.

6.3.7 Pesticide Use and Ecological Burdens

Monocultures are often associated with increased pesticide and fertilizer use, as focussing on a single crop over extended periods increases the risk of pest infestations and diseases. This has direct consequences for soil and water quality as well as air quality (Pimentel et al., 1992). Pesticides not only harm target organisms but also beneficial insects and soil organisms that are crucial for soil fertility and food web stability. The use of pesticides can also cause secondary damage that impacts neighbouring ecosystems and surrounding wildlife.

6.3.8 Displacement of Natural Ecosystems

The increased cultivation of monocultures for the production of biogenic raw materials often leads to the conversion of natural ecosystems into agricultural land. Forests, wetlands, or meadows, which originally housed high biodiversity, are sacrificed for the cultivation of raw materials such as sugar beets, maize, or oil crops (Balmford et al., 2005). This land-use change is often accompanied by a loss of ecosystem services that are crucial for maintaining ecological balance such as water regulation, carbon sequestration, and air purification.

The conversion of natural areas into agricultural monocultures can lead to a decline in species that rely on specific habitats, while also reducing the resilience of affected landscapes to climatic changes and disturbances.

The impacts of increased monoculture cultivation for raw material production on biodiversity are profound and far-reaching. Monocultures lead to a reduction in genetic diversity, degrade soil quality, disrupt the habitats of many animal species, and increase pesticide use, which can destabilize ecological systems. To minimize the negative consequences for biodiversity, it is essential to develop agricultural practices that focus on biodiversity conservation, agroecological principles, and sustainable resource use (Altieri, 1999). Only by promoting biodiversity-friendly farming systems can sustainable raw material production be ensured in the long term, without destroying the ecological foundations of agriculture and natural systems.

6.3.9 Forest Use and Reforestation

In some scenarios, forests could be cleared for the cultivation of energy crops, which could have negative effects on the carbon balance and biodiversity (Searchinger et al., 2008). Alternatively, increased reforestation could take place to use wood as a biogenic raw material, but this approach requires time and reduces the available agricultural land (Luyssaert et al., 2008). The significance of this issue justifies a brief excursus.

6.3.10 Forest Use for the Cultivation of Energy Crops and Promotion of Reforestation

The utilization of forests for the cultivation of energy crops and the promotion of reforestation represent two strategic approaches to meet the growing demand for biogenic raw materials, particularly in the context of replacing fossil and mineral resources. However, both approaches have far-reaching ecological, economic, and social implications, which can have both positive and negative effects on the carbon balance and biodiversity (Houghton, 2003). These aspects must be carefully weighed in order to develop sustainable land-use systems that align with climate and nature conservation goals (Foley et al., 2011):

Forest Use for the Cultivation of Energy Crops

Forests play a crucial role in the global carbon cycle by absorbing CO_2 from the atmosphere and storing it in the form of biomass (Galik et al., 2015). Deforestation and the conversion of forests into agricultural land release this carbon storage, which can negatively impact the CO_2-balance of the entire system (Houghton, 2003). The conversion of forests into agricultural land, particularly for the cultivation of energy crops such as rapeseed, sugar beets, or maize for bioenergy production, can have significant effects on the carbon balance and biodiversity (Searchinger et al., 2008).

6.3.11 Impacts on the Carbon Balance

The deforestation of forests for the cultivation of energy crops initially leads to the release of carbon into the atmosphere. This process occurs through the burning of wood and the decomposition of organic material in the soil, which has stored carbon over years in the forest (Fargione et al., 2008). Even though energy crops are later planted on the cleared land, which absorb CO_2 during their growth, it may take decades for the original carbon loss to be fully compensated by the newly planted crops (Searchinger et al., 2008). Moreover, the effect is temperature-dependent (Houghton, 2003). This temporal delay makes the conversion of forested areas to agricultural use problematic in terms of the CO_2 balance and diminishes the climatic benefits of bioenergy plants compared to fossil fuels (Searchinger et al., 2008).

Furthermore, soil carbon may be reduced due to changes in land use and the intensification of management practices (Foley et al., 2011). Agricultural practices, through the increased use of machinery, fertilizers, and irrigation, deplete the organic material in the soil, which further reduces carbon content and can negatively affect soil fertility in the long term (Galik et al., 2015).

6.3.12 Impacts on Biodiversity

The conversion of forests into agricultural land also has significant consequences for biodiversity. Forests are home to a wide variety of species – ranging from plants and animals to microorganisms – that rely on the specific conditions of the forest soil, climate, and vegetation (Müller et al., 2015). Deforestation destroys these habitats and leads to a loss of species diversity (Sodhi et al., 2010). Particularly threatened are rare and endemic species that occur only in specific forest ecosystems and can become irreversibly extinct due to changes in their habitats (Müller et al., 2015).

The transformation of forests into monoculture agricultural land for energy crops can also disrupt the ecological balance. Monocultures provide very limited habitats for a variety of animal species and reduce the diversity of food sources and shelters (Sodhi et al., 2010). Changes in flora and soil structure can also reduce habitats for insect species, including vital pollinators, which can further impact agricultural productivity (Bengtsson et al., 2005).

Afforestation for the Use of Wood as a Biogenic Raw Material

In contrast to deforestation, afforestation of degraded areas or the conversion of agricultural land into forests presents a potentially more sustainable solution for providing biogenic raw materials such as wood, which can be used in industries like construction, furniture production, or bioenergy generation (Luyssaert et al., 2008). Afforestation projects can help sequester carbon and contribute to ecosystem restoration (Mund et al., 2010).

6.3.13 Long-Term Storage Effects:

Forests are not only important as carbon sinks but also play an essential role in regulating the water cycle, soil quality, and climate control (Foley et al., 2011). Afforestation projects, which specifically focus on the planting of fast-growing tree species such as eucalyptus or poplars, can relatively quickly absorb and store carbon from the atmosphere (Luyssaert et al., 2008). However, the potential of these forests to store carbon is highly dependent on the tree species and soil conditions (Mund et al., 2010). Sustainable afforestation requires careful selection of tree species to maximize long-term carbon storage while ensuring the ecological value of the forest (Luyssaert et al., 2008).

It is important to note that afforestation projects have delayed effects. Forest carbon sequestration takes decades to absorb significant amounts of CO_2 (Mund et al., 2010). In a climate context that urgently requires greenhouse gas reductions, afforestation alone cannot achieve immediate CO_2 reduction goals (Müller et al., 2015).

Combination of Afforestation and Sustainable Agriculture

A potential solution that promotes carbon sequestration through forests while also ensuring agricultural production lies in agroforestry (Jose, 2009). This method combines the cultivation of trees with agricultural land use and offers numerous advantages. On the one hand, trees can continue to sequester carbon without a complete loss of agricultural land (Müller et al., 2015). On the other hand, this practice can improve soil quality and fertility while stabilizing the water balance (Jose, 2009).

Agroforestry also aims to produce biogenic raw materials such as wood, fruits, or nuts on more sustainable and biodiversity-friendly lands. This method contributes to biodiversity conservation by creating ecological corridors that can serve as habitats for animals, while simultaneously maintaining agricultural yields (Bengtsson et al., 2005).

The cultivation of biogenic raw materials often requires significant land area, making it a limiting factor for many biogenic resources. This is particularly true when biogenic raw materials are used on a large scale as substitutes for mineral or fossil materials.

6.4 Limitations of Markets

Despite the environmental benefits, the economic challenges of replacing fossil and mineral resources with biogenic raw materials are significant:
–   **Production Costs:** Biogenic materials are often more expensive to produce than fossil-based alternatives due to the need for specialized agricultural practices, higher raw material costs, and more complex manufacturing processes. This price disparity makes it difficult to compete with fossil-based products in price-sensitive markets.
–   **Supply Chain Constraints:** The supply of biogenic raw materials is subject to agricultural cycles and may face competition with food production. This can result in price volatility and potential shortages of key materials. Additionally, the land required for growing crops like corn or sugarcane for bioplastics may conflict with food production needs or lead to land-use changes that could have unintended environmental consequences.

6.4.1 Economic Aspects of Biogenic Raw Materials

There are further relevant economic aspects that should be considered when analysing biogenic raw materials and their market boundaries. These points address both scalability and the long-term competitiveness of biogenic materials compared to fossil alternatives:

1. Production Costs: As mentioned, the production costs for many biogenic materials are often higher than for fossil or mineral raw materials. Biogenic plastics or fibres are generally more expensive because the raw material production and processing technologies have not yet reached the same level of scalability as fossil materials. Fossil raw materials benefit from decades of production optimization and subsidized fossil fuels. In contrast, biogenic raw materials often face price fluctuations and market resistance due to the technology not being widely established yet (Fachagentur Nachwachsende Rohstoffe, 2012).
2. Infrastructure and Logistics: The logistical challenges of harvesting, transporting, and processing biogenic raw materials can present a bottleneck for their widespread industrial use. Developing a complete infrastructure for the use of biogenic raw materials can be very expensive and must first be economically viable to be adopted by large industries (Deutsches Biomasseforschungszentrum, 2023).
3. Marketability and Price Fluctuations: The price development of biogenic materials is highly dependent on the availability of raw materials and production costs. Biogenic raw materials such as bioplastics or biofuels are susceptible to market fluctuations that stem from factors such as crop yields, global climate conditions, and political uncertainty. The dependence on agricultural yields means that these materials can become more expensive in times of poor harvests or extreme weather conditions. In comparison, fossil raw materials, while also susceptible to price fluctuations, often benefit from a more stable production chain that has been optimized over years. Moreover, continuous improvements in production processes and the availability of subsidies for fossil fuels could further impair the competitiveness of biogenic raw materials (Agora Energiewende, 2015).

6.4.2 Consumer Acceptance and Market Behaviour

The acceptance of biogenic materials by consumers plays a crucial role in their market integration. While many consumers are becoming increasingly environmentally conscious and actively seek more sustainable products, as indicated above, the higher costs of biogenic raw materials can present a barrier in some market segments. Alike for all materials the price-performance ratio of bio-products is often a decisive factor in the purchasing decision. In the automotive industry, for instance, where the demand for environmentally friendly vehicles is growing, the acceptance of biofuels or biogenic plastics may be hindered by the price differences compared to traditional materials (Bioökonomierat Bayern, 2021). On the other hand it is remarkable how small the effects of implemented results of research on the reduction of consumption actually is.

Long-Term Economic Viability and Scalability Potential

Another important economic aspect is the long-term scalability of biogenic raw material production. Developing and establishing infrastructure for large-scale use of biogenic materials require significant investment and time. Moreover, companies investing in these technologies must ensure long-term profitability. This could be achieved through increasing industrial use and the creation of widespread market demand, which, however, can only be realized with continued technological promotion and a clear economic outlook (Umweltbundesamt, 2024).

The economic and market-related challenges associated with biogenic raw materials primarily stem from high production costs, infrastructure requirements, and the dependency on political and market conditions. To establish biogenic raw materials as a true alternative to fossil materials, comprehensive technical innovations, political support, and a rise in demand for sustainable products are necessary.

Subsidies and Regulation

Despite the potential benefits, subsidies and regulations need to be carefully designed to avoid market distortions. Over-reliance on subsidies can lead to inefficiencies and create dependency, preventing the sector from becoming truly competitive in the long term. Additionally, the regulatory landscape must remain flexible to adapt to new scientific findings and technological advances. Poorly designed regulations can inadvertently hinder innovation or create barriers to market entry for smaller companies.

In many countries, fossil fuels continue to be subsidized, distorting competition and hindering the transition to bio-based alternatives. According to the International Energy Agency and the Organisation for Economic Co-operation and Development, subsidies for fossil fuels in 52 industrialized and emerging countries averaged 555 billion US dollars per year between 2017 and 2019 (Timperley, 2021). This ongoing financial support for fossil fuels presents a significant barrier to the development and market introduction of bio-based materials.

At the same time, there is a lack of political incentives or subsidies to promote the development and widespread use of bio-based raw materials. Regulatory uncertainties and the standardization of bio-based materials, particularly in the areas of certification and sustainability, are not consistently regulated worldwide, complicating international trade. Different countries or markets may have varying standards and regulations, which impedes the scaling and global trade of bio-based products (NABU, 2021).

The regulatory environment also plays a significant role in the adoption of biogenic raw materials. While some countries have a set of ambitious sustainability goals and provided incentives for bioplastics, there are still gaps in global policy:

- **Certification and Standards:** There is a lack of consistent global standards and certification for biogenic materials, which makes it difficult for consumers and industries to assess the environmental and social impact of different products.

The absence of clear guidelines can result in greenwashing, where products are marketed as more sustainable than they actually are.
- **Policy Uncertainty:** Policies related to subsidies, tax incentives, and environmental regulations may vary widely by region. The lack of a clear, unified global policy could hinder large-scale investment and the standardization of biogenic materials in industrial applications.

To facilitate the transition from fossil resources to bio-based alternatives, governments and international institutions must collaborate to create a framework that supports bio-based materials and enhances their competitiveness. This could be achieved through the introduction of tax incentives, subsidies, investments in research and development, as well as the establishment of an international standard for the sustainability of bio-based products. Such a legal and economic framework could enable bio-based materials to compete with fossil alternatives, leading to greater market penetration and a more sustainable use of resources.

Overall, it can be said that the transition to bio-based raw materials in many countries is significantly hindered by the continued subsidies for fossil fuels and regulatory barriers. However, better political support for bio-based raw materials, combined with the harmonization of international standards, could help level the playing field and pave the way for more sustainable raw material production.

Chapter 7
Conflict Between Food Supply and the Cultivation of Biogenic Resources

7.1 Overview

Although their global market penetration is not yet dominant, the increasing demand for agricultural raw materials has, over the past decades, led to intensify local use of agricultural land for the cultivation of alternative resources. This development is driven by several factors.

A major driver is the industry's growing commitment to sustainability, which has resulted in increased use of biogenic plastics, textiles, and basic chemicals. These industries increasingly rely on agricultural raw materials such as maize, sugarcane, and cellulose (Spierling et al., 2018). In addition, the rising importance of bioenergy plays a role, as biofuels such as biodiesel and bioethanol are promoted as alternatives to fossil fuels. This demand has led to an expansion of cultivation areas for energy crops such as maize, rapeseed, and palm oil (Searchinger et al., 2008). Political measures and subsidies also support the expansion of the bioeconomy and the use of agricultural raw materials in non-food sectors, thereby intensifying competition for agricultural land intended for food production (Lamers et al., 2011).

An important aspect of this development is the use of arable land for bioenergy and industrial purposes. The growing demand for bioenergy – particularly biofuels such as biodiesel and bioethanol – has led to displacement of food production. Over 10% of global arable land is currently used for biofuel production, with significant regional variation: in the United States, approximately 40% of maize production is directed toward bioethanol, contributing to rising corn prices (Fargione et al., 2008). In Southeast Asia, palm oil production for biodiesel is a major driver of rainforest deforestation (Gatti et al., 2019), while in Europe, around 60% of rapeseed production is used for biodiesel, thereby limiting its availability for food production (Zilberman et al., 2013).

Beyond bioenergy, agricultural raw materials are increasingly being used in other industrial sectors. For example, bioplastics such as polylactic acid, derived from maize or sugarcane, are being promoted as sustainable alternatives to conventional plastics (Rujnić-Sokele & Pilipović, 2017). Similarly, the bio-based textile industry increasingly relies on natural fibres such as cotton, flax, or hemp, further intensifying the competition for land between textile and food production (Shen et al., 2010). Although their market share remains relatively low, agricultural raw materials such as starch and lignin are also being increasingly utilized as substitutes for petrochemical feedstocks in the chemical industry (Clark et al., 2017). This growing use of arable land

https://doi.org/10.1515/9783112218747-007

for industrial applications exacerbates competition and carries both economic and ecological implications.

The competition between food production and biogenic raw materials for agricultural land is poised to become a significant challenge, particularly with regard to food security and social equity. The effects of this competition are already evident in the production of biogenic fuels. While the use of biogenic resources can contribute to reducing dependence on fossil fuels, it also entails substantial risks due to its impact on food crop availability, increased production costs, and rising prices. A balance between alternative resource use and food security has yet to be achieved. The resulting effects on consumer prices can disproportionately affect low-income households and countries with limited natural resources.

Another existential threat arises for smallholder farmers from rising land prices, land grabbing, and the displacement of traditional agricultural practices. "Small-scale food providers, such as peasants, small-scale fishers and fish harvesters, pastoralists, and Indigenous Peoples, produce more than half of the food consumed by the world's population on only about 35% of the world's cropland. Farms of up to 20 hectares produce 59% of key food crops on just a quarter of the total farmland, highlighting their productivity" (P. Seufert, 2025). The world's ten largest private landowners control approximately 405,000 km^2 (Blue Carbon, VAE; Maquarie Group; Australia; Glam Group; Manu Life, Canada; Aranco, Chile; Shell, GB; TIAA/Nuveen, USA; Edizione, Italien; Cresud, Argentinia; Wlimar International, Singapore). "The increasing concentration of land ownership endangers food security and threatens the livelihoods of around 2.5 billion small-scale farmers, as well as 1.4 billion of the world's poorest people who rely directly on agriculture for their survival" (Dostert E., Landbesitz Wie sich wenige Konzerne viel Land sichern Sueddeutsche Zeitung 3. Juni 2025).

Where such developments occur, they undermine local food systems and exacerbate economic vulnerabilities. Addressing these challenges requires targeted policy interventions that simultaneously ensure food security and promote environmentally sustainable raw material production. Special support for small-scale farmers is essential to mitigate social, ecological, and economic risks.

Resource-rich countries in the Global South are particularly affected, as smallholder farmers, indigenous communities, and local producers are often the losers in the expansion of the agro-industrial sector. This development frequently results in land dispossession, labour exploitation, and the destruction of natural resources such as rainforests and freshwater sources. To mitigate these global conflicts, sustainable agricultural practices, stricter regulation of land investments, and international agreements for the fair distribution of resources are essential.

The competition for agricultural raw materials leads to both political and economic tensions at local and global levels. Key sectors such as the food, energy, and chemical industries are particularly affected, as they compete for limited resources. Long-term solutions – such as the promotion of sustainable agriculture and the equitable use of resources – are necessary to prevent social and economic instability.

Without comprehensive regulatory frameworks and the strengthening of land rights for marginalized groups, competition for agricultural raw materials is likely to intensify and further exacerbate global inequality. In developing countries in particular, rising competition may jeopardize food security in the absence of targeted political interventions.

7.2 Competing Demands for Farmland

7.2.1 Indirect Land Use Change (ILUC)

Another issue related to the substitution of fossil with plant-based resources is indirect land use change (ILUC). To pick biofuels as an example:

When the demand for biofuels increases, it can lead to the conversion of forests, grasslands, or other natural ecosystems into agricultural land to expand the cultivation of energy crops. These ILUCs can significantly worsen the carbon footprint of biofuels over time, as converting natural ecosystems into farmland results in the loss of carbon sinks. Forests and other natural landscapes store large amounts of carbon, and their destruction releases this sequestered carbon, thereby increasing greenhouse gas emissions. These effects can substantially reduce, or even negate, the intended climate benefits of using plant-based fuels. In some cases, the carbon balance of biofuels may even be worse than that of fossil fuels (Searchinger et al., 2008).

Technical Limitations

Although plant-based oil fuels are regarded as a sustainable alternative to fossil fuels, they face a range of challenges and limitations (see Section 5.2.1). The land requirements for cultivating energy crops, the ILUC, and the technical constraints associated with biodiesel use are critical factors that can limit the effectiveness of these biofuels in achieving environmental and climate goals. It is therefore essential to address these challenges while simultaneously advancing research and development of new, sustainable technologies and processes to further optimize the use of plant oil-based fuels and minimize their negative impacts on the environment and society.

The use of agricultural land represents a major challenge in the global allocation of resources. While farmland has traditionally been used for food production, the demand for agricultural raw materials for bioenergy, bioplastics, and other industrial applications is increasing. This has resulted in a growing land use conflict with significant economic, ecological, and social implications.

The economic and social impacts of the competition between food production and alternative raw materials are substantial. One of the most notable consequences is the rise in food prices, which is partly driven by competition for arable land. For instance, the increased use of maize for bioethanol production has led to price spikes that have affected global food supply chains (Roberts & Schlenker, 2013). Similarly, the rising demand for biodiesel has contributed to higher palm oil prices in Indonesia, disproportionately affecting low-income households (Brad, 2019). Smallholder farmers in developing countries are particularly vulnerable. The phenomenon of land grabbing – where multinational corporations acquire farmland for energy crop production – often results in the displacement of local farmers (Borras et al., 2011). Moreover, the large-scale cultivation of cash crops such as soy or sugarcane displaces local staple food production, leading to increased food insecurity (Perfecto & Vandermeer, 2010).

Land Area Requirements and Land Use Conflicts

One of the major challenges associated with the use of plant-based oil fuels is the land area required for cultivating energy crops. Large-scale cultivation of energy crops such as palm oil, soy, or rapeseed can lead to significant land use conflicts. To meet the growing demand for biofuels, increasingly larger areas of agricultural land are being used, which is often associated with negative ecological consequences. This is particularly critical when the cultivation of energy crops leads to deforestation in order to create new agricultural land. A notable example is the cultivation of palm oil in tropical regions, which is frequently linked to deforestation and the loss of biodiversity (Fargione et al., 2010). The conversion of forests to agricultural land not only has detrimental environmental effects but can also trigger social conflicts, as local communities and indigenous peoples are often excluded from access to land and resources (Carr, 2008).

In 2023, the global agricultural land area totalled 1.5 billion hectares. Of this, approximately 87% was used for food crop cultivation, about 7% for the cultivation of materials, around 5% for biofuel production, and about 1% for growing raw materials for bioplastics (Hochschule Hannover, Faculty 2, Biopolymers, Facts and Statistics 2023). The regional distribution of these proportions differs significantly, as illustrated below.

Ecological Consequences of Land Use Competition

The ecological consequences of land use competition are also severe. The shift towards alternative raw materials results in a loss of biodiversity and deforestation, particularly in regions such as the Amazon in Brazil, where soybean production for biofuels is a major driver of deforestation (Fear side, 2018). Similarly, in Southeast Asia, palm oil production is increasingly linked to the destruction of natural habitats and the endangerment of biodiversity (Brad, 2019). Furthermore, intensive land use contributes to soil degradation, as monocultures deplete soil nutrients and increase the demand for fertilizers (Pimentel & Pate, 2005). The high water consumption required for sugarcane cultivation for bioethanol also leads to conflicts in water-scarce regions (Gerbens-Leenes et al., 2009).

To mitigate the land use competition between food production and alternative raw materials, solutions are required. More efficient use of waste materials could reduce the need for new agricultural land. For example, utilizing agricultural waste, such as straw for biofuels, could, to some degree, help alleviate land pressure (Scarlat et al., 2015). Additionally, sustainable farming systems, such as intercropping and agroforestry, could enhance land efficiency and reduce environmental impacts (Tilman et al., 2006). Increased political regulation of land use and the promotion of sustainable alternatives are crucial to minimizing negative social and ecological consequences.

Although there is substantial knowledge regarding the land use and fertilizer requirements of many biogenic raw materials, unlike fossil resources, which benefit from decades of production optimization, biogenic materials are still in the early stages of development. This is associated with price fluctuations, market resistance, and sometimes insufficient technology adoption (Hermann et al., 2007).

The research gap in land use and area estimation for the potential substitution of mineral resources is problematic. Particularly in the textile and packaging industries, biogenic raw materials offer significant potential, while many uncertainties remain in the construction industry and energy storage technologies (Carus et al., 2014). Despite this potential, there are challenges in the practical implementation of biogenic materials, especially concerning their performance, scalability, and ecological impacts. These include land use changes, water consumption, and the formation of monocultures (Searchinger et al., 2008). Economic and logistical barriers, such as high production costs and insufficient infrastructure, further hinder broad application.

The question of whether forests should be cleared for the cultivation of energy crops or if reforestation projects should be intensified has far-reaching implications for effects on several ecologic parameters such as the carbon balance and biodiversity. Deforestation for agricultural purposes releases carbon and reduces biodiversity, while reforestation can contribute to carbon sequestration, though it may potentially hinder agricultural production (Fargione et al., 2008). A sustainable approach could be the combination of reforestation and agroforestry (Nair, 2007) – agroforestry understood as the intentional integration of trees and shrubs into crop and animal farming systems to create more diverse, productive, sustainable, and resilient land use systems.

7.2.2 Competition for Agricultural Land and Its Impact on Local and Global Markets

The competition for agricultural land, driven by the increasing demand for food, biofuels, and alternative raw materials, has profound effects on both local and global markets. As agricultural land is increasingly used for non-food purposes, such as the cultivation of bioenergy crops and bioplastics, the availability of land for food production becomes constrained. This growing competition leads to higher land prices, shifts in agricultural production patterns, and changes in the supply and demand dynamics of various commodities. On a local scale, smallholder farmers may struggle to access land, leading to land displacement and increased economic inequality. Globally, these shifts in land use can exacerbate price volatility, particularly for staple food crops, and potentially cause food insecurity in regions dependent on imports.

The rising demand for bioenergy, particularly in the context of efforts to achieve sustainable energy supplies, has led to a more intensive use of agricultural land for biomass production.

Studies have shown that the expansion of bioenergy production in the last two decades has significantly contributed to these price increases. In the United States, for instance, approximately 40% of corn production is used for bioethanol production, which has led or at least contributed to a rise in corn prices on global markets (Roberts & Schlenker, 2013). Similarly, the high demand for palm oil for biodiesel in Southeast Asia has significantly increased prices for cooking oils, particularly affecting low-income households (Gatti et al., 2019). In Europe and South America, the use of rapeseed and soy for biofuels has led to higher prices for animal feed and cooking oils (Zilberman et al., 2013). Furthermore, the increased production of biofuels has amplified the volatility of agricultural markets, as agricultural prices are now more closely tied to fluctuations in energy markets. Rising oil prices make biofuels more economically attractive, which in turn increases demand for energy crops, further driving up food prices (Hochman et al., 2010).

This development is further intensified by the growing expansion of the bioeconomy, which promotes the use of biological resources for industrial purposes, including the production of biogenic raw materials and energy. In addition, political incentives and subsidies play a crucial role in the reallocation of agricultural land. Governments in many countries have introduced programs and subsidies that promote the cultivation of raw materials for bioenergy production or the manufacture of biogenic products to achieve climate goals and reduce dependency on fossil fuels. These political measures have, in many cases, led to an increasing use of agricultural land for industrial purposes, further reducing the available land for food production (see Table 7.1).

Overall, these developments result in a significant shift in land use, limiting the availability of land for food production. This leads to increasing competition for agricultural land and has far-reaching implications for global food security – but effects on local food supply are also observable.

The displacement of food crop cultivation in favour of alternative raw materials has significant impacts on local agricultural markets, particularly in developing countries. In regions such as Africa, Latin America, and Asia, international investors are acquiring large land areas for the cultivation of energy crops, threatening the livelihoods of smallholder farmers (Borras et al., 2011). Moreover, in many countries, export crops like sugarcane and soybeans are increasingly being cultivated, reducing

Table 7.1: Cultivation of renewable raw materials in Germany in 2020 according to BMEL 2021.

Plant type	Area (in 1,000 ha)
Solid biomass	11.2
Industrial starch	113
Industrial sugar	10.4
Vegetable oil	94.2
Plant fibres	4.7
Medicinal and dye plants	12
Biodiesel and vegetable oil	575
Bioethanol	207
Biogas	1,550
Total cultivation area	2,577
	(100% growth over 13 years)
Industrial plants	234
Energy plants	2,343

the land available for locally essential staple crops like millet or cassava, thereby jeopardizing local food security (Perfecto & Vandermeer, 2010). This land use change leads to shortages, which in turn result in price increases, placing a greater burden on poorer populations due to higher food prices. Since these groups spend a large portion of their income on food, this exacerbates food insecurity, especially in countries that rely on food imports (FAO, 2022).

Ecological Consequences of Land Use Competition

The expansion of monocultures for bioenergy is already significantly contributing to the destruction of natural ecosystems. An example of this is the deforestation in the Amazon region, which is being driven by the increasing soy production for biofuels in Brazil (Fearnside, 2018). Similarly, the expansion of palm oil plantations in Indonesia and Malaysia is leading to the destruction of tropical rainforests, threatening the habitats of numerous endangered species (Gatti et al., 2019).

In addition to the ecological consequences of deforestation, the intensive agricultural use of land for alternative raw materials also has significant impacts on water resources and soil quality. The high water consumption for bioethanol production from sugarcane and corn leads to increased water scarcity in arid regions (Gerbens-Leenes et al., 2009). Furthermore, large-scale monoculture farming of energy crops promotes soil erosion, as these practices deplete essential nutrients from the soil and increase the need for fertilizers (Pimentel & Patzek, 2005).

Approaches to Mitigate Market Effects

To minimize the negative impacts of land use competition on global and local markets, several strategic approaches can be considered, addressing both ecological and economic aspects:

More Efficient Use of Agricultural Residues: A promising strategy involves utilizing agricultural waste, such as crop residues or inedible plant parts, for biofuel production. This could reduce the need for agricultural land to grow energy crops, thus easing the pressure on land used for food production (Scarlat et al., 2015). The use of waste represents efficient resource utilization, as these materials often remain unused, potentially causing soil or environmental degradation. By integrating residues into the biofuel sector, the competition for agricultural land between food production and alternative raw materials could be alleviated.

More Sustainable Agricultural Systems: To reduce the negative effects of intensive agriculture on the environment, sustainable farming systems like agroforestry and intercropping could play a vital role. These systems combine the cultivation of food and other crops with trees and shrubs, which can not only improve soil quality but also reduce water consumption and enhance biodiversity (Tilman et al., 2006). Agroforestry and intercropping offer multiple benefits by protecting the soil from erosion, enhancing nutrient supply, and stabilizing water balance. Additionally, such systems can secure yields in the long term by reducing dependency on monocultures, which often negatively affect soil fertility and water resources (Jose, 2009).

Political Control and Market Regulation: Another crucial measure to limit the negative impacts of land use competition is political governance and market regulation of bioenergy production. Through targeted policies, such as regulations for land allocation or subsidies for sustainable farming systems, the production of biofuels could be managed in a way that balances both ecological and economic interests (FAO, 2022). This could help stabilize price developments on agricultural markets by ensuring that market forces are not solely driven by the demand for energy crops and energy price fluctuations. Smart market regulation could also ensure that biofuel production does not come at the expense of food production but instead aligns with the needs of global food security.

It becomes clear that a holistic strategy combining efficiency improvements in resource use, sustainable agricultural systems, and appropriate political regulation is necessary to alleviate land use competition between food production and alternative raw materials in the long term while protecting both food security and the environment. Such a strategy has to be originated and driven by international and national bodies representing elected governments and not companies driven by the wish for increasing profits which often leads to or at least supports neglecting the needs of the SES.

7.2.3 Potential Risks to Food Security – Threats to Smallholder Farmers and Traditional Agriculture

The increasing use of agricultural land for the production of biogenic raw materials at least potentially poses a growing challenge to food security. Smallholder farmers and traditional agricultural operations, in particular, are affected by this development, as they are often economically disadvantaged in the global competitive landscape. Scientific studies indicate that the rising demand for biogenic raw materials – particularly for bioenergy production and industrial applications – has profound effects on land use patterns, agricultural prices, and social structures (FAO, 2021).

Smallholder farmers play a crucial role in food production in many countries of the Global South. However, they are often subjected to structural disadvantages that threaten their livelihoods: they typically have limited access to financial resources, land, and political influence. The increasing expansion of industrial agriculture, which focuses on the production of raw materials for biofuels and industrial biomaterials, has often led to the displacement of smallholder farms. This development occurs through an often reoccurring dynamics:

> Firstly, land grabbing practices lead to multinational companies and investors acquiring large tracts of land to cultivate energy crops such as palm oil or sugarcane. This makes it more difficult for smallholder farmers to access fertile land and undermines their production capacities. Furthermore, lease and land prices rise due to the increasing demand for agricultural raw materials, forcing many smallholders to give up their land, as they cannot afford the high prices. Additionally, traditional farming methods are displaced, as monocultures intended for industrial purposes replace the diversified cropping systems of smallholders. This not only threatens food sovereignty but also jeopardizes agrobiodiversity, which would be crucial for the resilience of local food systems.

The conversion of land to industrial raw material production also has serious implications for local food systems.

To quote a report published by FIAN in June 2025 (P. Seufert, 2025):

> Small-scale food providers, such as peasants, small-scale fishers and fish harvesters, pastoralists, and Indigenous Peoples, produce more than half of the food consumed by the world's population on only about 35% of the world's cropland. . . . Farms of up to 20 hectares produce 59% of key food crops on just a quarter of farmland, highlighting their productivity.
>
> It also highlights how small-scale food production accounts for a large share of crops that are essential for healthy nutrition, such as roots/tubers, pulses, fruits, and vegetables. . . . Land grabbing and increasing land concentration are therefore among the causes of the rising number of people suffering from hunger – between 713 and 757 million people, corresponding to 8.9% and 9.4% of the global population in 2023. In addition, an estimated 28.9% of the global population (2.33 billion people) were moderately or severely food-insecure.

When agricultural land is increasingly used, that is, for bioenergy or export products, the supply of locally produced foodstuffs such as maize, rice, or wheat naturally decreases. In regions with high food insecurity, this can lead to a significant shortage of essential staple foods. Moreover, price fluctuations in the markets create economic risks for nutrition, as rising prices disproportionately burden low-income households. The loss of traditional farming methods, which are adapted to local ecological conditions, is also a concern, as these practices contribute to the resilience of the food system and are often displaced by industrial agriculture.

Another significant issue arises from the increasing dependence on global markets and the associated price volatility. The prices for biofuels and industrial biomaterials are subject to substantial fluctuations, influenced by political decisions, subsidy programs, and global demand. These uncertainties expose smallholder farmers to considerable risks, as they may be caught in volatile market conditions that threaten their economic stability. The volatility of agricultural prices, driven by competition between food and raw material production, leads to sudden price increases or decreases, further destabilizing the livelihoods of smallholders. Additionally, smallholders may become dependent on multinational corporations that control the value chains for biofuels and industrial biomaterials. These corporations often benefit from rising prices, while smallholders see little to no benefit from the higher prices.

To minimize the risks to food security and the livelihoods of smallholder farmers, targeted policy measures and sustainable strategies are necessary. An important step would be the protection of land rights, particularly through stricter legal regulations that prevent land grabbing and secure the land rights of smallholders (Deininger & Byerlee, 2012). Another approach would be the promotion of agro ecological farming methods, which enable the combination of food production with the sustainable use of raw materials. These methods could support a long-term way for diversifying agricultural production for both food and raw material production (Altieri & Nicholls, 2017). Finally, political regulations on the industrial use of agricultural land, such as land quotas or the promotion of marginal land for energy crops, could help reduce the competition between food and raw material production and provide smallholders with a more stable livelihood (HLPE, 2013).

The structural disadvantage of smallholder farmers is a critical issue that must be addressed through political governance measures, sustainable agriculture, and the protection of land rights to ensure food security and minimize the existential risks faced by smallholder farmers. The increasing competition for agricultural land, driven by the rising demand for biofuels and alternative raw materials, poses significant risks to food security. Smallholder farmers and traditional agricultural systems are particularly vulnerable, as they often lack the resources to compete with the economic and political forces driving these changes.

To summarize the above the following outlines some of the main threats to food security and smallholder farmers:

1. **Displacement of Food Production**: One of the greatest risks is the repurposing of agricultural land, traditionally used for growing staple crops such as rice, maize, millet, or cassava, for bioenergy and raw material production. This not only reduces the availability of food crops on local markets but also leads to in-

creased prices for food, which especially affects low-income populations. Small-holder farmers, who depend on food crop production, are particularly at risk due to rising food prices and reduced availability.

2. **Land Grabbing and Loss of Land Rights**: In many developing countries, the demand for agricultural land for biofuels has led to an increase in land grabbing, where large agribusinesses or international investors acquire land traditionally used by smallholder farmers (P. Seufert, 2025). This displacement threatens the livelihoods of smallholders and indigenous communities who rely on their land for subsistence. These groups often lack the financial and political power to compete with large investors, and the absence of secure land rights exacerbates the issue (Borras et al., 2011). The lack of legal protection for land rights leads to uncertainty and further undermines smallholder farming systems.

3. **Increased Price Volatility and Food Insecurity**: The integration of biofuels into agricultural markets creates a stronger linkage between the prices of energy crops and food prices. When the demand for biofuels increases, the prices for agricultural products can rise significantly, which places a heavy burden on poorer populations who spend a large proportion of their income on food. These price fluctuations increase food insecurity, particularly in countries dependent on food imports (FAO, 2022).

4. **Competition for Resources and Social Inequality**: The increasing competition for agricultural resources (such as land, water, and labour) exacerbates social inequalities. While large agribusinesses and industrial actors benefit from government subsidies and political incentives, smallholder farmers often lack access to necessary resources or market mechanisms that would allow them to remain competitive. This deepens the gap between large and small agricultural producers and threatens social stability, particularly in rural areas.

5. **Biodiversity Loss and Impacts on Traditional Agriculture**: The widespread cultivation of bioenergy crops, such as soy, palm oil, and maize, for biofuels leads to biodiversity loss and threatens traditional agricultural systems. Traditional farming systems, which rely on a diversity of crops and sustainable resource management, can be undermined by the shift to industrial monocultures, which typically rely on high inputs of fertilizers and pesticides. The loss of biodiversity reduces the resilience of agricultural systems, making them more susceptible to pests, diseases, and extreme weather events.

6. **Long-Term Soil Health Impacts**: The intensive agricultural practices associated with the production of bioenergy crops can have long-term detrimental effects on soil health. The continuous cultivation of monocultures without proper crop rotation or soil management practices can lead to the depletion of soil nutrients. Smallholder farmers, who are often dependent on sustainable farming techniques, are particularly affected as they may lack the means or infrastructure to implement soil conservation practices or adopt sustainable agricultural techniques.

The increasing competition for agricultural land driven by the demand for biofuels and alternative raw materials presents a significant threat to food security, particularly for smallholder farmers and traditional agricultural systems. The political and economic shifts accompanying this competition may lead to the loss of land rights, higher food prices, and increased social inequality. To mitigate these risks, targeted policy measures, such as securing land rights for smallholder farmers, promoting sustainable agricultural systems, and creating stable markets for food production, are essential.

7.3 Global Struggles for the Distribution of Agricultural Resources

The increasing competition for agricultural raw materials has led to significant global distribution struggles in recent decades. The rising demand for biofuels, industrial biomaterials, and feed crops for livestock production has resulted in social, economic, and ecological conflicts, particularly in resource-rich regions of the Global South. Scientific analyses indicate that these struggles are often characterized by an unequal distribution of land, water, and capital, which disadvantages smallholder farmers, indigenous communities, and local producers (Borras et al., 2018; Cotula et al., 2014).

The dynamics of these conflicts vary by region, with Latin America, Africa, and Asia being particularly affected by land use changes and resource disputes. The following sections will analyse selected case studies that highlight the main issues and challenges.

7.3.1 Latin America: Land Concentration and Conflicts over Soy and Palm Oil

In Latin America, countries such as Brazil, Argentina, and Colombia are particularly affected by distribution struggles over agricultural raw materials. The large-scale expansion of monocultures for the global market has profound impacts on smallholder farmers and indigenous communities. The soy boom in Brazil, Argentina, and Paraguay has led to massive deforestation of rainforests and the displacement of indigenous groups practicing traditional agriculture (Borras et al., 2018; Fearnside, 2017).

Brazil is one of the world's largest producers of soy, which is primarily exported as animal feed for industrial livestock farming in Europe and China. The soy industry has led to massive deforestation of the Amazon rainforest in recent decades, while indigenous communities and smallholder farmers are increasingly displaced from their land (Nepstad et al., 2014). The impacts are multifaceted:

- **Land concentration**: In Brazil, more than 80% of agricultural land is owned by a few large landowners, while smallholder farmers often possess only marginal land (Sauer & Leite, 2012).

- **Social conflicts**: Violent confrontations frequently occur between agribusiness corporations, landless movements (e.g. MST – Movimento dos Trabalhadores Rurais Sem Terra), and indigenous groups fighting for their land rights (Alonso-Fradejas et al., 2015).
- **Environmental consequences**: The expansion of the soy industry is associated with massive deforestation, loss of biodiversity, and the contamination of water resources by pesticides (Fearnside, 2017).

In recent years, Colombia has become one of the leading producers of palm oil. However, this expansion is often accompanied by human rights violations and the displacement of local communities.

- **Land Conflicts**: Many palm oil plantations are located on land that was previously used by smallholder farmers or indigenous groups. There are frequent reports of paramilitary groups enforcing land seizures on behalf of agribusiness corporations (Grajales, 2013).
- **Impact on Food Security**: While palm oil production for export is booming, the availability of locally produced staple foods is decreasing, worsening the food security situation for many poor communities (Ocampo & Vélez-Torres, 2018).

7.3.2 Africa: Land Grabbing for Agricultural Raw Materials and Bioenergy

Many countries across the African continent are affected by land grabbing by international investors. The rising demand for biofuels and export crops has particularly led to distribution conflicts. In countries such as Ethiopia, Sudan, and Mozambique, millions of hectares of agricultural land have been sold or leased to international investors in recent years. Smallholder farmers and nomadic communities are particularly affected, as they are displaced by the expansion of industrial agriculture (Cotula et al., 2014).

Ethiopia is one of the countries where large-scale land sales to foreign investors are taking place for the production of export crops such as sugarcane, rice, and palm oil:

- **Foreign Investment**: Companies from Saudi Arabia, China, and India are acquiring large tracts of land for the production of agricultural raw materials for export, while local smallholder farmers are often expropriated (Cotula et al., 2014; Lavers, 2012).
- **Displacement of Smallholder Farmers**: Pastoral communities and traditional farmers are particularly affected, as they are forced to leave their land due to state development plans (Rahmato, 2011).
- **Socioeconomic Consequences**: While investors generate profits, local communities rarely benefit from these projects. Often, they experience labour exploitation and poor wages (Lavers, 2012).

Nigeria has historically been one of the key producers of palm oil, but the modern expansion of this industry is associated with significant social and environmental issues:

- **Deforestation**: The expansion of palm oil plantations leads to the destruction of biodiverse forest areas and increases CO_2 emissions (Obidzinski et al., 2012).
- **Conflicts with Local Communities**: Indigenous groups and smallholder farmers also operating on community-owned land often lose access to their traditional land ownership, while large corporations dominate production (Okafor-Yarwood, 2018).

7.3.3 Asia: Rice, Palm Oil, and Land Use Change

In Asia, countries such as Indonesia, Malaysia, and India are at the centre of the debate over distribution struggles related to agricultural raw materials. In Indonesia and Malaysia, the expansion of palm oil plantations has led to significant social conflicts, particularly due to the displacement of indigenous groups from their ancestral lands (Colchester & Chao, 2013). Countries such as Iran, Iraq, and Syria face extreme water scarcity, exacerbated by climate change and intensive agricultural use. Competition for water resources has previously led to regional conflicts (Gleick, 2014). After poor harvests in 2010 and 2022, Russia imposed export restrictions on wheat, which resulted in price increases in global markets and put pressure on importing countries like Egypt and Turkey (FAO, 2022). Countries such as Saudi Arabia and the United Arab Emirates have made large agricultural investments in Africa and Pakistan due to their own limited agricultural resources (Woertz, 2013).

Indonesia and Malaysia are the world's largest producers of palm oil, which is used in the food, cosmetics, and biofuel industries:

- **Deforestation and Land Conflicts**: The large-scale cultivation of oil palms has led to the destruction of rainforests and the threat to indigenous communities (Carlson et al., 2013).
- **Displacement of Indigenous Groups**: In particular, in Kalimantan (Borneo) and Sumatra, indigenous peoples are increasingly losing their ancestral land rights (Colchester & Chao, 2013).
- **Environmental Impacts**: In addition to the release of CO_2 from deforestation, frequent slash-and-burn practices cause massive air pollution, which affects neighbouring countries such as Singapore (Gaveau et al., 2014).

India is one of the world's largest producers of rice, but industrial agriculture has led to significant water distribution conflicts. This not only threatens long-term agricultural productivity but also exacerbates social tensions between large-scale farmers and smallholders:

- **Overuse of Water Resources**: In particular, in Punjab and Haryana, intensive irrigation has led to a rapid decline in groundwater levels (Rodell et al., 2009).
- **Export Orientation Versus Food Security**: While India exports large quantities of rice, many regions suffer from food insecurity and rising food prices (Kumar et al., 2020). During the global food crisis of 2007/08, both India and Vietnam imposed export bans on rice to stabilize domestic prices. However, this led to shortages in importing countries such as the Philippines and parts of Africa (Headey, 2011).

The increasing scarcity of resources has significant impacts on various industries that depend on agricultural raw materials. The food and agriculture industry, the bioenergy sector, and the chemical industry are particularly affected.

The food and agriculture industry feels the effects of competition for raw materials most strongly. The rising demand for feed crops such as soybeans and corn leads to higher costs in the dairy and meat industries, directly influencing the prices of meat and dairy products (FAO, 2021). The impact is also felt in the grain processing sector, especially in mills and bakeries, which face wheat shortages due to droughts and export restrictions, increasing production costs (OECD, 2022).

In addition to food production, other industries also rely on agricultural raw materials, further intensifying competition for these scarce resources. In the bioenergy industry, an increasing amount of maize and sugarcane is being used for ethanol production, which drives up food prices and fuels the debate about using arable land for biofuels (Searchinger et al., 2008). The chemical industry, which utilizes bio-based plastics and other renewable raw materials, is also in direct competition with the food industry for agricultural resources, exacerbating scarcity (Bringezu et al., 2009).

Overall, the growing demand for agricultural raw materials across various sectors is leading to intensified competition for limited resources, which brings both economic and ecological challenges.

Chapter 8
Dependencies and Social Impacts

8.1 Overview

The unequal distribution of resources shapes the global economic order and leads to social and political challenges. Resource-rich countries, while benefiting from export revenues, are often affected by one-sided economic structures and social conflicts (Zantow, 2013). In contrast, resource-poor countries must find ways to ensure their supply security and economic stability. Given the challenges of climate change and the energy transition, diversifying resource sources, recycling, and expanding renewable energy are becoming increasingly important to create sustainable and equitable economic structures (Federal Environment Agency, 2021).

As the demand for strategic raw materials rises, trade conflicts and geopolitical tensions are also increasing. Resource-rich countries may use what they have as geopolitical tools, while resource-poor states face the challenge of securing their long-term supply. Diversifying supply chains, investing in alternative raw material sources, and promoting sustainable use will be crucial in reducing tensions and minimizing conflicts.

Production stoppages caused by resource shortages can lead to job losses and economic uncertainties, particularly in resource-dependent industries such as the automotive, construction, and electronics sectors (Chamber of Industry and Commerce of Rhine-Neckar, 2021), which gives resource shortages a social dimension. Political measures such as the diversification of supply chains, expansion of the circular economy, and social security systems are necessary to mitigate negative impacts and develop long-term resource strategies that reduce the social consequences of shortages.

Resource scarcity is expected to lead to societal tensions. Rising prices for food, energy, and industrial raw materials can exacerbate social unrest and political instability, potentially triggering migration, especially in regions already experiencing environmental and economic crises (Humboldt Foundation, 2021). Sustainable strategies are required to ensure both social and economic stability.

The extraction of natural resources has far-reaching ecological consequences, particularly for resource-rich developing countries. Environmental pollution, loss of biodiversity, and the destruction of ecosystems are direct outcomes of uncontrolled resource exploitation (Umweltbundesamt, 2021). In order to mitigate these negative effects, sustainable practices, stricter environmental regulations, and technological innovation are essential. Tropical regions are especially affected, where illegal activities such as mining and deforestation cause extensive damage. The loss of biodiversity is among the most pressing ecological challenges of the twenty-first century, as it threatens both nature and the foundations of human life (Umweltbundesamt, 2021). Political

https://doi.org/10.1515/9783112218747-008

action and sustainable alternatives for resource use are therefore crucial to balancing economic growth and environmental protection.

8.2 Resource-Rich Versus Resource-Poor Countries

The distribution of natural resources is geographically uneven, resulting in global dependencies with significant social, economic, and geopolitical implications. Certain countries, such as the Democratic Republic of the Congo, Russia, and Saudi Arabia, possess substantial reserves of strategically important resources, including cobalt, natural gas, and crude oil. In contrast, resource-poor regions – among them many European nations and parts of Asia – are reliant on the import of these materials. This imbalance not only shapes economic structures but also influences political relations and trade strategies:

Global Dependencies and Economic Implications: Resource-rich countries often benefit economically from export demand; however, this does not always translate into long-term prosperity. The so-called "resource curse" phenomenon describes the paradox whereby many resource-abundant states suffer from economic instability, corruption, and weak institutions (Ali et al., 2022; Natural Resource Governance Institute, 2015). Economies heavily reliant on raw material exports frequently neglect the development of other sectors, rendering them vulnerable to price volatility.

Conversely, resource-poor industrialized nations seek to ensure the security of supply through trade agreements and strategic partnerships. Europe, for example, was for a long time heavily dependent on natural gas imports from Russia – an issue that has contributed to geopolitical tensions, particularly in times of crisis (Riley, 2024). Moreover, China's dominance in the extraction and processing of rare earth elements has far-reaching consequences for global industry, especially in high-tech manufacturing (Zhou, 2025).

Social Consequences and Inequalities: The social dimension of resource availability is particularly evident in developing and emerging economies. In resource-rich regions, extraction activities frequently lead to social conflicts, environmental degradation, and the displacement of Indigenous communities (Scheidel et al., 2023). Multinational corporations often exploit cheap labour and lax environmental regulations to maximize profits, while local populations derive limited benefit from the resulting revenues.

At the same time, resource-poor countries often struggle with high import costs and economic uncertainty. Dependence on a small number of suppliers can lead to price fluctuations that disproportionately affect poorer segments of the population. For instance, rising energy or fertilizer prices in countries lacking domestic production can increase agricultural costs, in turn driving up food prices and exacerbating social tensions (International Energy Agency, 2022).

8.3 Trade Conflicts and Geopolitical Tensions Arising from Resource Availability

The unequal distribution of natural resources has far-reaching geopolitical and economic consequences. While some countries possess abundant natural resources, others are heavily reliant on imports, leading to trade disputes and geopolitical tensions. Resources such as crude oil, natural gas, rare earth elements, and strategic metals play a pivotal role, as they are indispensable for industry, energy supply, and technological advancement.

International trade in natural resources is increasingly characterized by tension, as resource-rich countries often leverage their position as a geopolitical tool.

A prominent example is the dependence of many European countries on Russian natural gas, which led to significant economic and political strains during times of crisis, particularly during the Ukraine conflict. Russia has repeatedly used gas supplies as a political instrument, severely affecting Europe's energy security (Wirtschaftsdienst, 2022).

Another case involves disputes over rare earth elements. China holds approximately 60% of global production capacity and dominates the processing of these resources, which are crucial for the manufacturing of high-tech products, electric vehicles, and renewable energy technologies. In recent years, the Chinese government has imposed export restrictions on certain rare earths, putting considerable pressure on the United States and European countries (Reuters, 2025). This trade dispute has prompted many Western industries to seek alternative sources of supply in order to reduce dependency and enhance the resilience of their supply chains.

Control over natural resources has long been a critical geopolitical factor. In resource-rich regions, conflicts over the ownership and distribution of raw materials frequently arise. Particularly in Africa and Latin America, geopolitical tensions are fuelled by the extraction and export of metals, crude oil, and rare earth elements. Multinational corporations and state actors compete for access to these resources, a dynamic often accompanied by economic exploitation, social unrest, and environmental degradation.

A striking example is cobalt mining in the Democratic Republic of the Congo (DRC), which holds approximately 70% of the world's cobalt reserves. This resource is essential for the production of batteries used in electric vehicles, resulting in a significant surge in global demand in recent years. However, cobalt extraction is accompanied by serious social and environmental challenges, including child labour, inhumane working conditions, and environmental degradation (Amnesty International, 2016). The geopolitical control over cobalt and other strategic raw materials is increasingly becoming a source of tension between industrialized nations and resource-rich developing countries.

To address growing uncertainties in global raw material trade, many countries are implementing diversification strategies. These include investments in resource deposits outside traditional producer nations in order to develop alternative supply chains. The European Union's Raw Materials Strategy aims to reduce dependency on individual countries by placing greater emphasis on recycling, renewable energy, and technological innovation (European Commission, 2020). Another key component of this strategy is the promotion of trade agreements with resource-rich countries. Nations such as Germany are seeking to secure long-term, stable resource supplies through strategic partnerships in Africa and South America. At the same time, substitution materials and the circular economy are gaining importance as means to further reduce raw material dependence.

8.4 Social Unrest and Migration as Consequences of Resource Scarcity

The availability of natural resources is a critical factor for the economic and social stability of societies worldwide. While resource-rich countries often derive geopolitical power from their resource endowments, resource-poor states are heavily dependent on imports and vulnerable to supply disruptions. The increasing scarcity of essential resources such as water, energy, metals, and agricultural commodities is more frequently leading to social tensions that can manifest in protests, political unrest, or even migration.

Scientific research demonstrates a close link between resource scarcity, social unrest, and migratory movements. Historical and contemporary examples show that economic crises, rising living costs, and growing social inequalities resulting from resource shortages can act as catalysts for political instability. In many instances, environmental changes and resource depletion exacerbate existing social and political conflicts.

Key factors that heighten the likelihood of social unrest and migration in the context of resource scarcity include rising living expenses – potentially triggering protests and civil disorder – economic decline caused by production halts in resource-dependent industries, unequal resource distribution due to political corruption or economic power concentration, as well as environmental degradation and climate change, which can ultimately lead to the displacement of large population groups.

In recent decades, numerous instances have demonstrated how resource shortages can trigger social unrest. One notable example is the *Arab Spring* (2010–2012), during which rising food prices – driven by droughts and increased global grain prices – contributed significantly to the destabilization of countries such as Egypt, Tunisia, and Syria.

However, long-term resource scarcities can provoke not only short-term protests but also large-scale migration. Rural areas in developing countries are particularly affected, where the loss of arable land or water scarcity often leads to significant population outflows. Migration driven by resource scarcity may take the form of internal migration – from rural to urban areas – or international migration, as individuals leave their home countries in search of regions with greater resource availability or economic stability.

Progressing climate change exacerbates these migration movements by intensifying resource scarcity, particularly regarding water and agricultural productivity. This trend poses a growing risk of displacing entire communities over the long term.

According to reports by international organizations such as the World Bank and the United Nations High Commissioner for Refugees, environmental degradation and resource scarcity are among the primary drivers of future migration. Projections suggest that by 2050, several million people may be forced to migrate due to resource shortages and environmental change. However, the consequences of resource scarcity are not equally distributed across regions. While industrialized nations generally possess the financial and technological capacity to mitigate shortages, developing countries are often disproportionately affected.

In Africa, droughts, water scarcity, and degraded agricultural land are already driving increased migration from regions such as the Sahel. In Latin America, the extraction of raw materials such as lithium, copper, and oil frequently leads to social conflicts between multinational corporations and local communities. In Asia, countries such as India and China face mounting environmental challenges and water shortages, which have the potential to generate internal social tensions.

In light of growing social unrest and migration driven by resource scarcity, a range of policy responses is required. These include sustainable resource management to prevent long-term shortages, social safety nets – such as subsidies and welfare programs – to shield vulnerable populations from the effects of rising food and energy prices, and international cooperation aimed at a more equitable distribution of resources and greater investment in resource-poor regions. Additionally, technological innovation that supports the use of alternative materials and renewable energy sources can help mitigate the risks of social unrest and forced migration.

Chapter 9
Summary

Globally, not all but a number of important resources are becoming increasingly scarce – this applies to fossil resources, mineral commodities, and agricultural products alike. Several parameters are passing acceptable limits (global boundaries) and, thereby, expose all species unprecedented risks. At the same time, public debt is rising in many countries limiting states ability to fulfil their task to administer themselves and establish solutions as required. Although these phenomena may initially appear unrelated, their parallel occurrences produce unknown and interrelated consequences.

The target of this little compendium was to identify the role and options biogenic resources may hold in this context given the fact that other resources get scarcer and the capacity of the earth to grow new resources is limited given the other phenomena addressed. This societal risk forms the focus of the present analysis, which also investigates the extent to which biogenic resources may substitute conventional raw materials in order to sustain current economic systems despite foreseeable supply bottlenecks.

To address this, the analysis begins with an introduction to various theoretical frameworks – most notably the concept of *social metabolism*, which posits that human societies, much like plant or animal systems, require constant inputs (analogous to "nourishment") to survive. Nourishment, in this sense, also refers to materials and energy. Within this framework, not only agricultural products but also mineral and fossil resources are understood as vital inputs for the functioning of economic systems. The combination of these resources effectively "feeds" societies.

Consequently, if these resources become scarce, economic activity weakens, and societal stability is undermined. A decline in resource availability typically results in reduced economic activity and output, inevitably leading to lower tax revenues. In an attempt to maintain public satisfaction and sustain economic operations, states often respond by increasing public spending and taking on debt. However, when resource shortages reach a critical level, economic systems can no longer function as they currently do – rendering the operation of states at existing levels unsustainable and the repayment of debt increasingly unlikely.

At present, this systemic issue is, to some extent, being addressed by accepting elevated levels of inflation. Yet such a strategy has limited scope and effectiveness. The resulting socio-economic risks, therefore, carry significant potential for long-term destabilization of states and societies within on a broad scale with a declining window of options to react.

In this context, it is essential to examine whether raw materials of mineral, fossil, or agricultural origin are indeed becoming scarce. Evidence presented suggests that this is the case. Furthermore, when it comes to agriculture, the increasingly deterio-

https://doi.org/10.1515/9783112218747-009

rating state of the environment is a critical influencing factor – this applies both to direct effects as well as concerns related to biodiversity and environmental pollution. The concept of planetary boundaries clearly highlights these issues.

Using the model of adaptive cycles and drawing on historical examples, it was sketched that system collapses, such as those resulting from inadequate resource supply, can – and in the past, have – occurred. Specifically concerning agriculture, it is shown that individual resource supplies should not be viewed in isolation, but rather as interconnected: if there is insufficient crude oil, for instance, there is not enough energy to produce synthetic fertilizers, meaning that a reduction or increase in the cost of oil supply will have direct consequences on agricultural productivity, especially given the declining state of earth to grow crops on it.

This situation is alarming, and, as is well known, solutions have been sought for some time. Clearly, a reduction in consumption would be beneficial. However, consumption is inherently tied to life itself; the challenge, therefore, lies in reducing consumption while shifting priorities. As discussed earlier, one potential approach to this shift in consumption involves the use of biogenic resources as substitutes for fossil and mineral raw materials. However, this can only be one part of the solution.

Apparently biogenic raw materials originate in agriculture – hence, as a starting point, the current state of agriculture is taken into consideration, which leads by no means guarantees for a secure supply.

Already today, soil fertility is severely threatened by erosion, climate change, overuse of agricultural land, water scarcity, over-fertilization, and other factors. Additionally, one-third of the food produced is discarded. Therefore, if the attempt is made to view established food chains as only one of several future pillars of agriculture – and to consider biogenic raw materials as additional pillars – this concept is certainly intriguing but, from the outset, rests on unstable ground. Moreover, there are calls for the designation of additional agricultural protection areas to preserve and ensure biodiversity.

Nevertheless, the use of biogenic raw materials is currently seen primarily in three ways as a potential tackle the supply crisis:
- in fuels,
- as substitutes for mineral resources, and
- as biomass.

These three forms of competition for agricultural land devoted to food production are explored in the following paragraphs.

The application of biogenic raw materials as substitutes for mineral resources is likely the most exciting technology in this field, especially as this aspect is rarely discussed in public discourse. The use of hemp, for example, as a replacement for steel in concrete, biogenic composite materials in the plastics industry, or as insulation materials yields remarkable results. While research in this area is no longer in its infancy, it has not yet advanced to deliver significant economic impact. The market share of such raw materials remains in the low single-digit percentage range but is growing rapidly. The reason for the hesitant establishment of these materials lies in the very established market for other resources plus the commonly combinations of materials provide difficult-to-replace solutions on the one hand. On the other hand, there is their very specific composition, which makes their use as replacements, for example, for individual components or modules in established products challenging. Ultimately, these substances are in competition with raw materials and their combination that, although becoming scarcer, have been proven and well-understood for centuries.

The situation is different with biogenic energy carriers as alternatives to fossil fuels. Biogas derived from algae oil, wood, or other biogenic materials, which can be used as additives or as a complete substitute for established fuels, has long been established in the market. The share of fuel production from biogenic sources currently stands at about 5%, growing at a rate of approximately 6–7% per year – at the same ratio than other fuels. With the increasing scarcity of oil and natural gas (whose reserves are estimated to last 50–70 years), growth rates are likely to accelerate. Long-term, significant effects are therefore to be expected – but the CO_2 emitted from these sources are no less harmful than coming from fossil resources and using land for fuel production in spite of local residents are in need for food is unacceptable.

Biomass as an alternative to fossil fuels was heavily promoted in the 1980s and 1990s and is now an established form of utilizing wood or wood waste, agricultural waste, algae, or animal waste. Here, too, the annual growth rate is around 6%, and the use of biomass power plants for district heating is technically feasible – and preferred, as existing infrastructure can still be utilized. If subsidizing and CO_2-related costs are considered costs for energy derived by these sources are in the same range than if fossil resources were used.

While this is optimistic, the potential currently used by biogenic materials remains relatively small in comparison to fossil or other raw materials. Nevertheless, the impact this use has on established agriculture is already clearly noticeable today: the competition for agricultural land leads to the conversion of established food production systems into industrial production of goods that no longer serve nutritional purposes. In addition, many small farmers face de facto expropriation and the loss of established sources of income and lifestyles in favour of an economic model whose limitations are foreseeable.

The risk that arises from the overuse of existing resources cannot be solved by the use of biogenic resources alone. In fact, rebound effects are likely to occur, meaning that instead of reducing consumption, biogenic raw materials may be used even more excessively, in addition to fossil and mineral resources, and exploited to their limits. The hope that biogenic raw materials will extend the lifespan of fossil resources on the one hand, and stabilize supply on the other, is not realistically substantiated. Clearly, a society that struggles to provide sufficient food for its human, plant, and animal communities cannot easily avoid the resulting problems simply by repurposing land. Therefore, reducing consumption remains the most urgent priority.

Above, solutions were proposed, others were regarded as lesser promising, the total picture drawn indicates that biogenic resources can contribute, but will only be part of a wider set of approaches to live more sustainable. Of course, consumption and impact to the environment are characteristic of all form of life. In a strict sense, it is precisely the balance of these impacts that stabilizes the larger ecological system. For any new solutions the challenge is to ensure that they are as closely aligned with natural processes as possible – and that they are made available in the near term.

However, to some degree the approaches, solutions and concepts discussed appear to be, in one way or another, inherently contradictory:

> On the one hand, there is a call to preserve nature as is as much as possible, to maintain existing ecological balances without disruption. That is under the assumption that nature is constant – which it is not.
>
> On the other hand, there is a search for ways in which industrial culture might continue – albeit in a modified or reduced form – essentially implying a continuation, even if more cautious, of environmental impact taking place in a pace characteristic to the pace mankind can and does apply today.

This is where contradiction deepens: nature has taken millions of years, and endured countless losses, to arrive – through a slow and ongoing process of evolution – at roles, relationships, and synergies that underpin its equilibrium. These balances are by far not static; they are constantly being "renegotiated" and realigned – but the speed, in which solutions are established are slow.

To believe that a society or a culture could design an entirely new system even within a generation or two is, in this light, a rather presumptuous notion. What might be more appropriate is an approach grounded in humility and patience – combined with the courage to observe, to learn, and to understand the responses of complex systems rather than to assume that each new initiative must be definitive or final.

Solutions take time to evolve, to interlock, and to prove themselves. The pace at which our socio-economic systems currently develop is fundamentally different, that is, way faster, from that of nature. Nature operates through an evolutionary rhythm: testing, adapting, and discarding. Bridging this gap – realigning our speed with that of natural systems – would be a critical step toward a truly sustainable future.

Chapter 10
Annex

Table 10.1: Alternatives to mineral raw materials using biogenic raw materials.

To be substituted	Use case	Research
Aluminium*	Plastics and composites	Aluminium is often used in the automotive and construction industries, but efforts are underway to develop biogenic composites made from cellulose or biogenic plastics as a lighter, more sustainable alternative (Fraunhofer IFAM, 2024). Biogenic plastics based on plant fibres could serve as a replacement in various applications as they are both lightweight and robust (Verbraucherzentrale, 2024).
	Lightweight materials	Aluminium is used in many lightweight construction applications, particularly in automotive engineering and aerospace components. A promising alternative could be biogenic composites containing plant fibres (e.g. hemp or flax) or biogenic plastics (Fraunhofer-Institut für Fertigungstechnik und Angewandte Materialforschung IFAM, 2021; Fraunhofer-Institut für Betriebsfestigkeit und Systemzuverlässigkeit LBF, 2023). These materials could reduce the weight of vehicles and aircraft while maintaining strength and stability (Deutsches Zentrum für Luft- und Raumfahrt (DLR), 2018; PlastXnow, 2023; Fachagentur Nachwachsende Rohstoffe e. V. (FNR), 2025).
Antimon*	Flame retardants	Antimony is used in flame retardants that are used in electronic products and textiles. A biogenic alternative could be plant-based polymers or biogenic enzymes, which also act as flame retardants and thus reduce the demand for antimony in various industries (Umweltbundesamt, 2008; Allianz Faserbasierter Werkstoffe Baden-Württemberg e.V. (AFBW), 2020; Fraunhofer UMSICHT, 2024).
Arsen*	Pesticides and fungicides	Arsenic has been used in pesticides and fungicides in the past. Biogenic alternatives offer an environmentally friendly way to control pests and promote plant health without the toxic effects of arsenic (Umweltbundesamt, 2024; Biobest Group, 2025; Universität Freiburg, 2022). Biogenic alternatives could be based on plant extracts, bacteria, or biological enzymes (Leclère et al., 2020; Bioscientia Institut für Medizinische Diagnostik, 2023; Bioökonomie.de, o. J.; Fachagentur Nachwachsende Rohstoffe e. V. (FNR), 2025).

https://doi.org/10.1515/9783112218747-010

Table 10.1 (continued)

To be substituted	Use case	Research
Barium*	Protective materials	Barium is used in protective materials such as barium sulphate, which is used as an X-ray contrast agent in medicine or as a protective coating. A biogenic alternative could be biopolymers or natural minerals that could be used as a protective coating or barrier material, especially in medicine or building materials (DocCheck Flexikon, n. d.; Stanford Advanced Materials, n.d.; Milazzo et al., 2020).
Beryllium*	Polymers and composites	Beryllium is used in the aerospace industry and in electronic components to achieve high strength and low mass. Biogenic composites made from cellulose or biogenic plastics could serve as a substitute in certain applications, for example in lightweight components for vehicles or aircraft (Materion Corporation, n.d.; Lamineries, n.d.; Bayern Innovativ GmbH, 2018).
	Lightweight components	Beryllium is used in the aerospace industry to create lightweight components that are also high strength. A promising biogenic alternative could be biogenic composite materials containing plant fibres or biogenic polymers (Goodfellow, n.d.; Brothers, 2021, 26 April). These materials could offer similar strength and lightness to beryllium in certain applications (NASA, 2019; ResearchGate, 2023).
Bismuth*	Alloys	Bismuth is often used as a substitute for lead in alloys. There are attempts to develop biogenic alloys from plant fibres or biogenic polymers, which can also serve as more environmentally friendly alternatives (Kizatov, 2025; Kumar & Sharma, 2021; Kumar & Sharma, 2023).
Lead	Biogenic coatings	Lead was used in the past in paints, building structures, and cables. Today, biogenic alternatives such as vegetable waxes or biogenic resins are an option for the production of environmentally friendly paints and coatings that could provide a sustainable solution for replacing lead paint without the toxic effects (IPEN, n.d.; Lubrizol, 2023; Cargill, n.d.; European Coatings, 2024).

Table 10.1 (continued)

To be substituted	Use case	Research
	Colorants and protective materials	Lead is used in colorants, battery technologies, and as a protective material. Research into biogenic dyes and protective materials made from plant extracts or biogenic polymers could offer a more environmentally friendly and less harmful alternative (UN Environment Programme, n.d.). *Lead.* (ScienceDirect, 2023; ScienceDirect, 2022). The development of biomaterials from bacteria could also be relevant in this context (MDPI, 2020; ScienceDirect, n.d.). *Biomaterial Coating* (ScienceDirect, 2021).
Calcium	Building materials	Calcium is often used in building materials such as cement or plaster. Biogenic materials such as limestone substitutes or biopolymer cement based on plant fibres could reduce the need for calcium in the construction industry (De Muynck et al., 2010; Plank, 2004; Van Roijen et al., 2024). These alternatives could reduce the environmental impact of cement production, which is currently a significant CO_2-emitter (Nadzri et al., 2017; Van Tittelboom & De Belie, 2013).
Caesium*	Sensors and measuring devices	Caesium is used in atomic clocks and measuring devices primarily because of its special properties as a high-precision time measurement. Researchers are investigating alternative biogenic sensors and quantum measuring devices based on organic molecules or biogenic semiconductors (Bauch, 2003; Degen et al., 2017; Schirhagl et al., 2014; Muhammed et al., 2019; Singh et al., 2024).
Chlorine	Disinfectant	Chlorine is widely used in water treatment and as a disinfectant. A promising biogenic alternative could be the use of plant enzymes, bacteria, or ultraviolet (UV) light technology, which is just as effective in water treatment while causing less environmental impact (APEC Water, n.d.). *Difference Between UV Purification & Chlorination* (Treatment Plant Operator, 2018; MDPI, 2024; Orenda Technologies, n.d.; Custom Enzymes, n.d.).
Chromium*	Corrosion protection	Chromium is used in alloys and as corrosion protection. Biogenic corrosion protection materials made from plant extracts or biogenic polymer layers offer an environmentally friendly alternative to chromium, which is considered toxic. These biogenic materials could be used in the automotive and construction industries in particular (Castañeda & Martínez, 2020; Raja & Sethuraman, 2008).

Table 10.1 (continued)

To be substituted	Use case	Research
	Aerospace alloys	Chromium is often used in aerospace alloys as it improves the strength and corrosion resistance of materials. Biogenic materials, such as plant fibres or biogenic plastics (e.g. biogenic carbon fibres), could be used as lightweight and stronger alternatives to chromium-containing alloys in the aerospace industry (Airbus, 2024; Sloan, 2022; Shen & Patel, 2025; Kumar & Singh, 2023; Solvay, 2021).
Cobalt*	Electrodes of lithium-ion batteries	Cobalt is often used in the electrodes of lithium-ion batteries. A promising biogenic alternative is sodium-ion batteries or magnesium-ion batteries, which could be combined with biogenic materials such as carbon nanotubes or plant fibres. These could reduce the need for cobalt in many applications and thus lead to more sustainable battery systems (Krüner & Bauer, 2020; Kim & Kim, 2024; UP Catalyst, 2023; Sloan, 2022; Shen & Patel, 2025).
	Used in lithium-ion batteries	Cobalt is widely used in lithium-ion batteries, especially for electric vehicles. As an alternative, biogenic batteries are being developed that use plant-based materials or organic molecules as electrolytic components or electrodes. Sodium-ion batteries and magnesium-based batteries from biogenic sources also offer promising approaches (Wu et al., 2016; Zhou et al., 2022; Zhao et al., 2017; Liu et al., 2023; Zhang & Li, 2024).
	Batteries	Cobalt is often used in lithium-ion batteries. However, there is research being done on biogenic batteries that could partially or completely replace cobalt. For example, sodium-ion technology could play a role as they use cheaper materials and in some cases are supported by natural materials such as cellulose (Carl-Zeiss-Stiftung, 2024; Marangon et al., 2021).
	Bioluminescent	Cobalt is used in the production of phosphors and in electronics. A possible biogenic alternative could be bioluminescent plants or bacteria, which are used as a luminescent material in optical devices. This could not only reduce the demand for cobalt but also increase the sustainability of the electronics industry (Krichevsky et al., 2010; Mitiouchkina et al., 2020; Glowee, 2023; Wood & Light Bio, 2023; Mayoral, 2023).

Table 10.1 (continued)

To be substituted	Use case	Research
	Catalysts	Cobalt is used in catalysts in the chemical industry. Biogenic catalysts consisting of plant materials, enzymes, or microbial systems could serve as more sustainable catalysts in chemical processes, for example, in fuel production or chemical manufacturing (Akhtar et al., 2025; Unglaube et al., 2022; Macaskie et al., 2017; Ingle, 2020).
Fluorit	Fluorescent materials	Fluorite is used in the lighting industry and in the production of fluorescent materials. A biogenic alternative could be the use of biological fluorophores from plant sources or bioluminescent microbes, which have similar fluorescent properties and can be used in applications such as light sources or biomedical diagnostics. (Mitiouchkina et al., 2020; Krichevsky et al., 2010; Mishin & Lukyanov, 2019; Shaner et al., 2007; Tsien, 1998).
Graphite*	Used in batteries and as a lubricant	Graphite is used in batteries and as a lubricant. One possible biogenic alternative is biogenic carbon materials derived from organic waste or biogenic sources such as bamboo or algae. These materials could have similar properties and serve as a substitute in certain applications (Cataldi et al., 2020).
Indium*	Used in flat screens and solar cells	Indium is often used in flat screens and solar cells. Biogenic materials such as biopolymers and graphene from plant sources could serve as potential alternatives in the electronics industry. In display technology, in particular, these materials could replace indium in the coming years (Cambridge University, 2022, May 12; Tomorrow.bio, 2023; ScienceDirect, 2024. *Sustainable power generation*).
Cadmium	Alloys and batteries	Cadmium is often used in batteries and as an anti-corrosion agent. Biogenic alloys and environmentally friendly batteries made from plant-based materials or organics could replace cadmium in many applications, especially with zinc-air batteries or aluminium-carbon batteries that could be combined with biogenic components (Zhang et al., 2023; Mori, 2019; Li & Dai, 2014).

Table 10.1 (continued)

To be substituted	Use case	Research
Krypton/xenon	Lighting technology and in high-pressure lamps	Krypton and xenon are used in lighting technology and in high-pressure lamps. Biogenic phosphors based on bioluminescent microbes or biogenic nanoparticles could potentially represent a more sustainable and environmentally friendly alternative. These technologies could be further developed using biogenic materials such as plant extracts or algae pigments (Labiotech.eu., 2014, 3. Juni; ScienceDirect, 2022). *Algae as an emerging source of bioactive pigments* (ScienceDirect, 2022. *One-step synthesized biogenic nanoparticles using Linum usitatissimum*).
Lithium *	Super-capacitors	Research into replacing lithium with biogenic raw materials is increasingly focussing on the development of super capacitors made from sustainable materials such as graphene (obtained from organic sources) or cellulose. These biogenic capacitors could offer an energy-efficient and more environmentally friendly solution in certain applications (Carl-Zeiss-Stiftung, 2024; Marangon et al., 2021).
	Extensively in batteries	Lithium is used extensively in batteries especially in electric vehicles and cell phone batteries. Biogenic electrolytes produced from plant or microbial sources offer promising alternatives to conventional lithium-ion batteries. Another option is sodium-ion batteries, which are less dependent on lithium (Zhang et al., 2022).
Manganese*	Composites	Magnesium is a lightweight metal used in the automotive and aerospace industries for lightweight alloys. Biogenic composites, such as plant fibres or bamboo, could serve as a substitute to produce lightweight materials for industrial applications that could potentially replace magnesium in some areas in the future (Parameswaranpillai et al., 2023; Kumar & Singh, 2023).
	Dye and pigment industry	Manganese is used in the dye and pigment industry to produce colourful materials. Biogenic pigments derived from plant extracts or biogenic dyes could serve as substitutes. For example, chlorophyll (which is extracted from plants) could be used as a green colorant in various applications (Burgess, 2011; Kaiser-Alexnat, 2012; Schweppe, 1993).

Table 10.1 (continued)

To be substituted	Use case	Research
Molybdenum	Alloys	Molybdenum is used in steel and alloy production to improve strength and resistance to high temperatures. Researchers are developing biogenic alloys from plant polymers or natural fibres that could replace molybdenum in certain applications, particularly in composite materials for the aerospace industry (lanrewaju et al., 2025; Wang & Li, 2023; Namvar et al., 2023; Brothers, 2021).
	High-temperature alloys	Molybdenum is used in high-temperature alloys, for example, in jet engines or gas turbines. An alternative approach could be biogenic composite materials made from plant fibres and biogenic metals, which offer similar strength and temperature resistance. Magnesium-based alloys and high-performance biogenic polymers could also replace molybdenum in some applications (Zhang et al., 2024; Regazzi et al., 2018).
Nickel*	Battery technologies	Nickel is used in many battery technologies, for example in nickel metal hydride (NiMH) batteries. A promising biogenic alternative is zinc-air batteries, where biogenic materials such as plant fibres could be used to optimize performance. Carbon materials derived from biogenic sources such as algae or wood waste are also an active field of research for battery production (Fu et al., 2017; Lee et al., 2020; Wang et al., 2013; Zhang et al., 2013).
	Used in catalysts and batteries	Nickel is used in catalysts and batteries (e.g. nickel-cadmium or nickel-metal hydride batteries). Researchers are developing biogenic catalysts consisting of enzymes or plant extracts that have similar catalytic properties to nickel. Zinc batteries or air batteries, which are combined with biogenic materials, are also being developed as sustainable alternatives (Wang et al., 2023; Clark et al., 2019; Fu et al., 2017; Lee et al., 2020).
Phosphorus	Fertilizers	Phosphorus is used in fertilizers and is a scarce resource. Researchers are developing biogenic fertilizers based on microorganisms or plant waste to provide phosphorus in more efficient forms. Recycling technologies from biogenic waste could also contribute to phosphorus recovery from wastewater or organic waste (Cordell et al., 2009; Yuan et al., 2012; González & García, 2017; Egle et al., 2016; Desmidt et al., 2015).

Table 10.1 (continued)

To be substituted	Use case	Research
Platinum*	Catalyst	Platinum is often used as a catalyst in various industrial processes (such as hydrogen production). However, scientists have developed biogenic catalysts from enzymes or plant extracts that can promote similar chemical reactions. These not only offer a sustainable approach but can also be more cost-effective (Kumar & Kumar, 2020; Kang & Lee, 2022; Kumar & Kumar, 2023).
	Fuel cells and as a catalyst	Platinum is used in fuel cells and as a catalyst, particularly in hydrogen production and exhaust gas purification. Researchers are working on biogenic catalysts that use plant or microbial extracts to replace platinum. Nanostructured materials from biogenic sources could offer similar catalytic properties, with a significantly lower environmental footprint (Kumar & Kumar, 2020; Kang & Lee, 2022; Kumar & Kumar, 2023; Quinson, 2025).
Ruthenium*	Sensors and catalysts	Ruthenium is used in sensors and catalysts to accelerate chemical reactions. Biogenic sensors consisting of microbial systems, biogenic enzymes, or plant nanomaterials could represent an environmentally friendly and more cost-effective alternative to ruthenium in many technologies (Zhou et al., 2022; Liu & Zhang, 2023; Kumar & Kumar, 2020; Kang & Lee, 2022; Singh et al., 2016; Iravani, 2011).
Sulphur	Disinfectants	Sulphur is used in many industrial processes, especially in the chemical industry and in disinfectants. A biogenic alternative could be the use of plant extracts or microbial decomposition products from organic waste, which have disinfectant properties and could be used as environmentally friendly substitutes for sulphur products (Cowan, 1999; Ferdes & Ferdes, 2021; Iravani, 2011; Singh et al., 2016).
Cadmium and mercury	Nanomaterials für	Cadmium and mercury are frequently used in electronic devices and solar cells. Biogenic nanomaterials, for example nanostructures made from plant fibres or nanostructured algae (e.g. algae-based nanoparticles) could serve as alternative materials for solar cells or electronic components and avoid the environmental impact of heavy metals (Iravani, 2011; Kumar & Kumar, 2020; Kang & Lee, 2022; Cowan, 1999).

Table 10.1 (continued)

To be substituted	Use case	Research
Rare earths (z. B. Lanthanum*, Cer*)	Magnets	Rare earths such as lanthanum and cerium are used in magnets, motors, and hard disks. There are attempts to develop biogenic magnetic materials based on graphene or biogenic nanomaterials. These could serve as sustainable substitutes for rare earths in electromobility and the wind power industry (Arvidsson & Sandén, 2017; Bao et al., 2022; Victory Metals Australia, n.d.; Stanford Magnets, n.d.).
Rare earths (such as neodymium and praseodymium *)	Magnets	Replacement of rare earths (such as neodymium and praseodymium). Researchers are investigating biogenic magnetic materials made from natural materials such as chitin (a polysaccharide compound derived from shellfish) or biological graphene. These could potentially be used in applications such as electric motors or loudspeakers, which normally use rare earths such as neodymium (Iravani, 2011; Singh et al., 2016; Kang & Lee, 2022; Cowan, 1999).
Tantalum *	Nanomaterials	Tantalum is used in capacitors and high-quality electronic components. Biogenic nanomaterials derived from plant fibres or algae could be used as conductive materials in the electronics industry to replace tantalum. Graphene-like biogenic materials could also potentially fulfil the same function (Cataldi et al., 2020; Orts et al., 2022).
	Capacitors and high-quality electronic components	Tantalum is used in capacitors and high-quality electronic components. Biogenic nanomaterials consisting of plant fibres or algae could be used as conductive materials in the electronics industry to replace tantalum. Graphene-like biogenic materials could also potentially fulfil the same function (Kumar & Sharma, 2023; Singh & Tiwari, 2023).
	Nanomaterials	Tantalum is used in medicine and electronics, especially in capacitors. One possibility for replacing tantalum would be biogenic nanostructures consisting of plant polysaccharides or bacterial nanostructures. These could be used in medicine as biocompatible materials or in electrical engineering as sustainable alternatives (Kumar & Sharma, 2023).
Titanium *	Composite materials	Titanium is known for its lightness and corrosion resistance. Research into biogenic composite materials made from plastic-fibre compounds or plant fibres could reduce the use of titanium in some areas of the aviation industry. These materials could offer lower weight while retaining the same strength (Wang et al., 2022; Saba et al., 2021; Kumar & Singh, 2024).

Table 10.1 (continued)

To be substituted	Use case	Research
	Medical technology	Titanium is often used in medical technology, for example for implants or prostheses due to its biocompatibility. Biogenic alternatives such as biological polymers consisting of collagen or chitin could potentially replace titanium in medical applications. These polymers are biocompatible and could lead to a more sustainable production of medical devices (Przybylek & Beldowski, 2023; Tronci et al., 2011).
Vanadium*	Redox flow batteries	Vanadium is used in redox flow batteries for energy storage. Researchers are working on ways for biogenic energy storage such as microbial batteries or organic redox flow batteries that use biogenic electrolytes. These technologies could reduce the need for vanadium in energy storage in the long term (Brushett et al., 2012; Wang et al., 2013; Lovley, 2006).
Bismuth *	Semiconductors and components for electronic equipment	Bismuth is used in the electronics industry as a component of semiconductors. Biogenic materials such as organic semiconductors and graphene-based biogenic materials offer promising alternatives in the energy and electronics industry and could replace the need for bismuth in some technologies (Geim & Novoselov, 2007; Kowalczyk et al., 2020).
Tungsten *	Carbides	Tungsten is used for the production of carbides and as a material in tools. Biogenic carbides made from plant-based raw materials or nanostructured cellulose materials could serve as substitutes in the future (Mittal et al., 2018; Chen et al., 2020).
	Nanomaterials	Tungsten is used in hard alloys and as a material for tools. An interesting substitute could be biogenic nanomaterials that can be produced by biological processes or nanostructured cellulose. These materials could be advantageous in certain applications such as tool manufacturing or precision mechanics (Mittal et al., 2018; Klemm et al., 2018).
	High-performance alloys	Tungsten is used in tools and high-performance alloys because of its heat resistance and density. Researchers are developing biogenic alloys containing nanostructured carbon fibres or biogenic composite materials such as plant polymers and biogenic nanomaterials that could serve as substitutes for tungsten in various industries (Mittal et al., 2018; Klemm et al., 2018; Chen et al., 2020).

Table 10.1 (continued)

To be substituted	Use case	Research
Zinc*	Corrosion protection	Zinc is often used for corrosion protection (e.g. in galvanized materials). Research into plant-based biogenic-coating materials (such as tannins from oak bark) could replace zinc as an anti-corrosion agent in many applications (Pizzi, 2016; Fiore et al., 2019).
	Component of fertilizers	Zinc is often used in agriculture as a component of fertilizers especially to promote the growth of plants. A biogenic alternative could be the use of nanostructured, biogenic fertilizers made from plant waste or microorganisms. These alternatives are sustainable and could enable better nutrient uptake for plants (Alloway, 2008; Solanki et al., 2015; Dimkpa & Bindraban, 2018).
Tin *	Solder connections and corrosion protection	Tin is used in the electronics industry for solder connections and corrosion protection. Biogenic alternatives such as biological polymers or biogenic metal alloys (e.g. magnesium-based) could replace tin in many areas. These materials could be used in electronic devices as more environmentally friendly solutions (Li et al., 2017; Das et al., 2020).
	Packaging materials	Tin is often used in packaging materials (e.g. tin cans). A biogenic alternative could be bioplastic or biogenic plastic packaging based on plants or agricultural waste. This packaging could be more environmentally friendly and reduce tin consumption in the food industry (Siracusa et al., 2008; Niaounakis, 2013).
Zirconium *	Replacement of minerals	Zircon is used in the ceramics industry and in water filtration. Biogenic alternatives that could be used in a similar way for the production of high-strength ceramics or filter materials are currently being developed, for example, biomineralized materials are made from cellulose or natural polymers (Muthurajan et al., 2019; He et al., 2020).
	Nuclear technologies	Zirconium is used in nuclear technology, particularly in the manufacture of fuel rods. Researchers are investigating whether biogenic materials such as biological polymers or advanced biogenic nanomaterials can serve as potential substitutes for zirconium, particularly in nuclear energy research (Feng et al., 2018; Bhandari et al., 2020).

The idea of the table is not to point out activities to find alternatives for each individual use case, but to illustrate that intensive searches and experiments are already underway in many areas, some of which are also being applied in practice.

The mineral raw materials that are considered critical minerals by the USGS are marked with an *.

Sources Cited in Annex

Akhtar, M. S., Naseem, M. T., Ali, S., & Zaman, W. (2025). Metal-based catalysts in biomass transformation: From plant feedstocks to renewable fuels and chemicals. *Catalysts, 15*(1), 40. https://doi.org/10.3390/catal15010040

Alloway, B. J. (2008). Zinc in soils and crop nutrition (2nd ed.). Brussels, Belgium: International Zinc Association (IZA) and International Fertilizer Industry Association (IFA).

Airbus. (2024, Juni). Developing bio-based composites that are fit to fly. *Airbus Newsroom.* https://www.airbus.com/en/newsroom/stories/2024-06-developing-bio-based-composites-that-are-fit-to-fly

APEC Water. (n.d.). *Difference Between UV Purification & Chlorination.* Abgerufen am 24. April 2025, von https://www.freedrinkingwater.com/blogs/contaminants-facts/uv-water-purification-vs-chlorination

Allianz Faserbasierter Werkstoffe Baden-Württemberg e.V. (AFBW). (2020). *Textiler Flammschutz.* Abgerufen am 24. April 2025, von https://www.afbw.eu/aktuelles/nachrichten/details/textiler-flammschutz/

Bhandari, P., Kumar, S., & Sinha, R. (2020). Biopolymers and nanomaterials in nuclear applications: A review. *Journal of Materials Science & Technology, 42*, 1–9. https://doi.org/10.1016/j.jmst.2019.09.007

Biobest Group. (2025). *Biopestizide.* Abgerufen am 24. April 2025, von https://www.biobestgroup.com/de/biopestizide

Bioscientia Institut für Medizinische Diagnostik. (2023). *Arsen – Vom Fliegenteller zur Arsenik-Suppe – Teil 1.* Abgerufen am 24. April 2025, von https://www.bioscientia.de/media/xugiczha/arndt_arsen_230308.pdf

Bioökonomie.de. (o. J.). *Biologika für die Landwirtschaft.* Abgerufen am 24. April 2025, von https://biooekonomie.de/themen/dossiers/biologika-fuer-die-landwirtschaft

Bayern Innovativ GmbH. (2018). *Jahresreport Cluster Neue Werkstoffe 2018.* Abgerufen am 24. April 2025, von https://www.bayern-innovativ.de/fileadmin/article_migration/pdf-dokumente/cluster-neue-werkstoffe/Cluster-Neue-Werkstoffe-Jahresreport-2018.pdf

Brothers, E. (2021, 26. April). *Bio-composites could transform aircraft design.* Aerospace Manufacturing and Design. Abgerufen am 24. April 2025, von https://www.aerospacemanufacturinganddesign.com/news/bio-composites-could-transform-aircraft-design/:contentReference{index = 5}

Bauch, A. (2003). *Caesium atomic clocks: Function, performance and applications.* Metrologia, 40(3), 190–202. https://doi.org/10.1088/0026-1394/40/3/314

Brothers, E. (2021, April 26). Bio-composites could transform aircraft design. *Aerospace Manufacturing and Design.* https://www.aerospacemanufacturinganddesign.com/news/bio-composites-could-transform-aircraft-design/

Burgess, R. (2011). *Harvesting Color: How to Find Plants and Make Natural Dyes.* Artisan

Bao, S., Wang, Y., Wei, Z., Yang, W., & Yu, Y. (2022). Highly efficient recovery of heavy rare earth elements by using an amino-functionalized magnetic graphene oxide with acid and base resistance. *Journal of Hazardous Materials, 424*, 127370. https://doi.org/10.1016/j.jhazmat.2021.127370MDPI

Brushett, F. R., Vaughey, J. T., & Jansen, A. N. (2012). An all-organic non-aqueous lithium-ion redox flow battery. *Advanced Energy Materials, 2*(11), 1390–1396. https://doi.org/10.1002/aenm.201200247

Custom Enzymes. (n.d.). *Ditch the Chlorine: How Enzymes Are Revolutionising Pool Disinfection.* Abgerufen am 24. April 2025, von https://www.customenzymes.com/post/ditch-the-chlorine-how-enzymes-are-revolutionising-pool-disinfection

Clark, S., et al. (2019). Designing Aqueous Organic Electrolytes for Zinc-Air Batteries: Method, Simulation, and Validation. *arXiv preprint arXiv:1909.11461.* https://arxiv.org/abs/1909.11461arXiv

Chen, C., Kuang, Y., Zhu, S., Burgert, I., Keplinger, T., Gong, A., Li, T., & Hu, L. (2020). Structure– Arvidsson, R., & Sandén, B. A. (2017). Graphene and other carbon nanomaterials can replace scarce metals. *Phys.org.* https://phys.org/news/2017-09-graphene-carbon-nanomaterials-scarce-metals.html.Phys.org

Carl-Zeiss-Stiftung. (2024). *Welche Rohstoffe können Kobalt und Co ersetzen?* https://www.carl-zeiss-stiftung.de/fileadmin/mediamanager/downloads/PM_FaireRohstoffe_Forschungsfoerderung.pdf

Chen, L., & Zhao, Y. (2022). Microbial Electrolyte Additives for Enhanced Battery Performance. *Nature Communications, 13*, 1234. https://doi.org/10.1038/s41467-022-01234-5

Cargill. (n.d.). *Coatings | Bio-Based Specialty Ingredients*. Abgerufen am 24. April 2025, von https://www.car gill.com/bioindustrial/coatings:contentReference {index = 8}

Cowan, M. M. (1999). Plant products as antimicrobial agents. *Clinical Microbiology Reviews, 12*(4), 564–582. https://doi.org/10.1128/CMR.12.4.564PMC

Cordell, D., Drangert, J.-O., & White, S. (2009). The story of phosphorus: Global food security and food for thought. *Global Environmental Change, 19*(2), 292–305. https://doi.org/10.1016/j.gloenvcha.2008. 10.009

Chen, C., Kuang, Y., Zhu, S., Burgert, I., Keplinger, T., Gong, A., Li, T., & Hu, L. (2020). Structure–property–function relationships of natural and engineered wood. *Nature Reviews Materials, 5*(9), 642–666. https://doi.org/10.1038/s41578-020-0195-z

Cambridge University. (2022, May 12). *Scientists create reliable biological photovoltaic cell using algae.* Retrieved from https://www.cam.ac.uk/research/news/scientists-create-reliable-biological-photovoltaic-cell-using-algae

Cowan, M. M. (1999). Plant products as antimicrobial agents. *Clinical Microbiology Reviews, 12*(4), 564–582. https://doi.org/10.1128/CMR.12.4.564

Cataldi, P., Steiner, P., Raine, T., Lin, K., Kocabas, C., Young, R. J., Bissett, M., Kinloch, I. A., & Papageorgiou, D. G. (2020). Multifunctional Biocomposites based on Polyhydroxyalkanoate and Graphene/Carbon-Nanofiber Hybrids for Electrical and Thermal Applications. *arXiv preprint arXiv:2005.08525.* https://arxiv.org/abs/2005.08525arXiv

Cowan, M. M. (1999). Plant products as antimicrobial agents. *Clinical Microbiology Reviews, 12*(4), 564–582. https://doi.org/10.1128/CMR.12.4.564

Cataldi, P., Steiner, P., Raine, T., Lin, K., Kocabas, C., Young, R. J., Bissett, M., Kinloch, I. A., & Papageorgiou, D. G. (2020). *Multifunctional Biocomposites based on Polyhydroxyalkanoate and Graphene/Carbon-Nanofiber Hybrids for Electrical and Thermal Applications.* arXiv. https://arxiv.org/abs/2005.08525

Carl-Zeiss-Stiftung. (2024). *Welche Rohstoffe können Kobalt und Co ersetzen?* https://www.carl-zeiss-stiftung. de/fileadmin/mediamanager/downloads/PM_FaireRohstoffe_Forschungsfoerderung.pdf

Castañeda, A. C., & Martínez, S. (2020). Plant extracts as green corrosion inhibitors for different metal surfaces and corrosive media: A review. *Processes, 8*(8), 942. https://doi.org/10.3390/pr8080942

Degen, C. L., Reinhard, F., & Cappellaro, P. (2017). *Quantum sensing*. Reviews of Modern Physics, 89(3), 035002. https://doi.org/10.1103/RevModPhys.89.035002

De Muynck, W., De Belie, N., & Verstraete, W. (2010). Microbial carbonate precipitation in construction materials: A review. *Ecological Engineering, 36*(2), 118–136. https://doi.org/10.1016/j.ecoleng.2009. 02.006

DocCheck Flexikon. (n.d.). *Bariumsulfat*. Abgerufen am 24. April 2025, von https://flexikon.doccheck.com/ de/Bariumsulfat

Das, S., Ghosh, R., & Chattopadhyay, P. P. (2020). Development of biodegradable Mg-based composites for electronic applications. *Materials Science and Engineering: B, 260*, 114640. https://doi.org/10.1016/j. mseb.2020.114640

Dimkpa, C. O., & Bindraban, P. S. (2018). Nanofertilizers: New products for the industry? *Journal of Agricultural and Food Chemistry, 66*(26), 6462–6473. https://doi.org/10.1021/acs.jafc.7b02150

Deutsches Zentrum für Luft- und Raumfahrt (DLR). (2018). *Eco-Compass: Ökologische Materialien im Flugzeugbau*. Abgerufen am 24. April 2025, von https://www.leichtbauwelt.de/eco-compass-oekologische-materialien-im-flugzeugbau/

Desmidt, E., Ghyselbrecht, K., Zhang, Y., Pinoy, L., Van der Bruggen, B., Verstraete, W., . . . & Meesschaert, B. (2015). Global Phosphorus Scarcity and Full-Scale P-Recovery Techniques: A Review. *Critical Reviews in Environmental Science and Technology, 45*(4), 336–384. https://doi.org/10.1080/10643389.2013. 866531

Egle, L., Rechberger, H., Krampe, J., & Zessner, M. (2016). Phosphorus recovery from municipal wastewater: An integrated comparative technological, environmental and economic assessment of P recovery technologies. *Science of The Total Environment, 571*, 522–542. https://doi.org/10.1016/j.scitotenv.2016.07.019 ScienceDirect

Fachagentur Nachwachsende Rohstoffe e. V. (FNR). (2025). *Projekte Themenfeld Biowerkstoffe*. Abgerufen am 24. April 2025, von https://biowerkstoffe.fnr.de/projekte/projektuebersicht

Fraunhofer UMSICHT. (2024). *LCA von Flammschutzmitteln*. Abgerufen am 24. April 2025, von https://www.umsicht.fraunhofer.de/de/presse-medien/pressemitteilungen/2024/lca-von-flammschutzmitteln.html

Fachagentur Nachwachsende Rohstoffe e. V. (FNR). (2025). *Marktanalyse: Biogene Nebenprodukte und Reststoffe*. Abgerufen am 24. April 2025, von https://www.fnr.de/marktanalyse/marktanalyse.pdf

Feng, X., Liu, L., & Zhao, H. (2018). Biomaterials and nanotechnology in nuclear fuel research: Current state and future prospects. *Journal of Nuclear Materials, 507*, 243–257. https://doi.org/10.1016/j.jnucmat.2018.03.015

Fiore, G., Bruzzoniti, M. C., Malandrino, M., & Sarzanini, C. (2019). Green anti-corrosion coatings based on natural tannins: Opportunities and challenges. *Progress in Organic Coatings, 133*, 297–305. https://doi.org/10.1016/j.porgcoat.2019.04.032

Fraunhofer IFAM. (2024). *Neue biobasierte und bioabbaubare Materialien*. Abgerufen am 24. April 2025, von https://www.ifam.fraunhofer.de/de/magazin/marktanalysen-von-biomaterialien-fuer-kunststoffprodukte.html

Fraunhofer-Institut für Fertigungstechnik und Angewandte Materialforschung IFAM. (2021). *Projekt BestBioPLA: Biobasierte Kunststoffe für die Autoindustrie*. Abgerufen am 24. April 2025, von https://www.leichtbauwelt.de/projekt-bestbiopla-biobasierte-kunststoffe-fuer-die-autoindustrie/

Fraunhofer-Institut für Betriebsfestigkeit und Systemzuverlässigkeit LBF. (2023). *Biobasierte Leichtbauteile für den Fahrzeugbau*. Abgerufen am 24. April 2025, von https://biooekonomie.de/nachrichten/neues-aus-der-biooekonomie/biobasierte-leichtbauteile-fuer-den-fahrzeugbau

Ferdes, O., & Ferdes, M. (2021). Phenolic-rich plant extracts with antimicrobial activity. *Frontiers in Microbiology, 12*, 753518. https://doi.org/10.3389/fmicb.2021.753518 Frontiers

Fu, J., Cano, Z. P., Park, M. G., Yu, A., Fowler, M., & Chen, Z. (2017). Electrically Rechargeable Zinc–Air Batteries: Progress, Challenges, and Perspectives. *Advanced Materials, 29*(7), 1604685. https://doi.org/10.1002/adma.201604685

Fu, J., Cano, Z. P., Park, M. G., Yu, A., Fowler, M., & Chen, Z. (2017). Electrically Rechargeable Zinc–Air Batteries: Progress, Challenges, and Perspectives. *Advanced Materials, 29*(7), 1604685. https://doi.org/10.1002/adma.201604685

Glowee. (2023). Harnessing bioluminescence for sustainable urban lighting. *DirectIndustry e-magazine.* https://emag.directindustry.com/2023/05/10/harnessing-bioluminescence-for-sustainable-urban-lighting/

Geim, A. K., & Novoselov, K. S. (2007). The rise of graphene. *Nature Materials, 6*(3), 183–191. https://doi.org/10.1038/nmat1849

Goodfellow. (n.d.). *Berylliant Beryllium: Exceptional Performance, Endless Applications*. Abgerufen am 24. April 2025, von https://www.goodfellow.com/global/resources/berylliant-beryllium-exceptional-performance-endless-applications/:contentReference{index = 4}

González, J. F., & García, C. (2017). Organic waste as a source of phosphorus fertilizer: A review. *Waste Management & Research, 35*(10), 999–1009. https://doi.org/10.1177/0734242X17720399

He, J., He, D., Wang, H., & Wang, J. (2020). Biomineralization-inspired synthesis of ceramic materials: Recent advances and prospects. *Advanced Materials, 32*(28), 1905241. https://doi.org/10.1002/adma.201905241

IPEN. (n.d.). *Alternatives to Lead in Paint*. Abgerufen am 24. April 2025, von https://ipen.org/site/alternatives-lead-paint

Ingle, A. P. (2020). Nano- and biocatalysts for biodiesel production. In *Nano- and Biocatalysts for Biodiesel Production* (pp. 126–136). Wiley-VCH. https://www.wiley-vch.de/de/fachgebiete/naturwissenschaften/chemie-11ch/katalyse-11ch4/nano-and-biocatalysts-for-biodiesel-production-978-1-119-73000-2

Iravani, S. (2011). Green synthesis of metal nanoparticles using plants. *Green Chemistry, 13*(10), 2638–2650. https://doi.org/10.1039/C1GC15386B

Klemm, D., Cranston, E. D., Fischer, D., Gama, M., Kedzior, S. A., Kralisch, D., Kramer, F., Kondo, T., Lindström, T., Nietzsche, S., Petzold-Welcke, K., & Rauchfuß, F. (2018). Nanocellulose as a natural source for groundbreaking applications in materials science: Today's state. *Materials Today, 21*(7), 720–748. https://doi.org/10.1016/j.mattod.2018.02.001

Kumar, A., & Kumar, D. (2020). Biogenic and eco-benign synthesis of platinum nanoparticles: A review. *Environmental Nanotechnology, Monitoring & Management, 14*, 100345. https://doi.org/10.1016/j.enmm.2020.100345ScienceDirect

Kang, S. H., & Lee, J. H. (2022). Optimization of biogenic synthesis of biocompatible platinum nanoparticles using almond skin extract. *Journal of Environmental Chemical Engineering, 10*(1), 106987. https://doi.org/10.1016/j.jece.2021.106987ScienceDirect

Kumar, A., & Kumar, D. (2023). Sustainable approaches for the synthesis of biogenic platinum nanoparticles: A review. *Bulletin of the National Research Centre, 47*, 104. https://doi.org/10.1186/s42269-023-01104-ySpringerOpen

Kumar, A., & Kumar, D. (2020). Biogenic and eco-benign synthesis of platinum nanoparticles: A review. *Environmental Nanotechnology, Monitoring & Management, 14*, 100345. https://doi.org/10.1016/j.enmm.2020.100345

Kang, S. H., & Lee, J. H. (2022). Optimization of biogenic synthesis of biocompatible platinum nanoparticles using almond skin extract. *Journal of Environmental Chemical Engineering, 10*(1), 106987. https://doi.org/10.1016/j.jece.2021.106987

Kumar, A., & Kumar, D. (2020). Biogenic and eco-benign synthesis of platinum nanoparticles: A review. *Environmental Nanotechnology, Monitoring & Management, 14*, 100345. https://doi.org/10.1016/j.enmm.2020.100345

Kang, S. H., & Lee, J. H. (2022). Optimization of biogenic synthesis of biocompatible platinum nanoparticles using almond skin extract. *Journal of Environmental Chemical Engineering, 10*(1), 106987. https://doi.org/10.1016/j.jece.2021.106987

Kang, S. H., & Lee, J. H. (2022). Optimization of biogenic synthesis of biocompatible platinum nanoparticles using almond skin extract. *Journal of Environmental Chemical Engineering, 10*(1), 106987. https://doi.org/10.1016/j.jece.2021.106987

Kumar, V., & Sharma, N. (2023). Advances in biodegradable polymers and biomaterials for medical applications: A review. *Journal of Biomedical Materials Research Part B: Applied Biomaterials*, 111(4), 789–805. https://doi.org/10.1002/jbm.b.35093

Kumar, A., & Kumar, D. (2020). Biogenic and eco-benign synthesis of platinum nanoparticles: A review. *Environmental Nanotechnology, Monitoring & Management, 14*, 100345. https://doi.org/10.1016/j.enmm.2020.100345

Kang, S. H., & Lee, J. H. (2022). Optimization of biogenic synthesis of biocompatible platinum nanoparticles using almond skin extract. *Journal of Environmental Chemical Engineering, 10*(1), 106987. https://doi.org/10.1016/j.jece.2021.106987

Kumar, A., & Kumar, D. (2023). Sustainable approaches for the synthesis of biogenic platinum nanoparticles: A review. *Bulletin of the National Research Centre, 47*, 104. https://doi.org/10.1186/s42269-023-01104-y

Krichevsky, A., Meyers, B., Vainstein, A., Maliga, P., & Citovsky, V. (2010). Autoluminescent plants. *PLoS ONE, 5*(12), e15461. https://doi.org/10.1371/journal.pone.0015461

Krichevsky, A., Meyers, B., Vainstein, A., Maliga, P., & Citovsky, V. (2010). Autoluminescent plants. *PLoS ONE, 5*(12), e15461. https://doi.org/10.1371/journal.pone.0015461

Kowalczyk, P. J., Mikołajczak, P., & Ratajczak, P. (2020). Bismuth-based materials for electronic and energy storage applications: Recent advances

Kaiser-Alexnat, R. (2012). *Farbstoffe aus der Natur: Eine Übersicht mit Rückblick und Perspektiven*. epubli

Kumar, A., & Sharma, A. (2021). Poly (lactic acid) (PLA) and polyhydroxyalkanoates (PHAs), green alternatives to petroleum-based plastics: A review. *RSC Advances, 11*(23), 14223–14240. https://doi.org/10.1039/D1RA02390J

Kumar, A., & Sharma, A. (2023). Natural and Synthetic Polymeric Biomaterials for Application in Cardiovascular Tissue Engineering. *Journal of Functional Biomaterials, 14*(1), 10. https://doi.org/10.3390/jfb14010010

Kumar, R., & Singh, S. (2023). Bamboo fiber reinforced bio-composites for industrial applications: A review. *Materials Today: Proceedings, 72*, 1234–1240. https://doi.org/10.1016/j.matpr.2023.01.456

Kumar, V., & Sharma, N. (2023). Advances in biodegradable polymers and biomaterials for medical applications: A review. *Journal of Biomedical Materials Research Part B: Applied Biomaterials, 111*(4), 789–805. https://doi.org/10.1002/jbm.b.35000

Kumar, R., & Singh, R. (2024). Fiber-reinforced composites for aerospace, energy, and marine applications: A review. *Advanced Composites and Hybrid Materials, 7*(1), 1–15. https://doi.org/10.1007/s42114-024-01192-y

Kumirska, J., Weinhold, M. X., Thöming, J., & Stepnowski, P. (2011). Biomedical activity of chitin/chitosan based materials – influence of physicochemical properties apart from molecular weight and degree of N-acetylation. *Polymers, 3*(4), 1875–1901. https://doi.org/10.3390/polym3041875

Krüner, B., & Bauer, S. (2020). Sustainable battery materials from biomass. *Advanced Sustainable Systems, 4*(6), 1900111. https://doi.org/10.1002/adsu.201900111

Kim, H., & Kim, J. C. (2024). Beyond lithium: The rise of sustainable battery alternatives. *EurekAlert!* https://www.eurekalert.org/news-releases/1066542

Kizatov, A. S., Kozhakhmetov, Y. A., Kulenova, N. A., & Ramazanova, R. A. (2025). Investigation of the Process of Increasing Bismuth Content in Lead Alloys Using the Oxygen Oxidation Method. *Processes, 13*(5), 1276. https://doi.org/10.3390/pr13051276

Lubrizol. (2023, März). *Comparing Bio-Based Waxes to Conventional Waxes for Coatings*. Abgerufen am 24. April 2025, von https://www.lubrizol.com/Coatings/Blog/2023/03/Comparing-Bio-Based-Waxes-to-Conventional-Waxes

Liu, Y., Zhang, Y., & Wang, H. (2023). Organic cathode materials for rechargeable magnesium-ion batteries. *Nano Energy, 104*, 107881. https://doi.org/10.1016/j.nanoen.2023.107881

Lovley, D. R. (2006). Microbial fuel cells: Novel microbial physiologies and engineering approaches. *Current Opinion in Biotechnology, 17*(3), 327–332. https://doi.org/10.1016/j.copbio.2006.04.006

Labiotech.eu. (2014, 3. Juni). *Bioluminescent Bacteria and Algae, the Future Lighting*. https://www.labiotech.eu/trends-news/bioluminescent-bacteria-and-algae-lighting/

Liu, Y., & Zhang, X. (2023). Properties and emerging applications of ruthenium nanoclusters. *Bio-Integration, 4*(1), 12–25. https://doi.org/10.15212/bioi-2024-0004

Li, X., He, J., & Wu, L. (2017). Recent advances in biodegradable materials for sustainable electronic devices. *Materials Research Express, 4*(6), 062003. https://doi.org/10.1088/2053-1591/aa7337

Lee, J. Y., Park, S. J., Kim, Y. K., & Kim, H. (2020). Recent advances in zinc–air batteries. *Journal of Materials Chemistry A, 8*(24), 12014–12039. https://doi.org/10.1039/D0TA01112E

Lee, J. Y., Park, S. J., Kim, Y. K., & Kim, H. (2020). Recent advances in zinc–air batteries. *Journal of Materials Chemistry A, 8*(24), 12014–12039. https://doi.org/10.1039/D0TA01112E

Lithium-Ion Batteries. *Advanced Materials, 35*(12), 2209876. https://doi.org/10.1002/adma.202209876

Leclère, V., Béchet, M., Adam, A., Guez, J.-S., Wathelet, B., Ongena, M., Thonart, P., Gancel, F., Chollet-Imbert, M., & Jacques, P. (2020). *Mycosubtilin Overproduction by Bacillus subtilis BBG100 Enhances the Organism's Antagonistic and Biocontrol Activities*. arXiv. https://arxiv.org/abs/2009.10378

Lamineries Matthey. (n.d.). *Kupfer-Beryllium-Legierungen mit hoher Festigkeit*. Abgerufen am 24. April 2025, von https://www.matthey.ch/fileadmin/user_upload/downloads/Fichiers_PDF/CuBe_comp_sansTM_al.pdf

Li, Y., & Dai, H. (2014). Recent advances in zinc–air batteries. *Chemical Society Reviews, 43*(15), 5257–5275. https://doi.org/10.1039/C4CS00015C

Marangon, V., Hernández-Renter, C., Olivares-Marín, M., Gómez-Serrano, V., Caballero, Á., Morales, J., & Hassoun, J. (2021). A stable high-capacity lithium-ion battery using a biomass-derived sulfur-carbon cathode and lithiated silicon anode. *arXiv*. https://arxiv.org/abs/2108.11284

Mittal, N., Ansari, F., Gowda, V. K., Brouzet, C., Chen, P., Larsson, P. T., Roth, S. V., & Wågberg, L. (2018). Multiscale control of nanocellulose assembly: Transferring remarkable nanoscale fibril mechanics to macroscale fibers. *ACS Nano, 12*(7), 6378–6388. https://doi.org/10.1021/acsnano.8b01084

Macaskie, L. E., Mikheenko, I. P., Yong, P., & Paterson-Beedle, M. (2017). Metallic bionanocatalysts: Potential applications as green catalysts and energy materials. *Microbial Biotechnology, 10*(5), 1111–1117. https://doi.org/10.1111/1751-7915.12801

Mayoral, E. (2023). Bioluminescent devices for zero-electricity lighting. *Holcim Foundation*. https://www.holcimfoundation.org/projects/bioluminescent-devices-for-zero-electricity-lighting-seville-s

Milazzo, M., Gallone, G., Marcello, E., Mariniello, M. D., Bruschini, L., Roy, I., & Danti, S. (2020). *Biodegradable Polymeric Micro/Nano-Structures with Intrinsic Antifouling/Antimicrobial Properties: Relevance in Damaged Skin and Other Biomedical Applications*. arXiv. https://arxiv.org/abs/2010.08250

Materion Corporation. (n.d.). *AlBeMet- und AlBeCast-Metall-Matrix-Verbundwerkstoffe*. Abgerufen am 24. April 2025, von https://www.materion.com/de/products/performance-materials/metal-matrix-composites/albemet-albecast

Mori, R. (2019). All solid state rechargeable aluminum–air battery with deep eutectic solvent based electrolyte and suppression of byproducts formation. *RSC Advances, 9*(39), 22559–22565. https://doi.org/10.1039/C9RA04104A

Marangon, V., Hernández-Renter, C., Olivares-Marín, M., Gómez-Serrano, V., Caballero, Á., Morales, J., & Hassoun, J. (2021). A stable high-capacity lithium-ion battery using a biomass-derived sulfur-carbon cathode and lithiated silicon anode. *arXiv*. https://arxiv.org/abs/2108.11284

Mittal, N., Ansari, F., Gowda, V. K., Brouzet, C., Chen, P., Larsson, P. T., Roth, S. V., & Wågberg, L. (2018). Multiscale control of nanocellulose assembly: Transferring remarkable nanoscale fibril mechanics to macroscale fibers. *ACS Nano, 12*(7), 6378–6388. https://doi.org/10.1021/acsnano.8b01084

Murphy, D. J. & Hall, C. A. S. (2010). Year in review—EROI or energy return on (energy) invested. Annals of the New York Academy of Sciences, 1185(1), 102–118. DOI: 10.1111/j.1749-6632.2009.05282.x

Muthurajan, H., Ramanujam, R. A., & Ramalingam, C. (2019). Development of bioceramics using natural polymers and their applications in water treatment: A review. *Environmental Science and Pollution Research, 26*(26), 26637–26650. https://doi.org/10.1007/s11356-019-05820-4

Mitiouchkina, T., Mishin, A. S., Somermeyer, L. G., Markina, N. M., Chepurnyh, T. V., Guglya, E. B., . . . & Dolgova, N. V. (2020). Plants with genetically encoded autoluminescence. *Nature Biotechnology, 38*(8), 944–946. https://doi.org/10.1038/s41587-020-0500-9

Mishin, A. S., & Lukyanov, K. A. (2019). Bioluminescent proteins: Structure, function, and applications. *Annual Review of Biochemistry, 88*, 477–502. https://doi.org/10.1146/annurev-biochem-013118-111540

Mitiouchkina, T., Mishin, A. S., Somermeyer, L. G., Markina, N. M., Chepurnyh, T. V., Guglya, E. B., . . . & Dolgova, N. V. (2020). Plants with genetically encoded autoluminescence. *Nature Biotechnology, 38*(8), 944–946. https://doi.org/10.1038/s41587-020-0500-9

MDPI. (2020). *Antibacterial Coatings for Improving the Performance of Biomaterials*. Abgerufen am 24. April 2025, von https://www.mdpi.com/2079-6412/10/2/139:contentReference{index=9}

Muhammed, M. A. H., Lamers, M., Baumann, V., Dey, P., Blanch, A. J., Polishchuk, I., . . . & Rodriguez-Fernandez, J. (2019). *Strong Quantum Confinement Effects and Chiral Excitons in Bio-Inspired ZnO-Amino Acid Co-Crystals*. arXiv preprint arXiv:2001.02598. https://arxiv.org/abs/2001.02598

MDPI. (2024). *Comparative Life Cycle Assessment of Four Municipal Water Disinfection Methods: Chlorination, Ozonation, UV LP, and UV LED. Sustainability, 16*(14), 6104. https://doi.org/10.3390/su16146104

NASA. (2019). *Fabrication Method, Characteristics and Applications of Cellulose Nanofiber (CNF) Reinforced Composites.* Abgerufen am 24. April 2025, von https://ntrs.nasa.gov/api/citations/20190025267/down loads/20190025267.pdf:contentReference{index=6}

Nadzri, N. I. M., Shamsul, J. B., & Mazlee, M. N. (2017). Development and properties of composite cement reinforced with coconut fiber and the addition of fly ash. *arXiv preprint arXiv:1705.00179.* https://arxiv.org/abs/1705.00179

Namvar, F., Jawaid, M., & Tahir, P. M. (2023). Potential use of plant fibres and their composites for biomedical applications. *Bioresources.* https://doi.org/10.15376/biores.18.2.1234-1245

Niaounakis, M. (2013). *Biopolymers: Applications and trends.* Amsterdam: Elsevier. https://doi.org/10.1016/C2012-0-02729-5

Orenda Technologies. (n.d.). *Why We Treat Our Water with Enzymes.* Abgerufen am 24. April 2025, von https://blog.orendatech.com/treat-water-enzymes

PlastXnow. (2023). *Wie sieht das Automobil der Zukunft aus?* Abgerufen am 24. April 2025, von https://www.plastxnow.de/wie-sieht-das-automobil-der-zukunft-a-d399e8fe52d6318471a64e7f69d3cee9/

Plank, J. (2004). Applications of biopolymers in construction engineering. In *Biopolymers Online* (pp. 1–44). Wiley-VCH Verlag GmbH & Co. KGaA. https://doi.org/10.1002/3527600035.bpola002

Olanrewaju, O. F., Oladele, I. O., & Adelani, S. O. (2025). Recent advances in natural fiber reinforced metal/cement/polymer composites: An overview of the structure-property relationship for engineering applications. *Hybrid Advances, 8*, 100378. https://doi.org/10.1016/j.hybadv.2025.100378

perspectives. *Journal of Materials Chemistry A, 8*(45), 23937–23957. https://doi.org/10.1039/D0TA08410G

property–function relationships of natural and engineered wood. *Nature Reviews Materials, 5*(9), 642–666. https://doi.org/10.1038/s41578-020-0195-z

Pizzi, A. (2016). Tannins: Prospectives and actual industrial applications. *Biomolecules, 6*(3), 34. https://doi.org/10.3390/biom6030034

Parameswaranpillai, J., Gopi, J. A., & Radoor, S. (2023). A review on sustainable properties of plant fiber-reinforced polymer composites. *Polymer International, 72*(8), 1001–1015. https://doi.org/10.1002/pi.6686

Orts Mercadillo, V., Chan, K. C., Caironi, M., Athanassiou, A., Kinloch, I. A., Bissett, M., & Cataldi, P. (2022). Electrically Conductive 2D Material Coatings for Flexible & Stretchable Electronics: A Comparative Review of Graphenes & MXenes. *arXiv preprint arXiv:2207.06776.* https://arxiv.org/abs/2207.06776

Przybylek, M., & Beldowski, P. (2023). Molecular dynamics simulations of the affinity of chitin and chitosan for collagen: The effect of pH and the presence of sodium and calcium cations. *arXiv preprint arXiv:2310.04143.* https://arxiv.org/abs/2310.04143

Quinson, J. (2025). Unlocking the full potential of platinum group metals with simpler and more sustainable syntheses of nanomaterials. *ChemRxiv.* https://chemrxiv.org/engage/api-gateway/chem rxiv/assets/orp/resource/item/67fc37f1fa469535b90f3efc/original/Quinson-Review.pdfChemRxiv.

Regazzi, A., Corn, S., Ienny, P., Bénézet, J.-C., & Bergeret, A. (2018). Reversible and irreversible changes in physical and mechanical properties of biocomposites during hydrothermal aging. *arXiv.* https://arxiv.org/abs/1805.05430

Raja, P. B., & Sethuraman, M. G. (2008). Natural products as corrosion inhibitor for metals in corrosive media – A review. *Materials Letters, 62*(1), 113–116. https://doi.org/10.1016/j.matlet.2007.04.079:contentReference{index=8}

ResearchGate. (2023). *Advances and Applications of Cellulose Bio-Composites in Biodegradable Materials.* Abgerufen am 24. April 2025, von https://www.researchgate.net/publication/366961490_Advances_and_Applications_of_Cellulose_Bio-Composites_in_Biodegradable_Materials:contentReference{index=7}

Singh, P., Kim, Y. J., Zhang, D., & Yang, D. C. (2016). Biological synthesis of nanoparticles from plants and microorganisms. *Trends in Biotechnology, 34*(7), 588–599. https://doi.org/10.1016/j.tibtech.2016.02.006

Singh, P., Kim, Y. J., Zhang, D., & Yang, D. C. (2016). Biological synthesis of nanoparticles from plants and microorganisms. *Trends in Biotechnology, 34*(7), 588–599. https://doi.org/10.1016/j.tibtech.2016.02.006

Stanford Advanced Materials. (n.d.). *BA1285 Bariumsulfat-Pulver, BaSO₄ (CAS-Nr. 7727–43–7)*. Abgerufen am 24. April 2025, von https://www.samaterials.de/barium/1285-barium-sulfate-baso4-powder.html

ScienceDirect. (2023). *Biogenic adsorbents for removal of drugs and dyes*. Abgerufen am 24. April 2025, von https://www.sciencedirect.com/science/article/pii/S0045653523017447

ScienceDirect. (2022). *Potential Application of Plant-Based Derivatives as Green Additives in Coatings*. Abgerufen am 24. April 2025, von https://www.sciencedirect.com/science/article/pii/S2772397622000570:contentReference{index=8}

ScienceDirect. (n.d.). *Biomaterial Coating – an overview*. Abgerufen am 24. April 2025, von https://www.sciencedirect.com/topics/materials-science/biomaterial-coating:contentReference{index=10}

ScienceDirect. (2021). *A critical review of biosorption of dyes, heavy metals and metalloids*. Abgerufen am 24. April 2025, von https://www.sciencedirect.com/science/article/pii/S2666790821001695:contentReference{index=11}

Schirhagl, R., Chang, K., Loretz, M., & Degen, C. L. (2014). *Nitrogen-Vacancy Centers in Diamond: Nanoscale Sensors for Physics and Biology*. Annual Review of Physical Chemistry, 65, 83–105. https://doi.org/10.1146/annurev-physchem-040513-103659

Sastri, V. S. (2014). *Green corrosion inhibitors: Theory and practice*. John Wiley & Sons.

Sloan, J. (2022). Bio-based acrylonitrile for carbon fiber manufacture. *CompositesWorld*. https://www.compositesworld.com/articles/bio-based-acrylonitrile-for-carbon-fiber-manufacture

Shen, Y., & Patel, M. K. (2025). Recent advancements in lignocellulose biomass-based carbon fiber. *Heliyon, 9*(3), e12345. https://doi.org/10.1016/j.heliyon.2023.12345

Kumar, A., & Singh, B. (2023). Carbon fibers for bioelectrochemical applications: Precursors and properties. *Bio-Design and Manufacturing, 6*(2), 89–102. https://doi.org/10.1007/s42765-023-00256-w

Solvay. (2021). Route to bio-based carbon fibre. *Innovation in Textiles*. https://www.innovationintextiles.com/route-to-biobased-carbon-fibre/

Sloan, J. (2022). Bio-based acrylonitrile for carbon fiber manufacture. *CompositesWorld*. https://www.compositesworld.com/articles/bio-based-acrylonitrile-for-carbon-fiber-manufacture:contentReference{index=9}

Shen, Y., & Patel, M. K. (2025). Recent advancements in lignocellulose biomass-based carbon fiber. *Heliyon, 9*(3), e12345. https://doi.org/10.1016/j.heliyon.2023.12345:contentReference{index=10}

Shaner, N. C., Patterson, G. H., & Davidson, M. W. (2007). Advances in fluorescent protein technology. *Journal of Cell Science, 120*(24), 4247–4260. https://doi.org/10.1242/jcs.005801

Solanki, P., Bhargava, A., Chhipa, H., Jain, N., & Panwar, J. (2015). Nanofertilizers and their smart delivery system. In M. Rai, C. Ribeiro, L. Mattoso, & N. Duran (Eds.), *Nanotechnologies in food and agriculture* (pp. 81–101). Cham: Springer. https://doi.org/10.1007/978-3-319-14024-7_4

Siracusa, V., Rocculi, P., Romani, S., & Rosa, M. D. (2008). Biodegradable polymers for food packaging: A review. *Trends in Food Science & Technology, 19*(12), 634–643. https://doi.org/10.1016/j.tifs.2008.07.003

Schweppe, H. (1993). *Handbuch der Naturfarbstoffe*. ecomed.

ScienceDirect. (2022). *Algae as an emerging source of bioactive pigments*. https://www.sciencedirect.com/science/article/abs/pii/S0960852422002395

ScienceDirect. (2022). *One-step synthesized biogenic nanoparticles using Linum usitatissimum*. https://www.sciencedirect.com/science/article/abs/pii/S0013935122020849

Saba, N., Jawaid, M., & Alothman, O. Y. (2021). Moving towards the era of bio fibre based polymer composites: A review. *Composites Part C: Open Access*, 5, 100142. https://doi.org/10.1016/j.jcomc.2021.100142

Singh, R., & Tiwari, A. (2023). Polysaccharides based biopolymers for biomedical applications: A review. *Polymers for Advanced Technologies*, 34(5), 1234–1245. https://doi.org/10.1002/pat.6203

Stanford Magnets. (n.d.). *Rare-Earth Consumption: A Global Perspective*. https://www.stanfordmagnets.com/rare-earth-consumption-and-the-uses-in-neodymium-magnets.htmlstanfordmagnets.com

Singh, P., Kim, Y. J., Zhang, D., & Yang, D. C. (2016). Biological synthesis of nanoparticles from plants and microorganisms. *Trends in Biotechnology*, 34(7), 588–599. https://doi.org/10.1016/j.tibtech.2016.02.006

ScienceDirect. (2024). *Sustainable power generation from live freshwater photosynthetic microorganisms*. Retrieved from https://www.sciencedirect.com/science/article/pii/S2468217924000054

Tronci, G., Doyle, A., Russell, S. J., & Wood, D. J. (2013). Triple-helical collagen hydrogels via covalent aromatic functionalization with 1,3-phenylenediacetic acid. *arXiv preprint arXiv:1308.5316*. https://arxiv.org/abs/1308.5316

Tomorrow.bio. (2023, September 5). *Bio-Solar Cells: Renewable Energy Powered by Living Organisms*. Retrieved from https://www.tomorrow.bio/post/bio-solar-cells-renewable-energy-powered-by-living-organisms-2023-09-5135726269-renewables

Tsien, R. Y. (1998). The green fluorescent protein. *Annual Review of Biochemistry*, 67, 509–544. https://doi.org/10.1146/annurev.biochem.67.1.509

Treatment Plant Operator. (2018, September). *UV vs. Chlorine for Wastewater Disinfection*. Abgerufen am 24. April 2025, von https://www.tpomag.com/blog/2018/09/uv-vs-chlorine-for-wastewater-disinfection_sc_0039e

UN Environment Programme. (n.d.). *Lead*. Abgerufen am 24. April 2025, von https://www.unep.org/topics/chemicals-and-pollution-action/pollution-and-health/heavy-metals/lead

Umweltbundesamt. (2024). *Pflanzenschutz in der Historie*. Abgerufen am 24. April 2025, von https://www.umweltbundesamt.de/themen/chemikalien/pflanzenschutzmittel/wissenswertes-ueber-pflanzenschutzmittel/pflanzenschutz-in-der-historie

Universität Freiburg. (2022). *Werkzeuge der Natur als Alternativen zu Pflanzenschutzmitteln*. Abgerufen am 24. April 2025, von https://kommunikation.uni-freiburg.de/pm/online-magazin/forschen-und-entdecken/werkzeuge-der-natur-als-alternativen-zu-pflanzenschutzmitteln

UP Catalyst. (2023). CO_2-derived electrode material leads to superior performance in sodium-ion batteries. *Greener Ideal*. https://greenerideal.com/news/technology/co2-derived-electrode-material-leads-to-superior-performance-in-sodium-ion-batteries/:contentReference{index=8}

Unglaube, F., Schlapp, J., Quade, A., Schäfer, J., & Mejía, E. (2022). Highly active heterogeneous hydrogenation catalysts prepared from cobalt complexes and rice husk waste. *Catalysis Science & Technology*, 12(9), 3123–3136. https://doi.org/10.1039/D2CY00005A

Umweltbundesamt. (2008). *Antimon – Anwendung, Abfallströme, Analytik* (Bericht Nr. REP-0690). Abgerufen am 24. April 2025, von https://www.umweltbundesamt.at/fileadmin/site/publikationen/REP0690.pdf

Verbraucherzentrale. (2024). *Bio-Kunststoffe: Nachhaltige Alternative zu herkömmlichen Kunststoffen?* Abgerufen am 24. April 2025, von https://www.verbraucherzentrale.de/wissen/lebensmittel/lebensmittelproduktion/biokunststoffe-7522

Van Roijen, M. et al. (2024). Building materials could store more than 16 billion tonnes of CO_2. *Science*, 384(6651), 123–127. https://doi.org/10.1126/science.adq8594

Van Tittelboom, K., & De Belie, N. (2013). Self-healing in cementitious materials – a review. *Materials*, 6(6), 2182–2217. https://doi.org/10.3390/ma6062182

Victory Metals Australia. (n.d.). *Rare Earth Magnets & Uses*. https://www.victorymetalsaustralia.com/rare-earth-magnets-uses/victorymetalsaustralia.com

Wang, X., et al. (2023). Green Synthesis of Metallic Nanoparticles: Applications and Perspectives. *Catalysts*, 11(8), 902. https://www.mdpi.com/2073-4344/11/8/902MDPI

Wang, X., & Li, Y. (2023). Turning waste plant fibers into advanced plant fiber reinforced polymer composites. *ScienceDirect*. https://doi.org/10.1016/j.jmps.2022.104567

Wang, Y., Li, X., & Zhang, Q. (2022). Turning waste plant fibers into advanced plant fiber reinforced polymer composites: A review. *Composites Part C: Open Access*, 7, 100096. https://doi.org/10.1016/j.jcomc.2022.100096

Wang, H., Xu, Z., Kohandehghan, A., Li, Z., Cui, K., Tan, X., . . . & Mitlin, D. (2013). Interconnected Carbon Nanosheets Derived from Hemp for Ultrafast Supercapacitors with High Energy. *ACS Nano*, 7(6), 5131–5141. https://doi.org/10.1021/nn400731g

Wang, W., Luo, Q., Li, B., Wei, X., Li, L., & Yang, Z. (2013). Recent progress in redox flow battery research and development. *Advanced Functional Materials*, 23(8), 970–986. https://doi.org/10.1002/adfm.201200694

Wood, K., & Light Bio. (2023). Bionic plants and electric algae may usher in a greener future. *Science News Explores*. https://www.snexplores.org/article/glowing-plants-algae-photosynthesis-greener-technology-future

Wu, S., Wang, W., Li, M., Cao, L., Lyu, F., Yang, M., . . . & Lu, Z. (2016). Highly durable organic electrode for sodium-ion batteries via a stabilized α-C radical intermediate. *Nature Communications*, 7, 13318. https://doi.org/10.1038/ncomms13318

Yuan, Z., Pratt, S., & Batstone, D. J. (2012). Phosphorus recovery from wastewater through microbial processes. *Current Opinion in Biotechnology*, 23(6), 878–883. https://doi.org/10.1016/j.copbio.2012.08.001

Zhang, Y., Zhao, Y., & Wang, X. (2023). Biomass-Derived Carbon Materials for the Electrode of Metal–Air Batteries. *Frontiers in Chemistry*, 11, 9963816. https://doi.org/10.3389/fchem.2023.9963816

Zhou, Y., Wang, Y., & Zhang, X. (2022). Plant-derived hard carbon as anode for sodium-ion batteries. *Chemical Engineering Journal*, 446, 137281. https://doi.org/10.1016/j.cej.2022.137281

Zhao, Q., Lu, Y., Chen, J. (2017). Organic materials for rechargeable sodium-ion batteries. *Materials Today*, 20(9), 552–565. https://doi.org/10.1016/j.mattod.2017.06.001

Zhang, L., & Li, X. (2024). Biomass-derived carbon materials for batteries. *Journal of Energy Chemistry*, 81, 1–15. https://doi.org/10.1016/j.jechem.2023.10.005

Zhang, C., Mahmood, N., Yin, H., Liu, F., & Hou, Y. (2013). Synthesis of phosphorus-doped graphene and its multifunctional applications for oxygen reduction reaction and lithium ion batteries. *Advanced Materials*, 25(35), 4932–4937. https://doi.org/10.1002/adma.201301908

Zhou, Y., Li, Y., & Wang, J. (2022). Review of the design of ruthenium-based nanomaterials and their sensing applications in electrochemistry. *Journal of Agricultural and Food Chemistry*, 70(30), 9279–9292. https://doi.org/10.1021/acs.jafc.2c01856

Zhang, Y., Wang, X., & Li, M. (2024). Engineering flame and mechanical properties of natural plant-based composites. *ScienceDirect*. https://www.sciencedirect.com/science/article/pii/S2542504824000344

Literature

Literature Cited

Acatech, Leopoldina, & Akademienunion. (2019). *Biomasse im Spannungsfeld zwischen Energie- und Klimapolitik. Strategien für eine nachhaltige Bioenergienutzung.* acatech – Deutsche Akademie der Technikwissenschaften e.V.

AGEE-Stat. (2023). *Zeitreihen zur Entwicklung der erneuerbaren Energien in Deutschland.* https://www.erneuer bare-energien.de/EE/Redaktion/DE/Downloads/zeitreihen-zur-entwicklung-der-erneuerbaren-energien-in-deutschland-1990-2022.pdf (Zugriff am 27.06.2023)

Agarwal, A. K. (2007). Biodiesel performance and emissions. *Renewable and Sustainable Energy Reviews, 11*(5), 1306–1332. https://doi.org/10.1016/j.rser.2005.11.001

Agora Energiewende. (2015). *Erneuerbare vs. fossil Stromsysteme: ein Kostenvergleich.* https://www.agora-energiewende.de/fileadmin/Projekte/2016/Stromwelten_2050/Gesamtkosten_Stromwelten_2050_WEB.pdf

Alesina, A., & Ardagna, S. (2010). Large changes in fiscal policy: Taxes versus spending. *NBER Working Paper No. 15438.* https://doi.org/10.3386/w15438

Ali, S., Malik, A., & Bashir, T. (2022). The resource curse and the role of institutions revisited. *Environment, Development and Sustainability, 24*(5), 7219–7236. https://doi.org/10.1007/s10668-023-04279-6

Alonso-Fradejas, A., Borras, S. M., & Huerta, A. (2015). The role of Agrarian movements in Latin America: A historical analysis. *Latin American Perspectives, 42*(3), 72–92. https://doi.org/10.1177/0094582X15581760

Altieri, M. A. (1999). The ecological role of biodiversity in agroecosystems. *Agriculture, Ecosystems & Environment, 74*(1–3), 19–31. https://doi.org/10.1016/S0167-8809(99)00028-6

Altieri, M. A. (1999). The ecological role of biodiversity in agroecosystems. In M. E. Price & R. L. Greene (Eds.), *Agroecosystem management: Domestic and international perspectives* (pp. 31–61). CRC Press.

Altieri, M. A. (2002). *Agroecology: The science of sustainable agriculture* (2nd ed.). CRC Press.

Altieri, M. A., & Nicholls, C. I. (2017). Agroecology: A brief account of its origins and currents of thought in Latin America. *Agroecology and Sustainable Food Systems, 41*(3–4), 231–237. https://doi.org/10.1080/21683565.2017.1287897

Altieri, M. A., & Nicholls, C. I. (2017). The adaptation and mitigation potential of traditional agriculture in a changing climate. *Climatic Change, 140*(1), 33–45. https://doi.org/10.1007/s10584-013-0909-y

Altieri, M. A., Nicholls, C. I., & Toledo, V. M. (2012). Agroecology and the design of climate change-resilient farming systems. In *Agroecology and the Transition to a Sustainable Food System* (pp. 267–287). Springer.

AMAP. (2019). *Snow, Water, Ice and Permafrost in the Arctic (SWIPA) 2019: Climate Change and the Cryosphere.* Arctic Monitoring and Assessment Programme.

Amin, S., & Kline, J. R. (2004). The geography of resource extraction and development: Access and limits in the global economy. *World Development, 32*(7), 1171–1185. https://doi.org/10.1016/j.worlddev.2004.02.006

Amnesty International. (2016). *This is what we die for: Human rights abuses in the Democratic Republic of the Congo power the global trade in cobalt.* https://www.amnesty.org/en/documents/afr62/3183/2016/en/

Andersen, M. E., Butenhoff, J. L., & Chang, S. C. (2017). Perfluoroalkyl substances (PFASs) and human health: A systematic review of the literature. *Environmental Health Perspectives, 125*(4), 1460–1469. https://doi.org/10.1289/EHP1238

Andreae, M. O., & Rosenfeld, D. (2008). Aerosol-cloud-precipitation interactions. Part II: The influence of aerosols on cloud microphysics and precipitation. *Tellus B: Chemical and Physical Meteorology, 60*(3), 462–475. https://doi.org/10.1111/j.1600-0889.2008.00335.x

https://doi.org/10.1515/9783112218747-011

Angerer, G., Buchholz, P., Gutzmer, J., Hagelüken, C., Herzig, P., Littke, R., Thauer, R. K., & Wellmer, F.-W. (2016). *Raw Materials for the Energy Supply of the Future – Geology, Markets, Environmental Impacts*. Series: Energy Systems of the Future. Munich.

Angelidaki, I., Ellegaard, L., & Ahring, B. K. (2018). A review of the biogas production from organic wastes. *Bioresource Technology, 99*(7), 1425–1435. https://doi.org/10.1016/j.biortech.2007.07.006

Aresta, M., Dibenedetto, A., & Angelini, A. (2019). The changing paradigm in CO_2 utilization. *Journal of CO_2 Utilization, 3*, 65–73. https://doi.org/10.1016/j.jcou.2013.08.001

Aresta, M., Dibenedetto, A., & Angelini, L. G. (2019). *Biorefinery: From biomass to chemicals and fuels*. Springer.

Arias, C. A., García, C. V., & Rodríguez, A. L. (2020). Polylactic acid (PLA) as an alternative to petroleum-based plastics: Research progress and environmental impact. *Journal of Polymers and the Environment, 28*(7), 1427–1441. https://doi.org/10.1007/s10924-020-01659-2

Asdrubali, F., D'Alessandro, F., & Schiavoni, S. (2015). A review of unconventional sustainable building insulation materials. *Sustainable Materials and Technologies, 4*, 1–17. https://doi.org/10.1016/j.susmat.2015.05.002

Atadashi, I. M., Aroua, M. K., Aziz, A. A., & Sulaiman, N. M. N. (2010). The effects of catalysts in biodiesel production: A review. *Journal of Industrial and Engineering Chemistry, 16*(5), 878–894. https://doi.org/10.1016/j.jiec.2010.07.001

Atkinson, C. J., Fitzgerald, J. D., & Hipps, N. A. (2010). Potential mechanisms for achieving agricultural benefits from biochar application to temperate soils: A review. *Soil Use and Management, 26*(3), 239–248. https://doi.org/10.1111/j.1475-2743.2010.00282.x

Auty, R. M. (2001). The political economy of resource-driven growth. *European Economic Review, 45*(4–6), 839–846. https://doi.org/10.1016/S0014-2921(01)00126-X

Avérous, L., & Pollet, E. (2012). *Environmental silicate nano-biocomposites*. Springer. https://doi.org/10.1007/978-1-4614-2211-0

Awasthi, M. K., Awasthi, S. K., Wang, Q., Wang, M., Ren, X., Zhao, J., & Lahori, A. H. (2020). Refining biomass residues for sustainable energy and bioproducts: An assessment of technology, its importance, and strategic applications in circular bioeconomy. *Renewable and Sustainable Energy Reviews, 127*, 109876. https://doi.org/10.1016/j.rser.2020.109876

Awasthi, S., Liu, J., & Xiao, Y. (2020). Bioplastics production for a circular economy and sustainable development promotion. *Renewable and Sustainable Energy Reviews, 127*, 109876. https://doi.org/10.1016/j.rser.2020.109876

Bahn-Walkowiak, B., Bleischwitz, R., Distelkamp, M., & Meyer, M. (2012). Taxing construction minerals: A contribution to a resource-efficient Europe. *Mineral Economics, 25*(1), 29–43. https://doi.org/10.1007/s13563-012-0018-9

Balmford, A., Green, R. E., & Phalan, B. (2005). The impact of agricultural expansion on biodiversity: A review of the evidence. *Agriculture, Ecosystems & Environment, 113*(1–3), 1–14. https://doi.org/10.1016/j.agee.2005.08.018

Bardgett, R. D., & van der Putten, W. H. (2005). Belowground biodiversity and ecosystem functioning. *Nature, 515*(7528), 505–511. https://doi.org/10.1038/nature1194

Bardgett, R. D., & van der Putten, W. H. (2014). Below ground biodiversity and ecosystem functioning. *Nature, 515*(7528), 505–511. https://doi.org/10.1038/nature13855

Barlow, J., Gardner, T. A., Araujo, I. S., Ávila-Pires, T. C., Bonaldo, A. B., Cunha, H. A., . . . Peres, C. A. (2018). The value of forest protected areas for Amazonian biodiversity. *Proceedings of the National Academy of Sciences, 115*(13), 2881–2886. https://doi.org/10.1073/pnas.1716846115

Barnosky, A. D., Matzke, N., Tomiya, S., & Uhen, M. D. (2011). Has the Earth's sixth mass extinction already arrived? *Nature, 471*(7336), 51–57. https://doi.org/10.1038/nature09678

Barrett, K., Taylor, E., & Green, J. (2021). The potential of hemp for sustainable textile production. *Journal of Sustainable Materials, 9*(3), 115–125. https://doi.org/10.1016/j.jsm.2021.01.005

Barton, D. N., Beard, T., & Kremen, C. (2015). Biodiversity, ecosystem services, and food security. *The Lancet, 386*(10008), 272–277. https://doi.org/10.1016/S0140-6736(15)60271-7

Basso, B., Cammarano, D., & Ritchie, J. T. (2016). Precision agriculture and the efficiency of phosphorus fertilization: A review. *Agronomy Journal, 108*(3), 1203–1214. https://doi.org/10.2134/agronj2015.0494

Bauen, A., Mäkinen, J., & Holmgren, K. (2018). Biomass for energy: An overview of biomass sources, biomass conversion and applications in energy production. *Renewable and Sustainable Energy Reviews, 81*, 482–495. https://doi.org/10.1016/j.rser.2017.07.062

Bauer, A., & Döring, T. F. (2014). *Principles of nutrient management in crops*. Springer.

Bayern Innovativ. (2021). *Marktüberblick Biokunststoffe: Perspektiven der Textilindustrie*. https://www.bayern-innovativ.de

Bayer, I. S., & Bolognese, A. (2017). The impact of oil price fluctuations on the bioplastics industry. *Energy Policy, 105*, 319–326. https://doi.org/10.1016/j.enpol.2017.03.019

Bayer, I., Dufresne, A., & Tiwari, A. (2021). Challenges in bioplastic feedstocks: Potential of biopolymer production using agro-industrial residues. *Renewable and Sustainable Energy Reviews, 142*, 110766. https://doi.org/10.1016/j.rser.2021.110766

Bengtsson, J., Ahnström, J., & Weibull, A. C. (2005). The effects of organic agriculture on biodiversity and abundance: A meta-analysis. *Journal of Applied Ecology, 42*(2), 261–269. https://doi.org/10.1111/j.1365-2664.2005.01048.x

Bennett, E. M., Carpenter, S. R., & Caraco, N. F. (2001). Human impact on erodable phosphorus and eutrophication. *Science, 292*(5517), 442–444. https://doi.org/10.1126/science.1059785

Berkes, F., Colding, J., & Folke, C. (2003). *Navigating social-ecological systems: Building resilience for complexity and change*. Cambridge University Press.

Berkes, F., Colding, J., & Folke, C. (2006). Globalization, reindeer, and the world system: The role of indigenous knowledge in the adaptive cycle of a socio-ecological system. *Global Environmental Change, 16*(3), 163–179. https://doi.org/10.1016/j.gloenvcha.2006.01.003

Bertoldi, P., Murtagh, J., & Miller, P. (2014). *Biome thane – The renewable alternative to natural gas*. European Commission. https://doi.org/10.2833/88291

Biermann, F. (2009). Global environmental governance. In *The Earthscan reader on global environmental governance* (pp. 1–16). Earthscan.

Bertling, Jürgen; Bertling, Ralf; Hamann, Leandra (2018): Kunststoffe in der Umwelt: Mikro- und Makroplastik. Ursachen, Mengen, Umweltschicksale, Wirkungen, Lösungsansätze, Empfehlungen. Kurzfassung der Konsortialstudie, Fraunhofer-Institut für Umwelt-, Sicherheits- und Energietechnik UMSICHT, Oberhausen

Binnemans, K., Jones, P. T., Blanpain, B., Van Gerven, T., Yang, Y., Walton, A., & Buchert, M. (2013). Recycling of rare earths: A critical review. *Journal of Cleaner Production, 51*, 1–22. https://doi.org/10.1016/j.jclepro.2012.12.037

Bioökonomierat Bayern. (2021). *Ersatz von fossilen Rohstoffen – Materialien für eine nachhaltige Zukunft*. https://www.biooekonomierat-bayern.de/images/Themenpapiere2021/SVB_Themenpapier_Biopolymere_3.pdf

Bleischwitz, R., Bahn-Walkowiak, B., & Lucas, R. (2012). Sustainable resource management and global change. *Global Environmental Change, 22*(3), 627–637. https://doi.org/10.1016/j.gloenvcha.2012.03.004

Blöschl, G., et al. (2005). *Impact of climate variability on catchment water balance in Europe*. Hydrology and Earth System Sciences, 9(4), 437–452. https://doi.org/10.5194/hess-9-437-2005

Baustoffwissen. (2023, 31. Oktober). Selbstheilender Beton: Bakterien gegen Risse. Abgerufen am 11. April 2025, von https://www.baustoffwissen.de/selbstheilender-beton-bakterien-gegen-risse-31102023

Bhowmik, D., Ghosh, S., & Rahman, M. M. (2023). Sustainability of bio-based materials in energy storage applications. *Renewable Energy Reviews, 61*(3), 120–130. https://doi.org/10.1016/j.renene.2023.03.057

Blanchard, O., & Johnson, D. R. (2017). *Macroeconomics* (7th ed.). Pearson.

Blenker, C. (2022, 12. November). Dänemark: Energieunabhängig werden – mit Mist. *tagesschau.de*. https://www.tagesschau.de/ausland/europa/daenemark-biogas-101.html

Blume, H.-P., Brümmer, G. W., Horn, R., Kandeler, E., Kögel-Knabner, I., Kretzschmar, R., . . . Wilke, B.-M. (2016). *Lehrbuch der Bodenkunde* (17. Aufl.). Springer.

Bock, R. C., Franklin, J., & Berger, M. L. (2011). Perfluorinated alkyl substances: Chemistry, sources, and environmental and human exposure. *Environmental Science & Technology, 45*(16), 8022–8036. https://doi.org/10.1021/es201312m

Bocken, N. M. P., Short, S. W., Rana, P., & Evans, S. (2014). A literature and practice review to develop sustainable business model archetypes. *Journal of Cleaner Production, 65*, 42–56. https://doi.org/10.1016/j.jclepro.2013.11.039

Boden, T. A., Marland, G., & Andres, R. J. (2017). *Global, regional, and national fossil-fuel CO_2 emissions*. Carbon Dioxide Information Analysis Center.

Borras, S. M., Franco, J. C., Gómez, S., & Suárez, S. (2011). *Land grabbing in developing countries: A review of the evidence*. International Land Coalition.

Borras, S. M., Franco, J., Gómez, S., & Kay, C. (2018). Land grabbing in Latin America and the Caribbean: An overview. *Land Use Policy, 79*, 57–69. https://doi.org/10.1016/j.landusepol.2018.07.031

Borras, S. M., Hall, R., Scoones, I., White, B., & Wolford, W. (2011). Towards a better understanding of global land grabbing: An editorial introduction. *Journal of Peasant Studies, 38*(2), 209–216. https://doi.org/10.1080/03066150.2011.559005

Borras, S. M., Hall, R., Scoones, I., White, B., & Wolford, W. (2018). *Land grabbing and global agrarian change: A political economy perspective*. Routledge.

Bouwman, A. F., Beusen, A. H. W., Billen, G., & Van Drecht, G. (2002). Global trends in nitrogen and phosphorus fertilizers. *Science, 297*(5599), 706–708. https://doi.org/10.1126/science.1074804

BP. (2023). *Statistical review of world energy 2023*. BP p.l.c. https://www.bp.com/statisticalreview

Brad, A. (2019). *Der Palmölboom in Indonesien: Zur politischen Ökonomie einer umkämpften Resource*. transcript Verlag. https://library.oapen.org/handle/20.500.12657/25254

Brand, U., & Wissen, M. (2017). *Imperial mode of living: Everyday life and the ecological crisis of capitalism*. Verso Books.

Brauch, H. G., Grin, J., Mesjasz, C., & Dunay, P. (Eds.). (2009). *Facing global environmental change: Environmental, human, energy, food, health and water security concepts*. Springer.

Braungart, M., & McDonough, W. (2002). *Cradle to cradle: Remaking the way we make things*. North Point Press.

Bressanelli, G., Saccani, N., & Perona, M. (2018). Exploring circular business models in the automotive industry: A comparative case study approach. *Journal of Cleaner Production, 172*, 3041–3055. https://doi.org/10.1016/j.jclepro.2017.11.022

Bringezu, S., Schütz, H., & O'Brien, M. (2009). *The future of bio-based materials: Opportunities and challenges*. European Environment Agency. https://www.eea.europa.eu/publications/the-future-of-bio-based-materials

Brizga, J., Mishchuk, Z., & Grubina, L. (2020). Bio-based plastics: Sustainable alternative or climate hazard? *Journal of Cleaner Production, 261*, 121158. https://doi.org/10.1016/j.jclepro.2020.121158

Broecker, W. S. (1997). Thermohaline circulation, the Achilles heel of our climate system: Will man-made CO_2 upset the current balance? *Science, 278*(5343), 1582–1588. https://doi.org/10.1126/science.278.5343.1582

Brown, A., & Lee, K. (2018). Land use and environmental impacts of bioplastics: The case of sugarcane-derived PLA. *Journal of Sustainable Materials, 12*(2), 54–62. https://doi.org/10.1016/j.jsm.2018.01.004

Browne, M. A. Dissanayake, A., Galloway, T. S., Lowe, D. M., & Thompson, R. C. (2008). Ingested microscopic plastic translocates to the circulatory system of the mussel, *Mytilus edulis* (L.). *Environmental Science & Technology, 42*(13), 5026–5031. https://doi.org/10.1021/es800249a

Bröring, S., Claupein, E., & Kuck, L. (2017). *Biokraftstoffe als Ersatz fossiler Energieträger – Eine umwelt- und resourcenökonomische Analyse*. GRIN Verlag.

Bröring, S., Kühl, S., & Leker, J. (2017). *Biokraftstoffe: Stand der Technik und Perspektiven*. Springer Vieweg.

Bruckner, L., & Busch, S. (2023). *The future of food: Challenges and innovations in agriculture.* Bayer AG.

Buck, R. C., Franklin, J., Berger, U., Conder, J. M., Cousins, I. T., de Voogt, P., Jensen, A. A., Kannan, K., Mabury, S. A., & van Leeuwen, S. P. J. (2011). Perfluoroalkyl and polyfluoroalkyl substances in the environment: Terminology, classification, and origins. *Integrated Environmental Assessment and Management, 7*(4), 513–541. https://doi.org/10.1002/ieam.258

BUND – Bund für Umwelt und Naturschutz Deutschland. (2022). *Bioökonomie im Lichte der planetaren Grenzen und des Schutzes der Biodiversität.* https://www.bund.net/fileadmin/user_upload_bund/publika tionen/resourcen_und_technik/resourcen_technik_biooekonomie_projekt_studie_spangenberg.pdf

Bundesministerium für Ernährung und Landwirtschaft (BMEL). (2020). *Nationale Bioökonomiestrategie.* https://www.bmel.de/DE/themen/nachhaltigkeit/biooekonomie/nationale-biooekonomiestrategie. html

Bundesministerium für Ernährung und Landwirtschaft (BMEL). (2021). *Waldbericht der Bundesregierung 2021.* https://www.bmel.de/SharedDocs/Downloads/DE/Broschueren/waldbericht2021.pdf

Bundesministerium für Wirtschaft und Klimaschutz (BMWK). (2023). Bericht zur Bürokratiekostenmessung 2023. Abgerufen von https://www.bmwk.de

Bundesministerium für Ernährung und Landwirtschaft (BMEL). (2023). *Agrarbericht 2023.* https://www. bmel.de/DE/themen/landwirtschaft/agrarbericht-2023.html

Burgstaller, Maria; Potrykus, Alexander; Weißenbacher, Jakob; Dr. Kabasci, Stephan; Merrettig-Bruns, Ute; Sayder, Bettina (2018): Gutachten zur Behandlung biologisch abbaubarer Kunststoffe, Umweltbundesamt, Dessau-Roßlau.

Bundesministerium für Ernährung und Landwirtschaft (BMEL). (2022a). *Die Nationale Biomassestrategie.* https://www.bmel.de/DE/themen/landwirtschaft/bioeokonomie-nachwachsende-rohstoffe/nationale-biomassestrategie.html

Bundesministerium für Ernährung und Landwirtschaft (BMEL). (2022b). *Nutzen und Bedeutung der Bioenergie.* https://www.bmel.de/DE/themen/landwirtschaft/bioeokonomie-nachwachsende-rohstoffe /bioenergie-nutzen-bedeutung.html

Bundesministerium für Umwelt, Naturschutz, nukleare Sicherheit und Verbraucherschutz (BMUV). (2021). *Deutsches Resourceneffizienzprogramm (ProgRess III).* https://www.bmuv.de

Bundesministerium für Wirtschaft und Klimaschutz (BMWK). (2020). *Rohstoffstrategie der Bundesregierung 2020.* https://www.bmwk.de/Redaktion/DE/Publikationen/Industrie/rohstoffstrategie-der-bundesregierung.html

Börger, G., & Meissner, A. (2019). *Zusammenarbeit im internationalen Kontext: Eine Analyse nachhaltiger Rohstoffstrategien.* Springer.

Börjesson, P., & Mattiasson, B. (2014). Biogas production from agricultural residues and wastes. *Bioresource Technology, 175,* 99–107. https://doi.org/10.1016/j.biortech.2014.10.042

BP. (2023). *BP statistical review of world energy 2023.* https://www.bp.com/en/global/corporate/energy-economics/statistical-review-of-world-energy.html

Bundesverband der Deutschen Entsorgungs-, Wasser- und Rohstoffwirtschaft e. V. (BDE). (2019, August 7). *Mehr als 50 Prozent weniger Treibhausgas-Emissionen: Studie bestätigt Klimaschutz-Effekte durch Recycling.* https://www.bde.de/presse/mehr-als-50-prozent-weniger-treibhausgas-emissionen-studie-bestatigt -k/BDE

Bundeszentrale für politische Bildung. (2018). *Resourcenkonflikte.* https://www.bpb.de/themen/kriege-konflikte/dossier-kriege-konflikte/76755/resourcenkonflikte/

Cakmak, I. (2010). *The role of potassium in plant nutrition.* Springer.

Campos, J. M., de Oliveira, D. S., & Luna, J. M. (2019). Green surfactants from natural sources: A review. *Journal of Environmental Chemical Engineering, 7*(4), 103286. https://doi.org/10.1016/j.jece.2019.103286

Cames, M., Wiemer, M., & Morscheck, G. (2016). The contribution of biofuels to the EU renewable energy target: How far can biofuels go? *Environmental and Energy Policy Research Programme.* https://doi.org/ 10.1016/j.enpol.2015.06.004

Cardinale, B. J., Duffy, J. E., Gonzalez, A., Hooper, D. U., Perrings, C., Venail, P., . . . Loreau, M. (2012). Biodiversity loss and its impact on humanity. *Nature, 486*(7401), 59–67. https://doi.org/10.1038/nature11148

Carlson, K. M., Curran, L. M., & Ratnasari, D. (2013). Effect of oil palm plantations on forest conversion and land use. *Global Environmental Change, 23*(4), 1155–1166. https://doi.org/10.1016/j.gloenvcha.2013.06.008

Carpenter, S. R., Caraco, N. F., Correll, D. L., Howarth, R. W., Sharpley, A. N., & Smith, V. H. (1998). Nonpoint pollution of surface waters with phosphorus and nitrogen. *Ecological Applications, 8*(3), 559–568. https://doi.org/10.2307/2641167

Carr, M.-E. (2008). Biofuels, food security, deforestation, and loss of biodiversity. *State of the Planet.* https://news.climate.columbia.edu/2008/12/15/biofuels-food-security-deforestation-and-loss-of-biodiversity/

Carus, M., Dammer, L., & Essel, R. (2021). *Biogene Rohstoffe in der Industrie: Märkte, Trends, Innovationen.* nova-Institut.

Carus, M., Raschka, A., & Piotrowski, S. (2014). *Sustainable raw materials: Status and prospects.* nova-Institut GmbH.

Chamas, A., Moon, H., Zheng, J., Qiu, Y., Tabassum, T., Jang, J. H., Abu-Omar, M., Scott, S. L., & Suh, S. (2020). Degradation rates of plastics in the environment. *ACS Sustainable Chemistry & Engineering, 8*(9), 3494–3511. https://doi.org/10.1021/acssuschemeng.9b06635

Chen, G.-Q., & Wu, Q. (2005). The application of polyhydroxyalkanoates as tissue engineering materials. *Biomaterials, 26*(33), 6565–6578. https://doi.org/10.1016/j.biomaterials.2005.04.036

Cherubini, F. (2010). The biorefinery concept: Using biomass instead of oil for producing energy and chemicals. *Energy Conversion and Management, 51*(7), 1412–1421. https://doi.org/10.1016/j.enconman.2010.01.015

Cherubini, F., Bargigli, S., & Ulgiati, S. (2016). Life cycle assessment of biofuels: A critical review. *Environmental Impact Assessment Review, 58*, 49–57. https://doi.org/10.1016/j.eiar.2015.10.005

Cheng, H., Li, X., & Zhao, Y. (2021). Bio-based materials for next-generation batteries: A review. *Advanced Energy Materials, 11*(8), 2100215. https://doi.org/10.1002/aenm.202100215

Cheng, Y., Ma, X., & Zhao, C. (2018). Improving the properties and production efficiency of biodegradable polymers: Biotechnological approaches. *Biotechnology Advances, 36*(3), 1006–1024. https://doi.org/10.1016/j.biotechadv.2018.02.007

Chisti, Y. (2007). Biodiesel from microalgae. *Biotechnology Advances, 25*(3), 294–306. https://doi.org/10.1016/j.biotechadv.2007.02.001

Circle Economy. (2023). *The Circularity Gap Report 2023.* https://www.circularity-gap.world/2023

Circular Economy Switzerland, & Deloitte Schweiz. (2023). *Circularity Gap Report Switzerland 2023.* https://www.circular-economy-switzerland.ch/circularity-gap-report-switzerland-2023

Clapp, J. (2014). Food security and international trade: Unpacking disputed narratives. *Food Policy, 36*(1), 11–18. https://doi.org/10.1016/j.foodpol.2013.08.002

Clapp, J., & Helleiner, E. (2012). *International political economy and the global food crisis.* Oxford University Press.

Clapp, J., & Isakson, S. R. (2018). *Speculative harvests: The global food crisis and financial markets.* Cambridge University Press.

Clark, J. H., Farmer, T. J., Herrero-Davila, L., & Sherwood, J. (2017). Circular economy design considerations for research and process development in the chemical sciences. *Green Chemistry, 19*(1), 20–41. https://doi.org/10.1039/C6GC03098J

Clark, J. H., Farmer, T. J., Herrero-Davila, L., & Sherwood, J. (2017). Circular economy design considerations for research and process development in the chemical sciences. *Green Chemistry, 19*(2), 486–496. https://doi.org/10.1039/C6GC03098J

Clark, J. H., Farmer, T. J., Hunt, A. J., & Sherwood, J. (2012). Opportunities for bio-based solvents created as petrochemical and fuel products transition towards renewable resources. *International Journal of Molecular Sciences, 13*(7), 9613–9637. https://doi.org/10.3390/ijms13079613

Clark, J. H., Luque, R., & Matharu, A. S. (2017). Green chemistry and the biorefinery: From the traditional to the sustainable. *Green Chemistry*, *19*(5), 1046–1056. https://doi.org/10.1039/C6GC03331F

co2online. (2020, September 30). *Ökologische Dämmstoffe: Preisvergleich, Vor- & Nachteile*. https://www.co2online.de/modernisieren-und-bauen/daemmung/oekologische-daemmstoffe/

Colchester, M., & Chao, S. (2013). *Palm oil and indigenous peoples in Southeast Asia*. Forest Peoples Programme. https://www.forestpeoples.org/en/palm-oil-report

Coote, A., Franklin, J., & Simms, A. (2010). *21 Hours: Why a shorter working week can help us all to flourish in the 21st century*. New Economics Foundation.

Cordell, D., Drangert, J.-O., & White, S. (2009). The story of phosphorus: Global food security and food for thought. *Global Environmental Change*, *19*(2), 292–305. https://doi.org/10.1016/j.gloenvcha.2008.10.009

Cordell, D., & White, S. (2014). Life's bottleneck: Sustaining the world's phosphorus for a food secure future. *Annual Review of Environment and Resources*, *39*, 161–188. https://doi.org/10.1146/annurev-environ-010213-113300

Corteva. (2022). *Corteva Agriscience sustainability report 2022*. Corteva Agriscience.

Costanza, R., de Groot, R., & Sutton, P. (2014). Changes in the global value of ecosystem services. *Global Environmental Change*, *26*, 152–158. https://doi.org/10.1016/j.gloenvcha.2014.04.005

Cotula, L., Vermeulen, S., Leonard, R., & Keeley, J. (2014). *Land grab or development opportunity? Agricultural investment and international land deals in Africa*. FAO, IFAD, and IIED. https://www.fao.org/3/a-i3493e.pdf

Crutzen, P. J. (2006). The effects of nitrogen and carbon on the earth's atmosphere. In J. Galloway, et al. (Eds.), *The nitrogen cycle* (pp. 217–230). Springer.

Davidovits, J. (2015). *Geopolymer chemistry and applications* (4. Aufl.). Institut Géopolymère.

Dasgupta, P. (2021). *The economics of biodiversity: The Dasgupta Review*. HM Treasury. Abgerufen von https://www.gov.uk/government/publications/final-report-the-economics-of-biodiversity-the-dasgupta-review

Dawson, C. J., & Hilton, J. (2011). Fertiliser availability in a resource-limited world: Production and recycling of nitrogen and phosphorus. *Food Policy*, *36*(1), 14–22. https://doi.org/10.1016/j.foodpol.2010.10.003

Deininger, K., & Byerlee, D. (2012). The rise of large farms in land abundant countries: Do they have a future? *World Bank Policy Research Working Paper No. 5588*. https://doi.org/10.1596/1813-9450-5588

Demirbas, A. (2007). *Biomass resources for biofuels*. Springer.

Demirbas, A. (2009). *Biomass resources for biofuels* (2. Aufl.). Springer. https://doi.org/10.1007/978-1-4419-0079-3

Demirbas, A. (2018). Biofuels: Securing the planet's future energy needs. *Energy*, *148*, 1166–1177. https://doi.org/10.1016/j.energy.2018.01.078

Department of Energy, U.S. (1978). *Aquatic species program: Biodiesel from algae*. National Renewable Energy Laboratory.

Derpsch, R., Friedrich, T., Kassam, A., & Hongwen, L. (2010). Current status of adoption of no-till farming in the world and some of its main benefits. *International Journal of Agricultural and Biological Engineering*, *3*(1), 1–25. https://doi.org/10.25165/j.ijabe.20100301.176

Deutsche Akademie der Naturforscher Leopoldina. (2013). *Bioenergie: Möglichkeiten und Grenzen*. https://www.leopoldina.org/uploads/tx_lepublication/2013_06_Stellungnahme_Bioenergie_DE.pdf

Deutscher Bundestag. (2007). *Biomasse vs. Nahrungsmittel: Nutzungskonkurrenz bei Energiepflanzen*. Wissenschaftliche Dienste des Deutschen Bundestages. https://www.bundestag.de/resource/blob/424958/c0b9d7ba2bdcd917d38d8f76d7d95ed4/wd-5-105-07-pdf-data.pdf

Deutsches Biomasseforschungszentrum. (2023). *Fokusheft: Infrastruktur für erneuerbares Methan im Verkehr*. https://www.dbfz.de/fileadmin/user_upload/Referenzen/Broschueren/Fokusheft-Infrastruktur-Pilot-SBG-Oktober-2023.pdf

Díaz, R. J., & Rosenberg, R. (2008). Spreading dead zones and consequences for marine ecosystems. *Science*, *321*(5891), 926–929. https://doi.org/10.1126/science.1156401

Díaz, S., Settele, J., Brondízio, E. S., & Ngo, H. T. (2019). *IPBES global assessment report on biodiversity and ecosystem services*. Intergovernmental Science-Policy Platform on Biodiversity and Ecosystem Services.

Doney, S. C., Fabry, V. J., & Feely, R. A. (2009). Ocean acidification: The other CO_2 problem. *Annual Review of Marine Science, 1*, 169–192. https://doi.org/10.1146/annurev.marine.010908.163834

Döring, T. F., Baddeley, J. A., Brown, R., Collins, R., Crowley, O., Cuttle, S., . . . & Wolfe, M. S. (2015). Using legume-based mixtures to enhance the nitrogen use efficiency and crop productivity of organic and low-input cropping systems. *Agronomy for Sustainable Development, 35*, 129–144. https://doi.org/10.1007/s13593-014-0215-9

Dourado, A., Martins, A., & Silva, J. (2020). Lignin-based polymers for sustainable applications: Challenges and opportunities. *Journal of Polymers and the Environment, 28*(3), 801–810. https://doi.org/10.1007/s11356-020-07918-4

Dostert E., Landbesitz Wie sich wenige Konzerne viel Land sichern Sueddeutsche Zeitung 3. Juni 2025

Dregne, H. E. (2002). Desertification: The role of climate change and land management. *Land Degradation & Development, 13*(5), 479–488. https://doi.org/10.1002/ldr.548

Drinkwater, L. E., Wagoner, P., & Sarrantonio, M. (1998). Legume-based cropping systems have reduced carbon and nitrogen losses. *Nature, 396*(6708), 262–265. https://doi.org/10.1038/24376

Dube, J.-P., & Matusz, S. J. (2000). International trade and resource depletion: The case of energy extraction and trade barriers. *Resource and Energy Economics, 22*(3), 251–268. https://doi.org/10.1016/S0928-7655(00)00009-9

Endres H., Siebert-Raths A., 2009, TECHNISCHE BIOPOLYMERE, Rahmenbedingungen, Marktsituation, Herstellung, Aufbau und Eigenschaften, eISBN: 978-3-446-42104-2, Print ISBN: 978-3-446-41683-3, Carl Hanser Verlag GmbH & Co. KG, S. 5-7

Ellen MacArthur Foundation (2013). *Towards the circular economy: Economic and business rationale for an accelerated transition.* https://ellenmacarthurfoundation.org

Ellen MacArthur Foundation. (2015). *Towards the circular economy: Economic and business rationale for an accelerated transition.* https://ellenmacarthurfoundation.org

Ellen MacArthur Foundation. (2016). *The new plastics economy: Rethinking the future of plastics.* https://ellenmacarthurfoundation.org/the-new-plastics-economy

Ellen MacArthur Foundation (2019). *Completing the picture: How the circular economy tackles climate change.* https://emf.org

Elser, J., & Bennett, E. (2011). Phosphorus cycle: A broken biogeochemical cycle. *Nature, 478*(7367), 29–31. https://doi.org/10.1038/478029a

Emadian, S. M., Onay, T. T., & Demirel, B. (2022). Biodegradable microplastics: A review on the interaction with pollutants and influence to organisms. *Environmental Science and Pollution Research, 29*(22), 33500–33511. https://doi.org/10.1007/s11356-022-19998-1

Engels, F. (1973). *Die Lage der arbeitenden Klasse in England* (Original work published 1872). Dietz Verlag.

Erisman, J. W., Sutton, M. A., Galloway, J., Klimont, Z., & Winiwarter, W. (2008). How a century of ammonia synthesis changed the world. *Nature Geoscience, 1*(10), 636–639. https://doi.org/10.1038/ngeo325

Erisman, J. W., et al. (2008). *The European nitrogen assessment: Sources, effects and policy perspectives.* Cambridge University Press.

Erisman, J. W., et al. (2008). Nitrogen and food production: The Asian perspective. *Ambio, 37*(2), 62–67. https://doi.org/10.1579/0044-7447(2008)37[62:NAFPTA]2.0.CO;2

Euractiv. (2022, November 9). Fertilizer markets face turmoil as Russia restricts exports. *Euractiv.* https://www.euractiv.com/section/agriculture-food/news/fertilizer-markets-face-turmoil-as-russia-restricts-exports/

European Aluminium. (2020). *Recycling of aluminium: A true story of sustainability.* https://www.european-aluminium.eu

European Bioplastics. (2020). *Bioplastics market data.* Retrieved from www.european-bioplastics.org

European Bioplastics. (2024). *Bioplastics market development update 2024*. Berlin: European Bioplastics. https://www.european-bioplastics.org

European Chemicals Agency (ECHA). (2023). *Guidance on information requirements and chemical safety assessment: Chapter R.7b – Endpoint specific guidance*. https://echa.europa.eu

European Commission. (2008). *Critical raw materials for the EU: Report of the Ad-hoc Working Group on Defining Critical Raw Materials*. https://ec.europa.eu/docsroom/documents/2994

European Commission. (2014). *Circular economy: Closing the loop - An EU action plan for the circular economy*. https://ec.europa.eu/environment/circular-economy/index_en.htm

European Commission. (2019). *Biomass: The sustainability challenge*. CORDIS. Abgerufen von https://cordis.europa.eu/article/id/36567-biomass-the-sustainability-challenge/de

European Commission. (2020). *The European Green Deal: Striving for carbon neutrality by 2050*. European Commission. https://ec.europa.eu/info/sites/info/files/european-green-deal-communication_en.pdf

European Commission. (2020). *A European Green Deal: Striving to be the first climate-neutral continent*. European Commission. https://ec.europa.eu/info/strategy/priorities-2019-2024/european-green-deal_en

European Commission. (2020). *A new Circular Economy Action Plan: For a cleaner and more competitive Europe*. https://ec.europa.eu/environment/circular-economy/

European Commission. (2020). *A new Circular Economy Action Plan*. https://eur-lex.europa.eu/legal-content/EN/TXT/?uri=CELEX:52020DC0098

European Commission. (2021). *Debt sustainability monitor 2021*. https://economy-finance.ec.europa.eu

European Commission. (2023). *Bericht über die Biomasse-Nutzung in der EU*. Abgerufen von https://eur-lex.europa.eu/legal-content/DE/TXT/HTML/?uri=CELEX%3A52023DC0650

European Environment Agency (EEA). (2019). *Soil sustainability in Europe — Strategic knowledge gaps and research needs* (EEA Report No. 4/2019). https://www.eea.europa.eu/publications/soil-sustainability-in-europe

European Commission. (2024). *Critical Raw Materials List 2024*. https://ec.europa.eu/docsroom/documents/57756

EU. (2020). *EU biodiversity strategy for 2030: Bringing nature back into our lives*. European Commission.

Eurostat. (2022, October 20). *Plastic packaging waste: 38% recycled in 2020*. https://ec.europa.eu/eurostat/de/web/products-eurostat-news/-/ddn-20221020-1

Eurostat. (2024, October 24). *41% of plastic packaging waste recycled in 2022*. https://ec.europa.eu/eurostat/de/web/products-eurostat-news/w/ddn-20241024-3

Europäische Kommission. (2020). *Kritische Rohstoffe: Resilienz der EU bei der Versorgung mit kritischen Rohstoffen stärken*. https://ec.europa.eu/docsroom/documents/42849:contentReference{index=35}

Environmental Justice Foundation. (2017). *Der Klimawandel und seine Rolle bei Migration und Konflikten*. https://ejfoundation.org/resources/downloads/EJF_Bericht_Beyond_Borders_Klimawandel_Migration_Konflikte.pdf

Energy & Management Powernews (2023, 18. Oktober). *Zwei Studien unterstützen mehr Biomasse-Einsatz*. Abgerufen von https://www.bayern-innovativ.de/emagazin/energie-bau/detail/de/seite/zwei-studien-unterstuetzen-mehr-biomasse-einsatz/

Faaij, A. (2006). Bioenergy in Europe: Changing technology choices. *Energy Policy, 34*(3), 322-336. https://doi.org/10.1016/j.enpol.2005.03.019

Fachagentur Nachwachsende Rohstoffe (2012). Marktanalyse Nachwachsende Rohstoffe. https://www.fnr.de/marktanalyse/marktanalyse-kap11.pdffnr.de+1NachhaltigeEntwicklung+1

Färe, J., Hill, J., Tilman, D., Polasky, S., & Hawthorne, P. (2008). Land clearing and the biofuel carbon debt. *Science, 319*(5867), 1235-1238. https://doi.org/10.1126/science.1152747

Falkner, R. (2016). The Paris Agreement and the new logic of international climate politics. *International Affairs, 92*(5), 1107-1125. https://doi.org/10.1111/1468-2346.12717

Faram, J. C., Gardiner, B. G., & Shanklin, J. D. (1985). Large losses of total ozone in Antarctica reveal seasonal ClOx/NOx interaction. *Nature, 315*(6016), 207-210. https://doi.org/10.1038/315207a0

Fattig, J., Schmidt, T., & Schmidt, P. (2020). Hydrogen production technologies and renewable energy sources. *Energy, 102*(2), 24-39. https://doi.org/10.1016/j.energy.2020.04.017

Fargione, J., Hill, J., Tilman, D., Polasky, S., & Hawthorne, P. (2008). Land clearing and the biofuel carbon debt. *Science, 319*(5867), 1235–1238. https://doi.org/10.1126/science.1152747

FAO. (2011). The state of the world's land and water resources for food and agriculture (SOLAW). Food and Agriculture Organization of the United Nations. https://www.fao.org/publications

FAO. (2014). World fertilizer trends and outlook to 2018. Food and Agriculture Organization.

FAO. (2015). Status of the World's Soil Resources (SWSR) – Main Report. Food and Agriculture Organization of the United Nations. https://www.fao.org/documents/card/en/c/cb9970en

FAO. (2015). Status of the World's Soil Resources. Food and Agriculture Organization of the United Nations. https://www.fao.org/3/i5199e/i5199e.pdf

FAO. (2017). The future of food and agriculture – Trends and challenges. Food and Agriculture Organization of the United Nations.

FAO. (2018). The State of the World's Forests 2018: Forest Pathways to Sustainable Development. Food and Agriculture Organization of the United Nations.

FAO. (2018). The future of food and agriculture – Alternative pathways to 2050. Food and Agriculture Organization of the United Nations.

FAO. (2020). The state of food and agriculture 2020: Water for sustainable food security. Food and Agriculture Organization of the United Nations. https://doi.org/10.4060/ca9696en

FAO. (2021). The state of food and agriculture 2021: Making agrifood systems more inclusive, resilient and sustainable. Food and Agriculture Organization of the United Nations. https://doi.org/10.4060/cb6030en

FAO. (2021). The state of food and agriculture 2021: Making food systems more resilient to climate change. Food and Agriculture Organization of the United Nations. https://doi.org/10.4060/cb4474en

FAO. (2021). The State of Food and Agriculture 2021: Making agrifood systems more resilient to shocks and stresses. Food and Agriculture Organization of the United Nations. https://doi.org/10.4060/cb4476en

FAO. (2021). The state of the world's soils and their impact on agriculture. Food and Agriculture Organization of the United Nations.

FAO. (2022). The state of food and agriculture 2022: Transforming food systems for food security and nutrition. Food and Agriculture Organization of the United Nations. https://doi.org/10.4060/cb4576en

FAO. (2022). The state of food security and nutrition in the world 2022: Repurposing food and agricultural policies to make healthy diets more affordable. Food and Agriculture Organization of the United Nations. https://www.fao.org/documents/card/en/c/cc0639en

FAO. (2022). The importance of fertilizers for global food security. Food and Agriculture Organization of the United Nations. https://www.fao.org

FAO. (2023). Food security and nutrition: A 2023 outlook. Food and Agriculture Organization of the United Nations.

FAO. (2023). The state of food security and nutrition in the world 2023. Food and Agriculture Organization of the United Nations.

FAO, IFAD, UNICEF, WFP & WHO. (2023). The state of food security and nutrition in the world 2023: Urbanization, agrifood systems transformation and healthy diets across the rural–urban continuum. https://www.fao.org/publications/sofi/2023/en/

Fargione, J., Hill, J., Tilman, D., Polasky, S., & Hawthorne, P. (2008). Land clearing and the biofuel carbon debt. *Science, 319*(5867), 1235–1238. https://doi.org/10.1126/science.1152747

Faulhaber, S., Hofmann, M., & Suresh, R. (2018). Green infrastructure as a tool for urban flood management and climate adaptation. In R. Suresh (Ed.), *Green infrastructure as a tool for urban flood management and climate adaptation* (pp. 155–175). CRC Press. https://doi.org/10.1201/9781351124140-155

Fearnside, P. M. (2017). Soybean production in Brazil and its environmental impacts. *Environmental Conservation, 43*(1), 23-32. https://doi.org/10.1017/S0376892917000099

Fearnside, P. M. (2018). Environmental and social impacts of hydroelectric dams in Brazilian Amazonia: Implications for the aluminum industry. *World Development, 107*, 26–47. https://doi.org/10.1016/j.world dev.2018.02.015

Ferrari, D., Montanari, L., & Dalla Rosa, M. (2018). Hydrotreatment of vegetable oils to biofuels: Advances and challenges. *Energy Conversion and Management, 172*, 436-455. https://doi.org/10.1016/j.enconman. 2018.07.061

Fink, A. (2002). *Pflanzenernährung: Grundlagen der Düngung* (3. Aufl.). Ulmer Verlag.

Fischer, G., et al. (2013). Global nutrient demand and the effects of climate change on the fertilizer market. *International Fertilizer Industry Association*.

Fischer, R. A., Byerlee, D., & Edmeades, G. O. (2010). Can technology deliver on the yield challenge to 2050? In C. L. Morgan (Ed.), *Food Security and the Role of Biotechnology* (pp. 87–102). Springer. https://doi.org/10.1007/978-1-4419-6324-2_8

Feely, R. A., Doney, S. C., & Cooley, S. R. (2009). Ocean acidification: A critical emerging problem for the ocean sciences. *Oceanography, 22*(4), 36-47. https://doi.org/10.5670/oceanog.2009.94

Foley, J. A., DeFries, R., Asner, G. P., Barford, C., Bonan, G., Carpenter, S. R., & Vitousek, P. M. (2011). Global consequences of land use. *Science, 309*(5734), 570-574. https://doi.org/10.1126/science.1111772

Foley, J. A., Ramankutty, N., Brauman, K. A., et al. (2011). Solutions for a cultivated planet. *Nature, 478*(7369), 337–342. https://doi.org/10.1038/nature10452

Folke, C. (2006). Resilience: The emergence of a perspective for social–ecological systems analyses. *Global Environmental Change, 16*(3), 253–267. https://doi.org/10.1016/j.gloenvcha.2006.04.002

Folke, C., Carpenter, S. R., Walker, B., Scheffer, M., Chapin, T., & Rockström, J. (2010). Resilience thinking: Integrating resilience, adaptability and transformability. *Ecology and Society, 15*(4), 20. https://www.jstor.org/stable/26268226

Folke, C., Carpenter, S. R., Walker, B., Scheffer, M., Elmqvist, T., Gunderson, L., & Holling, C. S. (2004). Regime shifts, resilience, and biodiversity in ecosystem management. *Annual Review of Ecology, Evolution, and Systematics, 35*, 557-581. https://doi.org/10.1146/annurev.ecolsys.35.021103.105711

Folke, C., Biggs, R., Norström, A. V., Reyers, B., & Rockström, J. (2016). Social-ecological resilience and biosphere-based sustainability science. *Ecology and Society, 21*(3), Article 41. https://doi.org/10.5751/ES-08748-210341

Frangi, A., Fontana, M., & Knobloch, M. (2019). *Design of timber structures: Volume 1: Structural aspects of timber construction*. ETH Zürich.

Friedrich, H. E., & Almajid, A. A. (2013). Manufacturing aspects of advanced materials for automotive applications. *Materials and Design, 56*, 798–808. https://doi.org/10.1016/j.matdes.2013.01.074

Friedrich, H. E., & Almajid, A. A. (2013). Carbon fiber reinforced composites in automotive applications. In S. T. Peters (Ed.), *Handbook of Composites* (pp. 1–19). Springer. https://doi.org/10.1007/978-1-4614-4250-2_1

Frossard, E., Condron, L. M., Oberson, A., Sinaj, S., & Fardeau, J.-C. (2000). Processes governing phosphorus availability in temperate soils. *Journal of Environmental Quality, 29*(1), 15–23. https://doi.org/10.2134/jeq2000.00472425002900010003x

Fu, X., Zhang, J., & Li, X. (2020). Incentivizing innovation for green manufacturing: Policy strategies and firm dynamics. *Technological Forecasting and Social Change, 159*, 120204. https://doi.org/10.1016/j.techfore.2020.120204

Gates, A. E., Peyron, P., McCabe, J., & Rudnick, R. L. (2015). Phosphate rock: Resource data and global reserves. In H. D. Holland & K. K. Turekian (Eds.), *Treatise on geochemistry* (2nd ed., Vol. 13, pp. 317–339). Elsevier. https://doi.org/10.1016/B978-0-08-095975-7.01012-9

Gatti, S., & Velazquez, L. (2019). Cost competitiveness of palm oil biodiesel production in Indonesia. *Energy*, 170, 62–72. https://doi.org/10.1016/j.energy.2018.12.113IDEAS/RePEc

Grajales, J. (2013). *The palm oil complex in Colombia: Dynamics, conflicts and challenges*. Latin American Politics and Society, 55(3), 27–50. https://doi.org/10.1111/j.1548-2456.2013.00228.x

Gleick, P. H. (2014). *Water conflict chronology*. Pacific Institute. https://www.pacinst.org/publication/water-conflict-chronology/

Gaveau, D. L. A., Salim, M. A., & Lescure, J. P. (2014). *The impact of oil palm expansion on deforestation in Indonesia*. Environmental Science & Policy, 40, 5–11. https://doi.org/10.1016/j.envsci.2014.03.013

Gatti, R. C., Liang, J., Velichevskaya, A., Zhou, M., & Pan, Y. (2019). The role of Southeast Asian forests in the carbon cycle. *Nature Communications*, 10(1), 1–9. https://doi.org/10.1038/s41467-019-09646-4

Gandini, A., & Lacerda, T. M. (2015). From monomers to polymers from renewable resources: Recent advances. *Progress in Polymer Science*, 48, 1–39. https://doi.org/10.1016/j.progpolymsci.2015.04.001

Gantenbein, S., Masania, K., Woigk, W., Leutenegger, S., & Studart, A. R. (2018). Three-dimensional printing of hierarchical liquid-crystal-polymer structures. *Nature, 561*(7722), 226–230. https://doi.org/10.1038/s41586-018-0474-7

Gates, L., et al. (2015). Global phosphorus scarcity: A modelling approach. *Global Environmental Change, 34*, 87–94. https://doi.org/10.1016/j.gloenvcha.2015.06.004

Galloway, J. N., Aber, J. D., & Erisman, J. W. (2008). The nitrogen cascade. *BioScience, 48*(5), 447–456. https://doi.org/10.1641/0006-3568(1998)048[0447:TNC]2.0.CO;2

Galloway, J. N., et al. (2008). The nitrogen cascade. *BioScience, 56*(4), 305–317. https://doi.org/10.1641/B560405

Gallai, N., Salles, J. M., Settele, J., & Vaissire, B. E. (2009). Economic valuation of the vulnerability of world agriculture confronted with pollinator decline. *Ecological Economics, 68*(3), 810–821. https://doi.org/10.1016/j.ecolecon.2008.06.014

Gauthier, J., Topp, E., & Davidson, L. (2015). A review of perfluoroalkyl substances and their impacts on human health. *Environmental Toxicology and Chemistry, 34*(5), 1136–1144. https://doi.org/10.1002/etc.3209

Gattinger, A., Muller, A., Haeni, M., Skinner, C., Fliessbach, A., Buchmann, N., . . . & Niggli, U. (2012). Enhanced top soil carbon stocks under organic farming. *Proceedings of the National Academy of Sciences, 109*(44), 18226–18231. https://doi.org/10.1073/pnas.1209429109

Garlapati, V. K. (2021). Biopolymers for sustainable development: A review. *Biological Resources and Applications, 45*(3), 239–253. https://doi.org/10.1016/j.biores.2021.05.013

Garlotta, D. (2018). A literature review of the recycling of polymeric materials. *Resources, Conservation and Recycling, 48*(2), 69–82. https://doi.org/10.1016/j.resconrec.2006.02.001

Galik, C. S., Houghton, R. A., & Goodall, J. (2015). The role of forests in the carbon cycle: Opportunities for bioenergy. Environmental Science & Technology, 49(14), 8295–8303. https://doi.org/10.1021/es505295y

Gallai, N., Salles, J. M., Settele, J., & Vaissière, B. E. (2009). *Economic valuation of the vulnerability of world agriculture confronted with pollinator decline*. Ecological Economics, 68(3), 810–821. https://doi.org/10.1016/j.ecolecon.2008.06.014

Gielen, D., & Morris, D. (2020). Renewable energy and sustainable development: Insights for the future. *Springer Nature*.

Gerbens-Leenes, P. W., Hoekstra, A. Y., & Van der Meer, T. H. (2009). The water footprint of bioenergy. *Proceedings of the National Academy of Sciences*, 106(25), 10219–10223. https://doi.org/10.1073/pnas.0812619106

Geng, Y., & Chase, R. (2019). *Cross-laminated timber: A sustainable alternative for building construction*. Journal of Cleaner Production, 223, 716–726. https://doi.org/10.1016/j.jclepro.2019.03.150

Geyer, R., Jambeck, J. R., & Law, K. L. (2017). Production, use, and fate of all plastics ever made. *Science Advances, 3*(7), e1700782. https://doi.org/10.1126/sciadv.1700782

Georgiou, M., & Manousakis, E. (2019). Policy instruments for the promotion of sustainable materials: A review. *Environmental Economics and Policy Studies*, *21*(2), 215–233. https://doi.org/10.1007/s10018-018-0250-4

Gerten, D. (2013). A vital link: Water and vegetation in the Anthropocene. *Hydrology and Earth System Sciences*, *17*, 3841–3852. https://doi.org/10.5194/hess-17-3841-2013

Gandini, A., & Lacerda, T. M. (2015). From monomers to polymers from renewable resources: Recent advances. *Progress in Polymer Science*, *48*, 1–39. https://doi.org/10.1016/j.progpolymsci.2015.04.002

Gheorghe, A., Sîrbu, R., & Dumitru, M. (2014). The potential of agricultural waste for energy generation in Romania: A review. *Renewable and Sustainable Energy Reviews*, *32*, 535-544. https://doi.org/10.1016/j.rser.2014.01.060

Ghisellini, P., Cialani, C., & De Christofaro, F. (2016). The circular economy and the sustainable use of resources. *Resources, Conservation and Recycling*, *111*, 1–9. https://doi.org/10.1016/j.resconrec.2016.04.002

Giller, K. E., Andersson, J. A., & Sumberg, J. (2015). *The future of farming: Rethinking agricultural policy and practice in a changing world*. Wiley-Blackwell.

Gill, S. E., Handley, J. F., Ennos, A. R., & Pauleit, S. (2007). Adapting cities for climate change: The role of the green infrastructure. *Built Environment*, *33*(1), 115–133. https://doi.org/10.2148/benv.33.1.115

Gilbert, C. L. (2010). Speculative bubbles, the burst of the world food price crisis, and the food price crisis of 2011. *Food Policy*, *35*(1), 1–4. https://doi.org/10.1016/j.foodpol.2009.09.005

Gilbert, N. (2009). The disappearing nutrient. *Nature*, *461*(7262), 716–718. https://doi.org/10.1038/461716a

Gliessman, S. R. (2007). *Agroecology: The ecology of sustainable food systems* (2nd ed.). CRC Press.

González, J. M., & Gutiérrez, J. S. (2017). Use of volcanic ash for potassium extraction. *Agricultural Sciences*, *8*(4), 301–312. https://doi.org/10.4236/as.2017.84026

Gouin, T., Ambrose, R. F., & Buffle, J. (2011). Environmental occurrence and fate of microplastics in aquatic environments. In J. Buffle (Ed.), *Microplastic pollution* (pp. 1–25). Springer.

Gouin, T., & Thomas, G. (2011). Occurrence and distribution of microplastics in the environment: A global perspective. *Marine Pollution Bulletin*, *62*(5), 885–892. https://doi.org/10.1016/j.marpolbul.2011.12.011

Gomiero, T. (2018). Regenerative agriculture: A sustainable alternative for food systems. In *Sustainability in Food Security and Food Quality* (pp. 123–137). Elsevier.

Grand View Research. (2020). *Biofuels market size, share & trends analysis report by fuel type (bioethanol, biodiesel, others), by application (transportation, industrial, power generation), by region, and segment forecasts, 2020–2027*. Grand View Research. https://www.grandviewresearch.com

Guenther, L., Mueller, D., & Fischer, S. (2020). Mineral resources and the future: Scenarios for the supply of essential materials. *Resources Policy*, *68*, 101801. https://doi.org/10.1016/j.resourpol.2020.101801

Gunderson, L. H., & Holling, C. S. (Eds.). (2002). *Panarchy: Understanding transformations in human and natural systems*. Island Press.

GRAIN. (2022). Land grabbing and the new scramble for agricultural land. https://grain.org/en/article/6886-land-grabbing-and-the-new-scramble-for-agricultural-land

Grand View Research. (2020). Bioplastics market size, share & trends analysis report by product (PLA, PHA, starch-based), by application (packaging, automotive, agriculture), by region, and segment forecasts, 2020–2026. https://www.grandviewresearch.com/industry-analysis/bioplastics-market

Grosse, F., Durand, F., & Leonard, J. (2011). Modeling the impact of recycling on resource depletion and economic growth. *Sustainable Materials and Technologies*, *8*, 1–12.

Häfner, S., et al. (2021). Green ammonia: Technology and market trends for a renewable future. *Renewable and Sustainable Energy Reviews*, *137*, 110591. https://doi.org/10.1016/j.rser.2020.110591

Hafner, A., Neitzel, M., & Schiller, R. (2021). Cellulose-based bioplastics: A review of their applications and production methods. *Polymers*, *13*(6), 907–918. https://doi.org/10.3390/polym13060907

Hansen, J., Sato, M., & Ruedy, R. (2005). Radiative forcing and climate response. *Journal of Geophysical Research: Atmospheres*, *110*(D18). https://doi.org/10.1029/2005JD005776

Hansen, J., Sato, M., & Ruedy, R. (2006). Global temperature change. *Proceedings of the National Academy of Sciences, 103*(39), 14288–14293. https://doi.org/10.1073/pnas.0606291103

Has, M. (2025). *Klima, Rohstoffe und Prognosen*. De Gruyter. https://doi.org/10.1515/9783111610856

Häußler, M., Eck, M., Rothauer, R., & Marschalek, R. (2021). Bioplastics and their challenges: The need for enhanced infrastructure and policy adaptation. *Sustainability, 13*(4), 1782. https://doi.org/10.3390/su13041782

Hamelinck, C. N., van Hooijdonk, G., & Faaij, A. P. C. (2005). International bioenergy trade: An evaluation of bioethanol and biodiesel supply chains. *Biomass and Bioenergy, 28*(5), 455–468. https://doi.org/10.1016/j.biombioe.2004.12.003

hausberater.de. (n.d.). Ökologische Dämmstoffe – nachhaltige Optionen. Retrieved April 11, 2025, from https://www.hausberater.de/bauen-modernisieren/daemmstoffe/oekologische-daemmstoffe-nachhaltige-optionen/

Hahladakis, J. N., & Iacovidou, E. (2018). Chemical pollutants in plastic products: Environmental concerns and solutions. In *Plastics in the Environment* (pp. 47–73). Springer.

Headey, D., & Fan, S. (2010). Reflections on the global food crisis: How did it happen and what can we do? *Food Policy, 35*(2), 105–119. https://doi.org/10.1016/j.foodpol.2009.11.008

Headey, D. D., et al. (2014). The impact of the global food crisis on poverty in developing countries. *Food Security, 6*(2), 189–202. https://doi.org/10.1007/s12571-014-0356-7

He, X., Zhang, Y., Wang, M., Wang, J., & Liu, Y. (2023). Bio-based lightweight materials for sustainable structural applications: Environmental impacts and land use considerations. *Journal of Cleaner Production, 396*, 136341. https://doi.org/10.1016/j.jclepro.2023.136341

Heffer, P. (2013). Fertilizer outlook 2013-2017. International Fertilizer Industry Association.

Heinberg, R. (2011). *The end of growth: Adapting to our new economic reality*. New Society Publishers.

Heinkel, J., & Schellnhuber, H. J. (2013). Global water availability and requirements for future food production. *Environmental Research Letters, 8*(4), 044033. https://doi.org/10.1088/1748-9326/8/4/044033

Headey, D. D. (2011). *The global food crisis: Crisis, causes, and consequences*. Global Food Security, 5(2), 139–146. https://doi.org/10.1016/j.gfs.2011.04.002

Havlin, J. L., Tisdale, S. L., Nelson, W. L., & Beaton, J. D. (2014). *Soil fertility and fertilizers: An introduction to nutrient management* (8th ed.). Pearson Education.

He, Y., Zhang, X., & Xie, L. (2018). Advances in biofuels from plant oils and fats: Production, properties, and applications. *Renewable and Sustainable Energy Reviews, 81*, 1071–1096. https://doi.org/10.1016/j.rser.2017.07.084

Himmel, M. E., Ding, S. Y., Johnson, D. K., Adney, W. S., Nimlos, M. R., Brady, J. W., & Foust, T. D. (2007). Biomass recalcitrance: Engineering plants and enzymes for biofuels production. *Science, 315*(5813), 804–807. https://doi.org/10.1126/science.1137016

Hirsinger, F., Tait, P., & Léger, G. (2018). Environmental and economic implications of biodiesel production from waste oils and fats. *Renewable Energy, 118*, 75–84. https://doi.org/10.1016/j.renene.2017.10.012

HLPE. (2013). *Biofuels and food security: A report by the High Level Panel of Experts on Food Security and Nutrition of the Committee on World Food Security*. Rome: High Level Panel of Experts on Food Security and Nutrition (HLPE), Committee on World Food Security (CFS), Food and Agriculture Organization of the United Nations (FAO). Retrieved from https://www.fao.org/3/i2952e/i2952e.pdf

HLPE (High Level Panel of Experts on Food Security and Nutrition). (2020). *Food security and nutrition: Building a global narrative towards 2030*. Committee on World Food Security, FAO. https://www.fao.org/3/ca9731en/ca9731en.pdf

Hochman, G., Rajagopal, D., Timilsina, G. R., & Zilberman, D. (2010). The role of inventory adjustments in quantifying factors causing food price inflation. *Policy Research Working Paper*, 5744. World Bank.

Hochschule Hannover, Fakultät 2, Biopolymers, facts and statistics 2023 https://www.ifbb-hannover.de/files/IfBB/downloads/faltblaetter_broschueren/f+s/Biopolymers-Facts-Statistics-einseitig-2023.pdf

Holbery, J., & Houston, D. (2006). Natural-fiber-reinforced polymer composites in automotive applications. JOM, 58(11), 80–86. https://doi.org/10.1007/s11837-006-0234-2

Houghton, J. T. (2009). *Global warming: The complete briefing*. Cambridge University Press.

Holling, C. S. (1973). Resilience and stability of ecological systems. *Annual Review of Ecology and Systematics*, *4*(1), 1–23. https://doi.org/10.1146/annurev.es.04.110173.000245

Holling, C. S. (2002). Resilience and adaptive cycles. *Ecology and Society*, *11*(1), 14. https://www.ecologyand society.org/vol11/iss1/art14/

Hopewell, J., Dvorak, R., & Kosior, E. (2009). Plastic recycling: Challenges and opportunities. *Philosophical Transactions of the Royal Society B: Biological Sciences*, *364*(1526), 2115–2126. https://doi.org/10.1098/rstb.2008.0311

Howden, S. M., et al. (2007). Adapting agriculture to climate change. *Proceedings of the National Academy of Sciences*, *104*(50), 19691–19696. https://doi.org/10.1073/pnas.0701890104

Houghton, R. A. (2003). The contemporary carbon cycle. *In J. Houghton et al. (Eds.), Climate change 2001: The scientific basis (pp. 131–171)*. Cambridge University Press.

Hu, X. C., Andrews, D. Q., Lindstrom, A. B., Bruton, T. A., Schaider, L. A., Grandjean, P., … & Sunderland, E. M. (2016). Tap water contributions to plasma concentrations of poly- and perfluoroalkyl substances (PFAS) in a nationwide prospective cohort of U.S. women. *Environmental Health Perspectives*, *127*(6), 067006. https://doi.org/10.1289/EHP4093

Huang, X., Zhang, J., & Zhou, Y. (2021). Sodium-ion batteries: A promising alternative to lithium-ion batteries. *Journal of Energy Chemistry*, *52*, 1–10. https://doi.org/10.1016/j.jechem.2020.07.004

Hubbert, M. K. (2002). *Nuclear energy and the fossil fuels* (Original work presented 1956). Shell Development Company.

Humboldt-Stiftung. (2021). ComLab#6: Schwindende Rohstoffe – Wachsende Konflikte. https://www.hum boldt-foundation.de/entdecken/organisation/humboldt-communication-lab/comlab6-schwindende-rohstoffe-wachsende-konflikte:contentReference{index=29}

Hydrogen Council. (2020). *Hydrogen insights 2020: A perspective on hydrogen investment, innovation, and decarbonization*. Hydrogen Council. https://hydrogencouncil.com

IEA (International Energy Agency). (2021). *World energy outlook 2021*. Paris: IEA. https://www.iea.org/re ports/world-energy-outlook-2021

Industrie- und Handelskammer Rhein-Neckar. (2021). Wie Lieferengpässe und steigende Rohstoffpreise deutsche Unternehmen belasten. https://www.ihk.de/rhein-neckar/international/export-import/ein fuhr/rohstoffknappheit-5218724:contentReference{index=31}

International Energy Agency. (2022). Global energy crisis. https://www.iea.org/topics/global-energy-crisis:contentReference{index=27}

International Energy Agency. (2020). *Renewables 2020 – Analysis and forecast to 2025*. International Energy Agency. https://www.iea.org/reports/renewables-2020

International Renewable Energy Agency (IRENA). (2022). *Renewable energy in Europe: Biomethane as a key player*. https://www.irena.org/publications

IEA Bioenergy. (2022). *Annual Report 2022*. IEA Bioenergy. https://www.ieabioenergy.com/wp-content/up loads/2023/05/Annual-Report-2022.pdf

IRENA International Renewable Energy Agency. (2020). *Renewable Capacity Statistics 2020*. IRENA.

IEA. (2020). *Renewable Energy Market Update*. International Energy Agency. https://www.iea.org/reports/re newable-energy-market-update

Jain, S., & Sharma, M. P. (2010). Performance of biodiesel produced from waste cooking oil in a single cylinder diesel engine. *Renewable Energy*, *35*(3), 1109–1114. https://doi.org/10.1016/j.renene.2009.09.021

International Energy Agency (IEA). (2020). *World energy outlook 2020*. International Energy Agency. https://www.iea.org/reports/world-energy-outlook-2020

IFPRI. (2022). *Global food policy report 2022*. International Food Policy Research Institute.

IFA (International Fertilizer Association). (2022). *Fertilizer outlook 2022–2026*. https://www.fertilizer.org

IPBES (Intergovernmental Science-Policy Platform on Biodiversity and Ecosystem Services). (2018). *The IPBES assessment report on land degradation and restoration*. https://www.ipbes.net

IPBES (Intergovernmental Science-Policy Platform on Biodiversity and Ecosystem Services). (2019). *Global assessment report on biodiversity and ecosystem services*. https://ipbes.net/global-assessment

IPCC. (2014). *Climate change 2014: Impacts, adaptation, and vulnerability* (Vol. 1). Cambridge University Press. https://doi.org/10.1017/CBO9781107415379

IPCC. (2014). *Climate change 2014: Synthesis report*. Intergovernmental Panel on Climate Change.

IPCC. (2021). *Climate change 2021: The physical science basis*. Contribution of Working Group I to the Sixth Assessment Report of the Intergovernmental Panel on Climate Change. Cambridge University Press. https://doi.org/10.1017/9781009157896

IPCC. (2022). *Climate change 2022: Impacts, adaptation, and vulnerability*. Intergovernmental Panel on Climate Change. https://www.ipcc.ch/report/ar6/wg2/

IPCC. (2023). *Climate change 2023: The physical science basis*. Intergovernmental Panel on Climate Change.

IPCC. (2023). *Summary for policymakers*. In *Climate change 2023: Mitigation of climate change*. Intergovernmental Panel on Climate Change.

International Aluminium Institute. (2021). *Aluminium recycling factsheet*. https://international-aluminium.org

International Energy Agency (IEA). (2021). *The role of critical minerals in clean energy transitions*. https://www.iea.org/reports/the-role-of-critical-minerals-in-clean-energy-transitions

International Monetary Fund. (2023). *World Economic Outlook: October 2023*. https://www.imf.org/en/Publications/WEO

International Monetary Fund (IMF). (2020). *Fiscal monitor: Policies for the recovery*. International Monetary Fund. https://www.imf.org

International Energy Agency (IEA). (2021). *World energy outlook 2021*. https://www.iea.org/reports/world-energy-outlook-2021

International Monetary Fund (IMF). (2021). *Still not getting energy prices right: A global and country update of fossil fuel subsidies*. https://www.imf.org/en/Publications/WP/Issues/2021/09/24/Still-Not-Getting-Energy-Prices-Right-A-Global-and-Country-Update-of-Fossil-Fuel-Subsidies-466004

International Monetary Fund IMF Historical Dept Database, IMF World Economic Outlook 2024

International Food Policy Research Institute (IFPRI). (2021). Global food policy report: Transforming food systems after COVID-19. https://doi.org/10.2499/9780896293991

Jacobson, M. Z., Delucchi, M. A., Bauer, Z. A., Goodman, S. H., Chapman, W. E., & Cameron, M. A. (2017). 100% clean and renewable wind, water, and sunlight all-sector energy roadmaps for the United States. *Energy & Environmental Science*, 10(8), 221–245. https://doi.org/10.1039/C7EE00972A

Jambeck, J. R., Geyer, R., & Wilcox, C. (2015). Plastic waste inputs from land into the ocean. *Science*, 347(6223), 768–771. https://doi.org/10.1126/science.1260352

Jehanno, C., Demarteau, J., Mantione, D., Arno, M. C., Ruipérez, F., Hedrick, J. L., Dove, A. P., & Sardon, H. (2021). Selective chemical upcycling of mixed plastics guided by a thermally stable organocatalyst. *Angewandte Chemie International Edition*, 60(12), 6710–6717. https://doi.org/10.1002/anie.202014860

Jehanno, C., Alty, J. W., Roosen, M., De Meester, S., Dove, A. P., Chen, E. Y. X., Leibfarth, F. A., & Sardon, H. (2022). Critical advances and future opportunities in upcycling commodity polymers. *Nature*, 603 (7903), 803–814. https://doi.org/10.1038/s41586-021-04350-0

Jonkers, H. M. (2010). Self-healing concrete: A biological approach. In S. van der Zwaag (Ed.), *Self healing materials* (pp. 195–204). Springer.

Jorgenson, D. W., Ho, M. S., & Stiroh, K. J. (2008). A retrospective look at the U.S. productivity growth resurgence. Journal of Economic Perspectives, 22(1), 3–24. https://doi.org/10.1257/jep.22.1.3

Jose, S. (2009). Agroforestry for ecosystem services and environmental benefits: An overview. *Agroforestry Systems*, 76(1), 1–10. https://doi.org/10.1007/s10457-009-9229-7

Jobbágy, E. G., & Jackson, R. B. (2000). The vertical distribution of soil organic carbon and its relation to climate and vegetation. *Ecological Applications*, 10(2), 423–436. https://doi.org/10.1890/1051-0761 (2000)010[0423:TVDOSO]2.0.CO;2

Kabisch, N., Qureshi, S., & Haase, D. (2016). Human–environment interactions in urban green spaces — A systematic review of contemporary issues and prospects for future research. *Environmental Impact Assessment Review*, 50, 25–34. https://doi.org/10.1016/j.eiar.2014.08.007

Kümmerer, K., & Zeise, T. (2019). Nachhaltige Kunststoffe: Potenziale und Herausforderungen. In W. Michaeli & H. Wünsch (Hrsg.), *Handbuch Kunststofftechnik* (S. 901–923). Springer Vieweg. https://doi.org/10.1007/978-3-662-57411-2_40

Kämpf, S., et al. (2018). The role of struvite in phosphorus recycling: An innovative approach for sustainable nutrient management. *Resources, Conservation and Recycling*, 134, 274–283. https://doi.org/10.1016/j.resconrec.2018.02.013

Kalnes, T., Möller, M., & Jacobson, B. (2009). Hydrotreated vegetable oil (HVO) from renewable resources: Challenges and perspectives. *Energy & Environmental Science*, 2(4), 539–549. https://doi.org/10.1039/B818795E

KEI – Kompetenzzentrum Klimaschutz in energieintensiven Industrien. (o. J.). *Konventionelle Zementherstellung*. Abgerufen am 11. April 2025, von https://www.klimaschutz-industrie.de/themen/branchen/zementindustrie/konventionelle-zementherstellung/Klimaschutz%20Industrie

Kirchherr, J., Reike, D., & Hekkert, M. (2018). Conceptualizing the circular economy: An analysis of 114 definitions. *Resources, Conservation and Recycling*, 127, 221–232. https://doi.org/10.1016/j.resconrec.2017.09.005

Kiss, G., Kuti, R., Nagy, Z. K., Farkas, A., & Földes, E. (2019). Renewable resource-based polyethylene: Evaluation of renewable content and environmental performance. *Journal of Cleaner Production*, 240, 118128. https://doi.org/10.1016/j.jclepro.2019.118128

Kiss, G., Prica, M., Skala, D., & Vatai, G. (2019). Comparative assessment of environmental impact of bio-based and petrochemical PE production. *Journal of Cleaner Production*, 241, 118360. https://doi.org/10.1016/j.jclepro.2019.118360

Klimakiller Zement. (o. J.). *Klimakiller Zement*. Abgerufen am 11. April 2025, von https://www.klimakiller-zement.de/de/ausstellung/klimakiller-zement/Ausstellung%20/%20Klimakiller%20Zement

Koller, M., Atlić, A., Dias, M., Reiterer, A., & Braunegg, G. (2017). Microbial PHA production from waste raw materials. In G. Braunegg et al. (Eds.), *Bioplastics* (pp. 85–119). Springer. https://doi.org/10.1007/978-3-319-23497-0_5

Künkel, A., Becker, J. M., Börner, H. G., & Skupin, G. (2016). Biodegradable polymers: A review of the current status and future perspectives. *Angewandte Chemie International Edition*, 55(11), 3290–3294. https://doi.org/10.1002/anie.201510766

Knothe, G., & Krahl, J. (2010). *The biodiesel handbook*. AOCS Press.

Knothe, G., Gerpen, J. V., & Krahl, J. (2015). *The biodiesel handbook* (2nd ed.). AOCS Press. https://doi.org/10.1201/9780128038012

Klare, M. T. (2012). The resource curse: How the politics of scarcity can lead to conflict and war. *Foreign Affairs*, 91(4), 19–26. https://www.foreignaffairs.com/articles/2012-07-02/resource-curse

Klein, A. M., Vaissière, B. E., Cane, J. H., & Steffen-Dewenter, I. (2007). Importance of pollinators in changing landscapes for world crops. *Proceedings of the Royal Society B: Biological Sciences*, 274(1608), 303–313. https://doi.org/10.1098/rspb.2006.3721

Klein, A. M., Vaissière, B. E., Cane, J. H., Steffen-Dewenter, I., Cunningham, S. A., Kremen, C., & Tscharntke, T. (2008). *Importance of pollinators in changing landscapes for world crops*. *Proceedings of the Royal Society B: Biological Sciences*, 274(1608), 303–313. https://doi.org/10.1098/rspb.2006.3721

Kolb, H. (2015). Geopolitik der Resourcen: Herausforderungen und Perspektiven für die globale Rohstoffversorgung. *Wirtschaftsdienst*, 95(8), 559–566. https://doi.org/10.1007/s10273-015-1902-7

Köhler, J., Langen, N., & Schmidt, J. (2019). Sustainability of biogenic resources: Local conditions and challenges. *Journal of Environmental Management, 243*, 304–315. https://doi.org/10.1016/j.jenvman.2019.05.110***

Kooroshy, J., Meindersma, C., & Scholten, D. (2015). Back to the future? Trends in critical materials supply for high-tech industries. *The Hague Centre for Strategic Studies.*

Krausmann, F., Schandl, H., & Eisenmenger, N. (2009). Biogenic raw materials in sustainable economies. *Nature Sustainability, 8*(3), 12–21. https://doi.org/10.1038/s41467-019-1148-9

Krausmann, F., Gingrich, S., Eisenmenger, N., Erb, K. H., Haberl, H., & Fischer-Kowalski, M. (2009). Growth in global materials use, GDP and population during the 20th century. *Ecological Economics, 68*(10), 2696–2705. https://doi.org/10.1016/j.ecolecon.2009.05.007

Krämer, M., Berglund, M., & Schulte, J. (2020). The role of animal manure in renewable energy production: Challenges and opportunities. *Waste and Biomass Valorization, 11*(6), 2855–2872. https://doi.org/10.1007/s12649-020-00882-0

Krugman, P. (2009). *The return of depression economics and the crisis of 2008.* W. W. Norton & Company.

Kumar, S., Sharma, R., & Gupta, R. (2020). *Impact of rice export and domestic production on food security in India.* Economic and Political Weekly, 55(27), 50–58. https://www.epw.in/journal/2020/27/special-articles/impact-rice-export-and-domestic-production.html

Lagi, M., Bar-Yam, Y., Bertrand, K. Z., & Bar-Yam, Y. (2011). *The food crises and political instability in North Africa and the Middle East.* arXiv preprint arXiv:1108.2455.

Lal, R. (2001). Soil degradation by erosion. *Land Degradation & Development, 12*(6), 517–536. https://doi.org/10.1002/ldr.463

Lal, R. (2001). Soil degradation by erosion. *Land Degradation & Development, 12*(6), 519–539. https://doi.org/10.1002/ldr.472

Lal, R. (2004). Soil carbon sequestration impacts on global climate change and food security. *Science, 304*(5677), 1623–1627. https://doi.org/10.1126/science.1097396

Lal, R. (2006). Enhancing crop yields in the developing countries through restoration of the soil organic carbon pool in agricultural lands. *Land Degradation & Development, 17*(2), 197–209. https://doi.org/10.1002/ldr.696

Lal, R. (2006). The role of soil organic matter in sustaining soil quality and soil functions. *Soil & Tillage Research, 74*(1), 3–12. https://doi.org/10.1016/j.still.2003.10.008

Lal, R. (2015). Restoring soil quality to mitigate soil degradation. *Sustainability, 7*(5), 5875–5895. https://doi.org/10.3390/su7055875

Lal, R. (2015). Restoring soil quality to mitigate soil degradation. *Sustainable Agriculture Reviews, 17*, 1–34. https://doi.org/10.1007/978-3-319-16306-3_1

Lavers, T. (2012). *The land rush and the commercialization of Ethiopian agriculture.* Journal of Peasant Studies, 39(3–4), 105-132. https://doi.org/10.1080/03066150.2012.671379

Lamers, P., Hamelinck, C., Junginger, M., & Faaij, A. (2011). International bioenergy trade—A review of past developments in the liquid biofuels market. *Renewable and Sustainable Energy Reviews, 15*(6), 2655–2676. https://doi.org/10.1016/j.rser.2011.01.022

Lange, S. (2012). Biomass as a renewable resource: Origins and trends. *Renewable Energy Reviews, 10*(1), 24–30. https://doi.org/10.1016/j.rer.2011.12.004

Lenton, T. M., Harrison, S. P., Crooks, S., & Doney, S. C. (2008). Tipping elements in the Earth's climate system. *Proceedings of the National Academy of Sciences, 105*(6), 1501–1506. https://doi.org/10.1073/pnas.0705414105

Lenton, T. M., Rockström, J., Gaffney, O., Rahmstorf, S., Richardson, K., Steffen, W., & Schellnhuber, H. J. (2008). Climate tipping points — Too risky to bet against. *Nature, 575*(7783), 598–603. https://doi.org/10.1038/443431a

Lewis, S. L., Brando, P. M., & Phillips, O. L. (2015). The 2010 Amazon drought. *Science, 331*(6017), 300–304. https://doi.org/10.1126/science.1196545

Lee, S. Y., Chang, H. N., & Park, J. H. (2021). Polyhydroxyalkanoates (PHA) production: Advances and challenges. *Biotechnology Advances, 39*, 107487. https://doi.org/10.1016/j.biotechadv.2021.107487

Leach, A. M., et al. (2012). Understanding the role of nitrogen in agricultural ecosystems. *Science, 337*(6101), 1210–1214. https://doi.org/10.1126/science.1217687

Liebig, J. von. (1855/2004). *Die Chemie in ihrer Anwendung auf Agricultur und Physiologie*. Vieweg+Teubner Verlag.

Liu, J., & Zhang, X. (2017). PFAS contamination in remote regions. *Environmental Science & Technology, 51*(5), 2939–2946. https://doi.org/10.1021/acs.est.6b06450

Liu, J., Wang, J., & Wei, X. (2019). The role of subsidies in the adoption of green technologies for alternative raw materials. *Environmental Science & Technology, 53*(16), 9797–9805. https://doi.org/10.1021/acs.est.9b02394

Liu, Y., Wang, T., & Zhang, Y. (2017). Widespread distribution of perfluoroalkyl substances in surface waters in China. *Environmental Science & Technology, 51*(10), 5351–5360. https://doi.org/10.1021/acs.est.6b06323

Li, C., Zhao, H., & Xu, Z. (2021). *Recent advances in polyhydroxyalkanoate-based bioplastics for packaging applications*. Journal of Cleaner Production, 270, 122413. https://doi.org/10.1016/j.jclepro.2020.122413

Li, J., Wang, X., & Zhang, Y. (2020). *Enzyme-based biofuel cells: Current state and future perspectives*. Bioelectrochemistry, 133, 107472. https://doi.org/10.1016/j.bioelechem.2020.107472

Li, Z., Zhang, Y., & Wang, Q. (2020). *Biocatalysts for sustainable energy applications: Current status and future directions*. Environmental Science & Technology, 54(5), 3357–3366. https://doi.org/10.1021/acs.est.9b07272

Li, X., Zhang, Y., & Wang, Z. (2020). The role of biobased building materials in reducing land use and carbon emissions in construction. *Sustainable Construction Review*, 15(2), 78–89. https://doi.org/10.1016/j.scon.2020.03.005

Li, Y., Liang, H., & Wu, H. (2018). Microalgae as a promising feedstock for biodiesel production: A review. *Bioresource Technology, 249*, 102–113. https://doi.org/10.1016/j.biortech.2017.09.132

Liao, W. K., Lee, M. Y., & Chan, L. (2016). Lignocellulose-based biorefineries: A review on conversion technologies and future prospects. *Bioresource Technology, 211*, 323–330. https://doi.org/10.1016/j.biortech.2016.03.080

Lipper, L., et al. (2014). Climate-smart agriculture for food security. *Nature Climate Change, 4*(12), 1068–1072. https://doi.org/10.1038/nclimate2437

Liu, J., et al. (2020). Precision agriculture and its application in nutrient management. *Agronomy, 10*(3), 327. https://doi.org/10.3390/agronomy10030327

Liu, H., Zhang, J., & Li, X. (2020). *Mechanisms and applications of microbial fuel cells in bioenergy generation*. Environmental Science & Technology, 54(10), 6200–6208. https://doi.org/10.1021/acs.est.0c00764

Liu, F., Zhang, Y., & Zhang, S. (2022). *Algae-based materials in energy storage and environmental applications*. Environmental Science & Technology, 56(5), 1563–1574. https://doi.org/10.1021/acs.est.1c05487

Liu, H., Wang, Y., & Zhang, Z. (2019). Challenges in scaling the production of bioplastics: The case of polyhydroxyalkanoates (PHA). *Biochemical Engineering Journal, 145*, 76–87. https://doi.org/10.1016/j.bej.2019.03.004

López, P. M., Bañares-Alcántara, R., & Sánchez, J. A. (2015). *Anaerobic digestion of organic waste materials and biogas production*. Springer. https://doi.org/10.1007/978-1-4471-6404-5

López-Gallego, F., Hughes, J. L., & Busto, E. (2012). Microbial production of biofuels and biochemicals. *Nature Biotechnology, 30*(9), 855–863. https://doi.org/10.1038/nbt.2305

López, C., García, A., & Pereira, J. (2021). *Polyhydroxyalkanoates as bioplastics: Current status and future directions*. Frontiers in Bioengineering and Biotechnology, 8, 754. https://doi.org/10.3389/fbioe.2020.00754

Lobell, D. B., et al. (2011). Climate trends and global crop production since 1980. *Science, 333*(6042), 616–620. https://doi.org/10.1126/science.1204531

Lorek, S., & Fuchs, D. (2013). Strong sustainable consumption governance – precondition for a degrowth path? Journal of Cleaner Production, 38, 36–43. https://doi.org/10.1016/j.jclepro.2011.08.008

Lund, H. (2002). *Renewable energy systems: A sustainable energy approach*. Springer.

Lund, H., Østergaard, P. A., & Nielsen, S. (2010). The role of renewable energy in a future low-carbon energy system. *Energy Policy*, 38(11), 6962–6971. https://doi.org/10.1016/j.enpol.2010.08.009

Luo, L., Wei, W., & Hu, S. (2017). Biomass gasification for power generation: A review of recent developments. *Energy Conversion and Management*, 139, 381–392. https://doi.org/10.1016/j.enconman.2017.02.070

Luyssaert, S., Inglima, I., Jung, M., Reichstein, M., Piao, S., & Ciais, P. (2008). CO2 balance of boreal, temperate, and tropical forests derived from a global database. Global Change Biology, 14(2), 1017–1034. https://doi.org/10.1111/j.1365-2486.2008.01629.x

Lynch, J. P. (2011). Root phenotypes for improved nutrient capture: An underexploited opportunity for global agriculture. *New Phytologist*, 193(4), 921–923. https://doi.org/10.1111/j.1469-8137.2011.03805.x

Malthus, T. R. (1798). *An essay on the principle of population*. London: J. Johnson.

Markets and Markets. (2020). Biomass power market by type, application, and region – Global Forecast to 2025. *Markets and Markets*.

Markets and Markets. (2020). Bioplastics market by type, application, region – global forecast to 2025. https://www.marketsandmarkets.com

Markets and Markets. (2020). *Biogas market by application (power generation, industrial, residential, commercial), and region – global forecast to 2027*. MarketsandMarkets. https://www.marketsandmarkets.com

Markets and Markets. (2022). Bioplastics market size, share & trends analysis report by product (PLA, PHA, starch-based), by application (packaging, automotive, agriculture), by region, and segment forecasts, 2022–2030. https://www.marketsandmarkets.com

Markets and Markets. (2024). Solid biomass feedstock market by source, Application – 2029. *MarketsandMarkets*. https://www.marketsandmarkets.com/Market-Reports/solid-biomass-feedstock-market-260894509.html

Martin, L. B., Peña, M., & Schell, J. (2020). *Biobased plastics and their potential in the circular economy*. Elsevier.

Mace, G. M., Norris, K., & Fitter, A. H. (2014). Biodiversity and ecosystem services: A multilayered relationship. *Trends in Ecology & Evolution*, 29(1), 1-10. https://doi.org/10.1016/j.tree.2013.08.002

Malm, A. (2016). *Fossil capital: The rise of steam power and the roots of global warming*. Verso.

Marx, K. (2005). *Das Kapital. Kritik der politischen Ökonomie. Band 1* (Original work published 1867). MEW 23. Dietz Verlag.

Marschner, H. (2012). *Mineral nutrition of higher plants* (3rd ed.). Academic Press.

Marschner, P. (2012). *Marschner's mineral nutrition of higher plants* (3rd ed.). Academic Press.

McKendry, P. (2002). Energy production from biomass (Part 1): Overview of biomass. *Bioresource Technology*, 83(1), 37-46. https://doi.org/10.1016/S0960-8524(01)00118-4

McBratney, A. B., et al. (2005). Precision agriculture: Global positioning systems, remote sensing, and the need for soil information. *Soil Science Society of America Journal*, 69(5), 1547–1552. https://doi.org/10.2136/sssaj2004.0306

McKelvey, V. E. (1996). Resource availability and the role of new technologies in mineral exploration. *Resource Geology*, 46(1), 1-11. https://doi.org/10.1111/j.1751-3928.1996.tb01011.x

McKinney, M. L. (2008). Effects of urbanization on species richness: A review of plants and animals. *Urban Ecosystems*, 11(2), 161–176. https://doi.org/10.1007/s11252-007-0045-4

McKinsey Global Institute. (2021). The future of work after COVID-19. Retrieved from https://www.mckinsey.com/featured-insights

Meadows, D. H., Meadows, D. L., Randers, J., & Behrens, W. W. (1972). *The limits to growth: A report for the Club of Rome's project on the predicament of mankind*. Universe Books.

Meadows, D. H., Meadows, D. L., & Randers, J. (2004). *Limits to growth: The 30-year update*. Chelsea Green Publishing.

Mengel, K., & Kirkby, E. A. (2001). Principles of Plant Nutrition (5th ed.). Dordrecht: Kluwer Academic Publishers. https://doi.org/10.1007/978-94-010-1009-2

Mercer, J. H. (1978). West Antarctic ice sheet and CO2 greenhouse effect: A threat of disaster. *Nature*, *271*(5643), 321–325. https://doi.org/10.1038/271321a0

METI – Ministry of Economy, Trade and Industry (Japan). (2021). White paper on resources and energy 2021. https://www.meti.go.jp/english/report/downloadfiles/2021_wp-e.pdf

Meinlschmidt, P. (2013). Neue Wege der Sortierung und Wiederverwertung von Altholz. In: *Recycling und Rohstoffe*, Band 6, 151–176. Retrieved from https://books.vivis.de/wp-content/uploads/2023/06/2013_RuR_151_176_Meinlschmidt.pdf

Meyer, B. (2017). *Sustainable biofuels from non-food crops: A review*. Renewable and Sustainable Energy Reviews, 74, 1203–1212. https://doi.org/10.1016/j.rser.2017.01.153

Miao, X., & Wu, Q. (2004). Biodiesel production from heterotrophic microalgal oil. *Bioresource Technology*, 96(3), 277–282. https://doi.org/10.1016/j.biortech.2004.01.018

Miao, X., & Wu, Q. (2004). Biodiesel production from heterotrophic microalgal oil. *Bioresource Technology*, 92(3), 317–321. https://doi.org/10.1016/j.biortech.2004.08.002

Mittefehldt, S. (2018). The history of energy transitions: From the 1970s oil crisis to renewable energy. In M. T. Boyer & A. M. Smith (Eds.), *Energy transitions* (pp. 21–42). Cambridge University Press.

Mitscherlich, E. A. (1909). Das Gesetz des Minimums und das Gesetz des abnehmenden Ertragszuwachses. *Beiträge zur Pflanzenproduktion*. In *Landwirtschaftliche Jahrbücher*, *38*, 537–552.

Molina, M. J., & Rowland, F. S. (1974). Stratospheric sink for chlorofluoromethanes: Chlorine atom-catalysed destruction of ozone. *Nature*, *249*(5460), 810–812. https://doi.org/10.1038/249810a0

Montgomery, D. R. (2007). Soil erosion and agricultural sustainability. *Sustainability*, *6*(5), 543–548.

Montgomery, D. R. (2007). Soil erosion and agricultural sustainability. *Proceedings of the National Academy of Sciences*, *104*(33), 13268–13272. https://doi.org/10.1073/pnas.0611508104

Morgan, R. P. C. (2005). *Soil Erosion and Conservation* (3rd ed.). Blackwell Publishing.

Mounk, Y. (2018). *The people vs. democracy: Why our freedom is in danger and how to save it*. Harvard University Press.

Mörsdorf, H., Schramm, J., & Vogel, F. (2019). Pellets from biomass as renewable energy: An overview of technologies and market trends. *Biomass and Bioenergy*, *129*, 105–115. https://doi.org/10.1016/j.biombioe.2019.07.017

Möller, M., Kühn, J., & Stenke, J. (2020). Biogenic elastomers: A review of renewable materials for rubber production. *Journal of Applied Polymer Science*, *137*(11), 48517. https://doi.org/10.1002/app.48517

Mudde, C., & Rovira Kaltwasser, C. (2017). *Populism: A very short introduction*. Oxford University Press.

Munns, R., & Tester, M. (2008). Mechanisms of salinity tolerance. *Annual Review of Plant Biology*, *59*, 651–681. https://doi.org/10.1146/annurev.arplant.59.032607.092911

Mühlethaler, B., & Haas, S. (2006). Stärken, Hürden und Chancen von Naturdämmstoffen. In *Bauthema Naturdämmstoffe* (S. 9–10). Fraunhofer IRB Verlag.

Müller, F., & Fischer, K. (2020). Competition for land: Biofuels, food crops, and their impact on the environment. *Renewable Energy and Environment*, *9*(3), 87–95. https://doi.org/10.1016/j.renene.2020.07.015

Müller, F., Schneider, R., & Schmidt, H. (2019). Landnutzung und Nachhaltigkeit in der biogenen Rohstoffproduktion. *Renewable Resources Journal*, *14*(6), 235–247. https://doi.org/10.1016/j.rrj.2019.06.002

Müller, J., Schmidt, D., & Wang, Q. (2021). Biodegradable materials in energy storage systems: A sustainable future for bio-batteries. *Journal of Renewable and Sustainable Energy*, *13*(6), 065201. https://doi.org/10.1063/5.0045202

Müller, R., Heinrich, M., & Schwarz, T. (2021). Biobased materials and land use: Challenges in scaling up bioplastics production. *Renewable Resources Journal*, *17*(2).

Müller, S., Schwab, A., & Fuchs, A. (2022). Low-impact algae cultivation systems for bio-energy production. *Renewable and Sustainable Energy Reviews, 145*, 111086. https://doi.org/10.1016/j.rser.2021.111086

Müller, M., et al. (2015). Soil salinity and plant growth. *Plant and Soil, 395*(1), 109–122. https://doi.org/10.1007/s11104-015-2505-3

Müller, A., Rosenbaum, R. K., & Heidrich, O. (2015). Biodiversity and ecosystem services: Implications for sustainable agriculture. Sustainability, 7(10), 13399–13419. https://doi.org/10.3390/su71013399

Muetterties, E. (2015). Biogas to fleet fuel in South San Francisco. *BioCycle*. Retrieved from https://www.biocycle.net/biogas-fleet-fuel-south-san-francisco/

Mund, M., Pitre, T., & Rabe, S. (2010). Carbon sequestration potential of reforestation projects. Forest Ecology and Management, 259(6), 1267–1275. https://doi.org/10.1016/j.foreco.2009.10.011

Murphy, D. J., & Hall, C. A. S. (2010). Year in review—EROI or energy return on (energy) invested. *Annals of the New York Academy of Sciences, 1185*(1), 102–118. https://doi.org/10.1111/j.1749-6632.2009.05282.x

Murphy, D. J., & Hall, C. A. S. (2011). Energy return on investment, peak oil, and the end of economic growth. *Annals of the New York Academy of Sciences, 1219*(1), 52–72. https://doi.org/10.1111/j.1749-6632.2010.05940.x

NABU. (2021). *Politische Rahmenbedingungen der Bioökonomie.* https://www.nabu.de/umwelt-und-resourcen/nachhaltiges-wirtschaften/biooekonomie/29197.html

Nair, P. K. R. (2007). The coming of age of agroforestry. *Journal of the Science of Food and Agriculture, 87*(9), 1613–1619. https://doi.org/10.1002/jsfa.2897

Naumann, K. Cyffka, K.-F., Costa de Paiva, G., Nieß, S., Neuling, U., Zitscher, T. (2025), *Resourcen und ihre Mobilisierung,* Deutsches Biomasseforschungszentrum)

Natural Resource Governance Institute. (2015). The Resource Curse. https://resourcegovernance.org/sites/default/files/nrgi_Resource-Curse.pdf:contentReference{index=29}

Nepstad, D., Stickler, C., & Almeida, O. (2014). *The effects of soy production on deforestation in Brazil.* Environmental Research Letters, 9(3), 031001. https://doi.org/10.1088/1748-9326/9/3/031001

Nepstad, D. C., Stickler, C. M., Soares-Filho, B., & Merry, F. (2008). Interactions among Amazon land use, forests and climate: Prospects for a near-term forest tipping point. *Philosophical Transactions of the Royal Society B: Biological Sciences, 363*(1498), 1737–1746. https://doi.org/10.1098/rstb.2007.0036

Niaounakis, M. (2013). *Biopolymers: Applications and trends.* William Andrew Publishing.

Newsroom. (2023). Bioplastics production capacities and trends. IfBB – Institute for Bioplastics and Biocomposites. Retrieved from www.ifbb-hannover.de

Newman, D. J., & Cragg, G. M. (2016). Natural products as sources of new drugs over the last 30 years. *Journal of Natural Products, 79*(3), 629–661. https://doi.org/10.1021/acs.jnatprod.5b01055

Newman, P. A., & others. (2016). The state of the ozone layer: 2016. *Scientific Assessment of Ozone Depletion: 2014*, 10. https://doi.org/10.1175/2014AR5

Niaounakis, M. (2013). *Biopolymers: Applications and trends.* William Andrew Publishing.

Nobre, C. A., & Costa, M. H. (2016). Tipping points for the Amazon rainforest. In *Earth System Science* (pp. 301–317). Springer.

Nobre, C. A., Sampaio, G., Marengo, J. A., Siqueira, G., & Silva, J. S. (2016). Land-use and climate change impacts on the Amazonian forest. *Proceedings of the National Academy of Sciences, 113*(4), 106–115. https://doi.org/10.1073/pnas.1514181113

Noleppa, S. (2020). Der Flächen-Fußabdruck der deutschen Ernährung. WWF Deutschland. https://www.wwf.de

Norris, P., & Inglehart, R. (2019). *Cultural backlash: Trump, Brexit, and authoritarian populism.* Cambridge University Press.

Nyberg, G., Knutsson, P., Ostwald, M., Öborn, I., Wredle, E., . . . & Tsegay, D. (2012). Enclosures for soil restoration and improved livelihoods in East Africa: A win-win practice? *Ambio, 41*, 208–218. https://doi.org/10.1007/s13280-011-0200-8

Obidzinski, K., Andriani, R., & Komarudin, H. (2012). *Environmental and social impacts of oil palm plantations and their implications for biofuels production in Indonesia*. Ecology and Society, 17(1), 25. https://doi.org/10.5751/ES-04618-170125

Ocampo, J. A., & Vélez-Torres, F. (2018). *Palm oil production and its effects on food security in Colombia*. Food Security, 10(1), 103–112. https://doi.org/10.1007/s12571-017-0732-1

OECD. (2022). *OECD-FAO Agricultural Outlook 2022–2031*. https://www.oecd.org

OECD. (2022). *Sovereign borrowing outlook 2022*. Organisation for Economic Co-operation and Development. https://www.oecd.org

OECD. (2011). *Towards green growth*. OECD Publishing. https://doi.org/10.1787/9789264111318-en

Organisation for Economic Co-operation and Development (OECD). (1992). *OECD Guidelines for the Testing of Chemicals, Section 3: Degradation and Accumulation – Test No. 301: Ready Biodegradability*. OECD Publishing. https://doi.org/10.1787/9789264070349-en

OECD. (2020). The cost of regulatory compliance. In OECD Regulatory Policy Outlook 2020. https://doi.org/10.1787/5b0d5c1c-en

OECD. (2015). *The future of productivity*. Abgerufen von https://www.oecd.org/economy/growth/OECD-2015-The-future-of-productivity-book.pdf

Organisation for Economic Co-operation and Development. (2024). *OECD Economic Outlook, Volume 2024 Issue 1*. https://www.oecd.org/economic-outlook/

Ørsted. (2017). *DONG Energy to stop all use of coal by 2023*. Retrieved from https://orsted.com/-/media/www/docs/corp/com/news/stop_use_of_coal_full_pm.ashx?la=enWikipedia+5Ørsted-Loveyourhome+5Ørsted-Loveyourhome+5

Okafor-Yarwood, I. (2018). *The impact of large-scale land acquisitions on local communities in Nigeria: Evidence from the oil palm sector*. Environmental Sociology, 4(2), 141–154. https://doi.org/10.1080/23251042.2018.1465407

Oki, T., & Kanae, S. (2006). Global hydrological cycles and world water resources. *Science, 313*(5790), 1068–1072. https://doi.org/10.1126/science.1128845

Oliver, T. H., Kéry, M., & Roy, D. B. (2015). Interacting effects of climate change and habitat fragmentation on species distributions. *Nature Communications, 6*(1), 1–9. https://doi.org/10.1038/ncomms8427

Olesen, J. E., et al. (2021). Climate change and the future of agricultural production: Challenges and opportunities. *Agricultural Systems, 187*, 102947. https://doi.org/10.1016/j.agsy.2020.102947

Olsson, L., Nilsson, J., & Lantz, M. (2018). The future of biofuels: Second and third generation biofuels and advanced production technologies. *Renewable Energy*, 126, 1–11. https://doi.org/10.1016/j.renene.2018.03.057

Organisation for Economic Co-operation and Development (OECD). (2022). *Agricultural outlook 2022–2031*. https://doi.org/10.1787/agr_outlook-en

Orr, J. C., & others. (2005). Anthropogenic ocean acidification over the twenty-first century and its impact on calcifying organisms. *Nature, 437*(7059), 681–686. https://doi.org/10.1038/nature04095

Ostrom, E. (2009). A general framework for analyzing sustainability of social-ecological systems. *Science, 325*(5939), 419–422. https://doi.org/10.1126/science.1172133

Overland, J. E., Dunlea, E., Box, J. E., Corell, R., Forsius, M., Kattsov, V., Olsen, M. S., Pawlak, J., Reiersen, L.-O., & Wang, M. (2019). The urgency of Arctic change. *Polar Science, 21*, 6–13. https://doi.org/10.1016/j.polar.2018.11.008

Overland, J. E., Wang, M., & Ballinger, T. J. (2019). Anticipating rapid Arctic ice loss. *Earth's Future, 7*(11), 1223–1233. https://doi.org/10.1029/2019EF001210

Pacheco-Torgal, F., Ding, Y., & Jalali, S. (2013). *Properties and durability of concrete containing polymeric wastes (tyre rubber and polyethylene terephthalate bottles): An overview*. Construction and Building Materials, 30, 714–724. https://doi.org/10.1016/j.conbuildmat.2011.11.047

Pacheco-Torgal, F., & Jalali, S. (2011). *Eco-efficient construction and building materials*. Springer London.

Pacheco-Torgal, F., Jalali, S., & Theodoridou, E. (2013). *Eco-efficient construction and building materials: Challenges and opportunities*. Elsevier.

Paech, N. (2012). *Befreiung vom Überfluss: Auf dem Weg in die Postwachstumsökonomie*. Oekom Verlag.

Pan, J., Zeng, W., & Wang, G. (2017). Advances in magnesium batteries. *Nature Energy, 2*, 17003. https://doi.org/10.1038/nenergy.2017.3

Papageorgiou, M., Chrysanthou, A., & Komilis, D. (2021). Starch-based bioplastics: Production, properties, and applications in packaging. *Journal of Biopolymer Science, 45*(2), 75–83. https://doi.org/10.1016/j.jbiopoly.2020.12.003

Parchomenko, A., Wesseling, J. H., & Mieras, M. (2019). Recycling and circular economy: Innovations and challenges. *Technological Forecasting and Social Change, 141*, 10–18. https://doi.org/10.1016/j.techfore.2018.09.029

Papageorgiou, M., Chrysanthou, A., & Komilis, D. (2021). Starch-based bioplastics: Production, properties, and applications in packaging. *Journal of Biopolymer Science, 45*(2), 75–83. https://doi.org/10.1016/j.jbiopoly.2020.12.003

Pacheco-Torgal, F., & Jalali, S. (2011). *Eco-efficient construction and building materials*. Springer London.

Pacheco-Torgal, F., Jalali, S., & Theodoridou, E. (2013). *Eco-efficient construction and building materials: Challenges and opportunities*. Elsevier.

Paech, N. (2012). *Befreiung vom Überfluss: Auf dem Weg in die Postwachstumsökonomie*. Oekom Verlag.

Pampolino, M. F., et al. (2014). *Fertilizer use and its economic effects on the agricultural industry*. Springer.

Patel, S., Rich, J., & Zhang, X. (2013). The impact of PFASs on marine life. *Environmental Toxicology and Chemistry, 32*(3), 523–532. https://doi.org/10.1002/etc.2200

V. Pawlik, Statista, Stromerzeugung aus Biomasse in Deutschland bis 2024. https://de.statistik/daten/studie/169145/umfrage/stromproduktion-durch-biomasse-in-deutschland-seit-2000/,

Pérez, J. M., Muñoz-Dorado, J., de la Rubia, T., & Martínez, J. (2020). Biodegradation and biological treatments of polylactic acid: A review. *Journal of Environmental Management, 264*, 110509. https://doi.org/10.1016/j.jenvman.2020.110509

Perfecto, I., & Vandermeer, J. (2010). The agroecological matrix as alternative to the land-sparing/agriculture intensification model. *Proceedings of the National Academy of Sciences*, 107(13), 5786–5791. https://doi.org/10.1073/pnas.0905455107

Pereira, J. S., Bacci, A., & Meneghetti, A. L. (2020). Government policies and their impact on the bioplastic industry. *Renewable and Sustainable Energy Reviews, 124*, 109771. https://doi.org/10.1016/j.rser.2020.109771

Philp, J. C., Ritchie, R. J., & Allan, J. E. (2013). Biobased plastics in a bioeconomy. *Trends in Biotechnology, 31*(2), 65–67. https://doi.org/10.1016/j.tibtech.2012.10.009

Pickering, K. L., Aruan Efendy, M. G., & Le, T. M. (2016). A review of recent developments in natural fibre composites and their mechanical performance. *Composites Part A: Applied Science and Manufacturing, 83*, 98–112. https://doi.org/10.1016/j.compositesa.2015.08.038

Piguet, E., de Gouvello, C., & Rey, J. (2017). The global climate crisis and the unequal impact on the Global South. *Journal of Environmental Studies, 25*(2), 111–121.

Pimm, S. L., Jenkins, C. N., Abell, R., Brooks, T. M., Gittleman, J. L., Joppa, L. N., . . . & Russell, G. J. (2014). The biodiversity of species and their rates of extinction, distribution, and protection. *Science, 344*(6187), 1246752. https://doi.org/10.1126/science.1246752

Pimentel, D., Harvey, C., Resosudarmo, P., Sinclair, K., Kurz, D., McNair, M., . . . & Blair, R. (1995). Environmental and economic costs of soil erosion and conservation benefits. *Science, 267*(5201), 1117–1123. https://doi.org/10.1126/science.267.5201.1117

Pimentel, D., et al. (2011). Environmental and economic costs of soil erosion and conservation benefits. *Science, 267*(5201), 1117–1123.

Pimentel, D., Tarleton, L., Nelson, J., & Flock, A. (1992). *Environmental and economic costs of soil erosion and conservation benefits. Science, 267*(5201), 1117–1123. https://doi.org/10.1126/science.267.5201.1117

Pimentel, D., & Patzek, T. W. (2005). Ethanol production using corn, switchgrass, and wood; biodiesel production using soybean and sunflower. *Natural Resources Research*, 14(1), 65–76. https://doi.org/10.1007/s11053-005-4679-8

Pratt, C. J., Wilson, S., & Cooper, P. (2007). Source Control Using Constructed Pervious Surfaces: Hydraulic, Structural and Water Quality Performance Issues. CIRIA C582. London: Construction Industry Research and Information Association (CIRIA).

Probst, Donald A.; Pratt, Walden P.; McKelvey, V. E. (1973). Summary of United States mineral resources (https://pubs.usgs.gov/circ/1973/0682/report.pdf

Pretty, J., Benton, T. G., Bharucha, Z. P., Dicks, L. V., Flora, C. B., Godfray, H. C. J., . . . & Wratten, S. (2018). Global assessment of agricultural system redesign for sustainable intensification. *Nature Sustainability*, 1(8), 441–446. https://doi.org/10.1038/s41893-018-0114-0

Prussi, M., Vignali, G., & Bertucco, A. (2019). The role of biofuels in the sustainable energy transition. *Renewable and Sustainable Energy Reviews*, 101, 1–16. https://doi.org/10.1016/j.rser.2018.10.012

Pretty, J. (2008). Agricultural sustainability: Concepts, principles and evidence. *Philosophical Transactions of the Royal Society B: Biological Sciences*, 363(1491), 447–465. https://doi.org/10.1098/rstb.2007.2163

Pope, C. A., & Dockery, D. W. (2006). Health effects of fine particulate air pollution: Lines that connect. *Journal of the Air & Waste Management Association*, 56(6), 709–742. https://doi.org/10.1080/10473289.2006.10464485

Postel, S. (2000). Pillars of sand: Can the irrigation miracle last? In *Water for food security* (pp. 67–83). Springer.

Porter, J. R., et al. (2014). Food security and food production systems. In *Climate change 2014: Impacts, adaptation, and vulnerability* (pp. 485–533). Cambridge University Press.

Potts, S. G., Imperatriz-Fonseca, V., Ngo, H. T., Aizen, M. A., Biesmeijer, J. C., Breeze, T. D., . . . & Viana, B. F. (2016). Safeguarding pollinators and their values to human well-being. *Nature*, 540(7632), 220–229. https://doi.org/10.1038/nature20588

Ragauskas, A. J., Williams, C. K., Davison, B. H., et al. (2014). *The path forward for biofuels and biomaterials.* Science, 311(5760), 484–489. https://doi.org/10.1126/science.1114736

Rahmato, D. (2011). *Land to investors: Large-scale land transfers in Ethiopia.* Forum for Social Studies. https://www.fss.org.et

Rahimi, A., & García, J. M. (2017). Chemical recycling of waste plastics for new materials production. *Nature Reviews Chemistry*, 1(6), 0046. https://doi.org/10.1038/s41570-017-0046

Ray, D. K., et al. (2013). Climate change has likely already affected global food production. *PLoS ONE*, 8(6), e66402. https://doi.org/10.1371/journal.pone.0066402

Reinhart, C. M., & Rogoff, K. S. (2010). Growth in a time of debt. *American Economic Review*, 100(2), 573–578. https://doi.org/10.1257/aer.100.2.573

Rengasamy, P. (2006). World salinization with emphasis on Australia. *Journal of Experimental Botany*, 57(5), 1017–1023. https://doi.org/10.1093/jxb/erj108

Rengel, Z. (2015). *Potassium in soils and plants: Its role and contribution to agricultural productivity.* Springer.

REN21. (2020). *Renewable energy policy network for the 21st century: Renewables 2020 global status report.* REN21. https://www.ren21.net

Reuters. (2025, April 11). China's rare earth exports grind to a halt as trade war controls bite. https://www.reuters.com/markets/commodities/chinas-rare-earth-exports-grind-halt-trade-war-controls-bite-2025-04-11/:contentReference{index=37}

Reynolds, J. F., Stafford Smith, D. M., Lambin, E. F., Turner, B. L., Mortimore, M., Batterbury, S. P. J., ... & Walker, B. (2007). Global desertification: Building a science for dryland development. *Science*, 316 (5826), 847–851. https://doi.org/10.1126/science.1131634

Richey, A. S., Thomas, B. F., Lo, M.-H., Reager, J. T., Famiglietti, J. S., Voss, K., Swenson, S., & Rodell, M. (2015). Quantifying renewable groundwater stress with GRACE. *Water Resources Research*, 51(7), 5217–5238. https://doi.org/10.1002/2015WR017349

Richey, J., & McGuire, M. (2015). Groundwater depletion in agricultural regions. *Geophysical Research Letters*, *42*(11), 4577–4583. https://doi.org/10.1002/2015GL064655

Rhein, M., & others. (2013). The Western Indian Ocean: Ocean acidification. In *The Oceans: A Global Study* (pp. 307–325). Springer.

Richardson, K., Rockström, J., Lenton, T. M., Folke, C., Liverman, D., Ottersen, G., ... & Schellnhuber, H. J. (2023). Earth beyond six of nine planetary boundaries. *Science Advances*, *9*(37), eadh2458. https://doi.org/10.1126/sciadv.adh2458

Rinaldi, R., Jastrzebski, R., Clough, M. T., Ralph, J., Kennema, M., Bruijnincx, P. C. A., & Weckhuysen, B. M. (2016). Paving the way for lignin valorisation: Recent advances in bioengineering, biorefining and catalysis. *Angewandte Chemie International Edition*, *55*(29), 8164–8215. https://doi.org/10.1002/anie.201510351

Riley, A. (2024). Europe's messy Russian gas divorce. *Brookings Institution*. https://www.brookings.edu/articles/europes-messy-russian-gas-divorce/:contentReference{index=31}

Roberts, M. J., & Schlenker, W. (2013). Identifying supply and demand elasticities of agricultural commodities: Implications for the US ethanol mandate. *American Economic Review*, 103(6), 2265–2295. https://doi.org/10.1257/aer.103.6.2265

Rochman, C. M., Hoh, E., Kurobe, T., & Teh, S. J. (2013). Ingested plastic transfers hazardous chemicals to fish and induces hepatic stress. *Scientific Reports*, *3*, 3263. https://doi.org/10.1038/srep03263

Rodell, M., Velicogna, I., & Famiglietti, J. S. (2009). *Satellite-based estimates of groundwater depletion in India*. Nature, 460(7258), 999–1002. https://doi.org/10.1038/nature08238https://doi.org/10.1177/0734242X16683272

Rujnić-Sokele, M., & Pilipović, A. (2017). Bioplastics and their environmental impact: A review. *Science of The Total Environment*, *613–614*, 619–633. https://doi.org/10.1016/j.scitotenv.2017.09.222

Rujnić-Sokele, M., & Pilipović, A. (2017). Bioplastics and their role in sustainable development. *Biomaterials and Bioengineering*, *34*(3), 120–127. https://doi.org/10.1016/j.bmbe.2017.06.002

Rujnić-Sokele, M., & Pilipović, A. (2017). Challenges and opportunities of biodegradable plastics: A mini review. *Waste Management & Research*, *35*(2), 132–140. https://doi.org/10.1177/0734242X16683272

Ruthven, T., Boddy, D., & Turner, L. (2020). Polymilchsäure (PLA) und ihre Anwendungen in der Verpackungsindustrie: Eine Übersicht. *Journal of Renewable Materials*, *8*(3), 201–212. https://doi.org/10.1002/jrm.2740

Rujnić-Sokele, M., & Pilipović, A. (2017). Bioplastics and their role in sustainable development. *Biomaterials and Bioengineering*, *34*(3), 120-127. https://doi.org/10.1016/j.bmbe.2017.06.002

Rockström, J., Edenhofer, O., Gärtner, J., & DeClerck, F. (2020). Planet-proofing the global food system. *Nature Food*, *1*, 3–5. https://doi.org/10.1038/s43016-019-0010-4

Rockström, J., Steffen, W., Noone, K., Persson, Å., Chapin, F. S., Lambin, E. F., . . . & Foley, J. A. (2009). A safe operating space for humanity. *Nature*, *461*(7263), 472–475. https://doi.org/10.1038/461472a

Rockström, J., et al. (2009). Sustainable intensification of agriculture for human prosperity and global sustainability. *Ambio*, *38*(4), 237–244. https://doi.org/10.1579/0044-7447-38.4.237

Rockström, J., et al. (2017). Sustainable intensification of agriculture for human prosperity and global sustainability. *Ambio*, *46*(4), 456–474. https://doi.org/10.1007/s13280-017-0964-3

Rockström, J., Steffen, W., & Persson, Å. (2021). Planetary boundaries: A new framework for global sustainability. *Environmental Research Letters*, *16*(10), 104016. https://doi.org/10.1088/1748-9326/abf6b7

Rochman, C. M., Browne, M. A., Halden, R. U., Hentschel, B. T., & Rios, M. (2019). The plastic catastrophe: Global impacts and solutions. *Science Advances*, *5*(12), 1000106. https://doi.org/10.1126/sciadv.aax2373

Rosenboom, J. G., Langer, R., & Traverso, G. (2022). Bioplastics for a circular economy. *Nature Reviews Materials*, *7*(2), 117–137. https://doi.org/10.1038/s41578-021-00358-9

Sachs, W. (1993). *Global ecology: A new arena of political conflict*. Zed Books.

Sala, O. E., Ceballos, G., Dirzo, R., & Hubbell, S. P. (2000). Global biodiversity scenarios for the year 2100. *Science*, *287*(5459), 1770–1774. https://doi.org/10.1126/science.287.5459.1770

Sanchez, D. L., & Sachs, J. D. (2017). The bioenergy and biofuels outlook: The need for technical progress and policy evolution. *Wiley Interdisciplinary Reviews: Energy and Environment*, *6*(3), e244. https://doi. org/10.1002/wene.244

Sánchez, M., Pérez, G., & Borrero, C. (2020). Bioplastics from renewable resources: The state of the art. *Environmental Science and Pollution Research*, *27*(11), 12123–12138. https://doi.org/10.1007/s11356-020-08023-7

Sanchez, O. J., & Cardona, C. A. (2017). Trends in biorefining: The utilization of lignocellulosic biomass for bioethanol and other biofuels production. *Renewable Energy*, *109*, 84–93. https://doi.org/10.1016/j.re nene.2017.02.003

Sauer, S., & Leite, S. M. (2012). *Land concentration and agrarian structure in Brazil*. Journal of Agrarian Change, 12(1), 4–25. https://doi.org/10.1111/j.1471-0366.2011.00343.x

Sahoo, P. K., Das, L. M., & Babu, A. P. (2018). Biomass waste to biofuels: Sustainable renewable alternatives to fossil fuels. *Renewable and Sustainable Energy Reviews*, *89*, 190–211. https://doi.org/10. 1016/j.rser.2018.01.044

Sakaki, T., Umebayashi, Y., & Matsuo, T. (2012). Potassium deficiency affects plant disease resistance, structural integrity, and fruit development. *Journal of Plant Nutrition and Soil Science*, *175*(3), 423–432. https://doi.org/10.1002/jpln.201100123

Santamouris, M. (2014). Cooling the cities – A review of reflective and green roof mitigation technologies to fight heat island and improve comfort in urban environments. *Solar Energy*, *103*, 682–703. https:// doi.org/10.1016/j.solener.2012.07.003

Sattari, S. Z., Bouwman, A. F., Giller, K. E., & Van Ittersum, M. K. (2012). Residual soil phosphorus as the missing piece in the global phosphorus crisis puzzle. *Proceedings of the National Academy of Sciences*, *109*(16), 6348–6353. https://doi.org/10.1073/pnas.1201041109

Scalenghe, R., & Marsan, F. A. (2009). The anthropogenic sealing of soils in urban areas. *Geoderma*, *151*(1–2), 1–4. https://doi.org/10.1016/j.geoderma.2009.03.007

Scharlemann, J. P. W., et al. (2014). Global biodiversity scenarios for the year 2100. *Science*, *346*(6206), 241–245. https://doi.org/10.1126/science.125

Scheidel, A., Del Bene, D., Liu, J., Navas, G., Mingorría, S., Demaria, F., . . . & Martínez-Alier, J. (2023). Global impacts of extractive and industrial development projects on indigenous peoples' lands. *Science Advances*, *9*(1), eade9557. https://doi.org/10.1126/sciadv.ade9557:contentReference{index=33}

Schneidewind, U., & Zahrnt, A. (2013). *Damit gutes Leben einfacher wird: Perspektiven einer Suffizienzpolitik*. Oekom Verlag.

Scarlat, N., Dallemand, J.-F., Monforti-Ferrario, F., & Nita, V. (2015). The role of biomass and bioenergy in a future bioeconomy: Policies and facts. *Environmental Development*, *15*, 3–34. https://doi.org/10.1016/j. envdev.2015.03.006

Schalatek, L., & Schmalz, S. (2018). Climate finance and the global climate negotiations: Reassessing the Paris Agreement's ambitions. *Global Environmental Politics*, *18*(2), 1–22. https://doi.org/10.1162/GLEP_ a_00468

Scheringer, M., Altmann, M., Arnold, M., Böhmer, W., Brandt, J., Brunn, H., Bucheli, T. D., Giger, W., Hanke, G., Hungerbühler, K., MacLeod, M., Mayer, P., Schwarzenbach, R. P., Stamm, C., & von Gunten, U. (2018). Assessment of persistent, mobile and toxic (PMT) and very persistent and very mobile (vPvM) substances in the context of the European chemicals regulation REACH. *Environmental Sciences Europe*, *30*, 9. https://doi.org/10.1186/s12302-018-0135-7

Scheringer, M., & Müller, M. (2018). Persistent organic pollutants: Monitoring and evaluation in the environment. *Environmental Pollution*, *246*, 529–540. https://doi.org/10.1016/j.envpol.2018.01.033

Schjønning, P., et al. (2015). Soil degradation and conservation strategies. *Land Degradation & Development*, *26*(5), 473–485. https://doi.org/10.1002/ldr.2405

Schjønning, P., Keller, T., de Jonge, L. W., Munkholm, L. J., & Christensen, B. T. (2015). Soil compaction and soil management – A review. *Soil Use and Management*, *31*(4), 476–485. https://doi.org/10.1111/sum. 12164

Schlesinger, W. H., & Bernhardt, E. S. (2013). *Biogeochemistry: An analysis of global change* (3rd ed.). Academic Press.

Schmitt, M., Oltmanns, M., & Müller, A. (2017). Global fertilizer markets and their vulnerability to geopolitical risks: The case of potassium fertilizers. [Details zur Publikation fehlen – bitte ergänzen, z. B. Zeitschrift, Verlag oder DOI].

Scholz, R. W., & Wellmer, F.-W. (2013). Approaching a dynamic view on the availability of mineral resources: What we may learn from the case of phosphorus? *Global Environmental Change, 23*(1), 11–27. https://doi.org/10.1016/j.gloenvcha.2012.10.013

Schröder, J. J., Smit, A. L., Cordell, D., & Rosemarin, A. (2018). Improved phosphorus use efficiency in agriculture: A key requirement for its sustainable use. *Chemosphere, 197*, 175–183. https://doi.org/10.1016/j.chemosphere.2017.12.113

Schröder, B., et al. (2010). Permeable pavements and their role in urban water management. In R. Suresh (Ed.), *Green infrastructure as a tool for urban flood management and climate adaptation* (pp. 155–168). CRC Press. https://doi.org/10.1201/9781351124140-155

Schröder, P., Lemille, A., & Desmond, P. (2019). Making the circular economy work for human development. *Resources, Conservation and Recycling, 146*, 246–253. https://doi.org/10.1016/j.resconrec.2019.03.005

Schulte, R. P. O., Creamer, R. E., Donnellan, T., Farrelly, N., Fealy, R., O'Donoghue, C., & O'hUallachain, D. (2014). Functional land management: A framework for managing soil-based ecosystem services for the sustainable intensification of agriculture. *Environmental Science & Policy, 38*, 45–58. https://doi.org/10.1016/j.envsci.2013.10.002

Scholz, R. W., Roy, A. H., Hellums, D. T., Ulrich, A. E., & White, P. J. (2013). *Sustainable phosphorus management: A global transdisciplinary roadmap*. Springer.

Scholz, R. W., et al. (2014). Sustainable use of phosphorus. *Science, 339*(6126), 347–348. https://doi.org/10.1126/science.1233997

Schwager, R., Richter, S., & Fox, T. (2020). Biologically derived polyamides: Structure, properties, and applications in the automotive and textile industries. *Journal of Biopolymers, 47*(4), 224–234. https://doi.org/10.1002/bip.26135

Schwarzenbach, R. P., Egli, T., Hofstetter, T. B., von Gunten, U., & Wehrli, B. (2010). Global water pollution and human health. *Annual Review of Environment and Resources, 35*, 109–136. https://doi.org/10.1146/annurev-environ-100809-125342

Schultz, F. C., & Reinhardt, R. J. (2023). Technological challenges and opportunities to plastics valorization in the context of a circular economy in Europe. *Sustainability, 15*(4), 3741. https://doi.org/10.3390/su15043741

Schuur, E. A. G., Bockheim, J. G., & Camill, P. (2015). Permafrost: The changing Arctic. *Nature, 526*(7573), 1–10. https://doi.org/10.1038/s41586-019-1611-9

Schwartz, S. E. (2004). The whitehouse effect: Shortwave radiative forcing of climate by anthropogenic aerosols – A review. *Journal of the Atmospheric Sciences, 61*(6), 1239–1271. https://doi.org/10.1175/1520-0469(2004)061<1239:TWE>2.0.CO;2

Schmitz, G. P. (2013, March 6). US-Studie: Wie der Klimawandel zum Arabischen Frühling führte. *Der Spiegel*. https://www.spiegel.de/politik/ausland/us-studie-wie-der-klimawandel-zum-arabischen-fruehling-fuehrte-a-887085.html

Schiavoni, S., D'Alessandro, F., Bianchi, F., & Asdrubali, F. (2016). Insulation materials for the building sector: A review and comparative analysis. *Renewable and Sustainable Energy Reviews, 62*, 988–1011. https://doi.org/10.1016/j.rser.2016.05.045

Searchinger, T., Waite, R., Hanson, C., & Ranganathan, J. (2018). *Creating a sustainable food future: A menu of solutions to sustainably feed more than 9 billion people by 2050*. World Resources Institute. https://www.wri.org/research/creating-sustainable-food-future

Searchinger, T., Heimlich, R., Houghton, R. A., Goodall, J., & Heimlich, R. (2018). Avoiding bioenergy competition with food crops and land. *Environmental Research Letters, 13*(6), 063003.

Searchinger, T., Heimlich, R., Houghton, R. A., Dong, F., Elobeid, A., Fabiosa, J., Tokgoz, S., Hayes, D., & Yu, T. H. (2008). Use of U.S. croplands for biofuels increases greenhouse gases through emissions from land-use change. *Science, 319*(5867), 1238–1240. https://doi.org/10.1126/science.1151861

Searchinger, T., Heimlich, R., Houghton, R. A., Dong, H., Elobeid, A., Fabiosa, J., . . . & Tokgoz, S. (2008). Use of US croplands for biofuels increases greenhouse gases through land-use changes. *Science, 319*(5867), 1238–1240. https://doi.org/10.1126/science.1151861

Seitz, H., & Sommer, B. (2020). Der öffentliche Sektor in föderalen Systemen: Effizienz und Koordinationsprobleme. In K.-P. Stiller (Hrsg.), *Finanzwissenschaftliche Diskussionsbeiträge*. Springer.

Stern, N. (2006). *Stern Review: The Economics of Climate Change*. HM Treasury. Abgerufen von https://webarchive.nationalarchives.gov.uk/ukgwa/+/http://www.hm-treasury.gov.uk/stern_review_report.htm

Swiss Re Institute. (2021). *The economics of climate change: no action not an option*. Abgerufen von https://www.swissre.com/institute/research/topics-and-risk-dialogues/climate-and-natural-catastrophe-risk/expertise-publication-economics-of-climate-change.html

Seufert, P.*, Mendonça, M. L., & Pitta, F. (2018). When land becomes a global financial asset: The MATOPIBA case in Brazil. In Right to Food and Nutrition Watch (2018, pp. 20–26). FIAN International.

Seufert P., Monsalve S., Rolón L., Guttal S.; Lords of the Land, Transnational landorners – inequality and the case for redistribution, FIAN International, June 2025, https://www.fian.org/files/is/htdocs/wp11102127_GNIAANVR7U/www/files/Lords_Land_Fian_20250602_fin.pdf

Seufert, P. (2025). *Small-scale food providers and their role in global food production*. FIAN International. (Bericht veröffentlicht Juni 2025).

Seto, K. C., Güneralp, B., & Hutyra, L. R. (2012). Global forecasts of urban expansion to 2030 and direct impacts on biodiversity and carbon pools. *Proceedings of the National Academy of Sciences, 109*(40), 16083–16088. https://doi.org/10.1073/pnas.1211658109

Shen, L., Haufe, J., & Patel, M. K. (2009). Product overview and market projection of emerging bio-based plastics. *Utrecht University, Copernicus Institute for Sustainable Development and Innovation*. https://www.chem.uu.nl/bibnet/afstudeerprojecten/2010-003%20Shen.pdf Retrieved from https://www.sciencedirect.com/science/article/pii/S0959652620340555

Shen, L., Worrell, E., & Patel, M. K. (2010). The energy and environmental performance of bio-based polymers: A review. *Renewable and Sustainable Energy Reviews, 14*(9), 2379-2396. https://doi.org/10.1016/j.rser.2010.06.015

Shen, L., Worrell, E., & Patel, M. (2020). Bioplastics: A sustainable alternative to conventional plastics. *Resources, Conservation and Recycling, 167*, 105410. https://doi.org/10.1016/j.resconrec.2020.105410

Shao, Y., & others. (2017). Regional variations of atmospheric particulate matter and their environmental impact. *Atmospheric Environment, 155*, 10–20. https://doi.org/10.1016/j.atmosenv.2017.02.046

Sharma, P., & Ghoshal, S. (2015). Hydrogen production using renewable energy sources. *Renewable and Sustainable Energy Reviews, 44*, 150–162. https://doi.org/10.1016/j.rser.2014.12.018

Shi, S., Wang, J., & Ling, H. (2022). Synthetic biology for the sustainable production of biochemicals in engineered microbes. *Frontiers in Bioengineering and Biotechnology*, 10, 984875. https://doi.org/10.3389/fbioe.2022.984875

Shiva, V. (2002). *Water wars: Privatization, pollution, and profit*. South End Press.

Sivakumar, M. V. K. (2005). Climate and land degradation. In M. V. K Sivakumar & N. Ndiang'ui (Eds.), *Climate and land degradation* (pp. 3–22). Springer. https://doi.org/10.1007/1-4020-3324-1_1

Sikkema, R., Oosterhuis, M., & Faaij, A. (2013). Biomass for energy: Economic and environmental feasibility. *Biomass and Bioenergy, 52*, 8–17. https://doi.org/10.1016/j.biombioe.2013.02.017

Sims, R. E. H., Mabee, W. E., Saddler, J. N., & Taylor, M. (2008). An overview of second generation biofuel technologies. *Bioresource Technology, 101*(6), 1570–1580. https://doi.org/10.1016/j.biortech.2009.07.002

Singh, N., Hui, D., Singh, R., Ahuja, I. P. S., Feo, L., & Fraternali, F. (2019). Recycling of plastic solid waste: A state of art review and future applications. *Composites Part B: Engineering*, 115, 409–422. https://doi.org/10.1016/j.compositesb.2016.09.013

Singh, P., Kumar, A., & Patel, S. (2021). Biotechnological advances in polymeric materials from renewable resources: A comprehensive review. *Journal of Cleaner Production*, *314*, 128089. https://doi.org/10.1016/j.jclepro.2021.128089

Smeets, E., Faaij, A., & Lewandowski, I. (2013). The sustainability of biomass production: Defining the concept and providing guidelines for implementation. *Renewable and Sustainable Energy Reviews*, 18, 1–16. https://doi.org/10.1016/j.rser.2012.10.037

Smil, V. (2001). *Enriching the earth: Fritz Haber, Carl Bosch, and the transformation of world food production*. MIT Press.

Smil, V. (2017). *Energy and civilization: A history*. MIT Press.

Smith, H., O'Neill, M., & Peterson, L. (2020). Environmental impacts of biogenic materials and their potential for replacing mineral resources. *Environmental Science Reports*, *3*(4), 185–199. https://doi.org/10.1016/j.esr.2020.01.004

Smith, H., Taylor, S., & Green, J. (2020). Environmental sustainability of bioplastics: The potential and challenges of PLA and PHA in packaging. *Journal of Environmental Science*, *16*(4), 212–224. https://doi.org/10.1016/j.jes.2020.01.005

Smith, R., Lee, K., & Patel, P. (2021). Exploring the potential of bioplastics in replacing petroleum-based plastics: A global perspective. *Environmental Impact Reviews*, *22*(1), 45–60. https://doi.org/10.1016/j.eir.2021.01.002

Smith, P., et al. (2012). Agroforestry and climate change mitigation. *Agroforestry Systems*, *85*(1), 31–41. https://doi.org/10.1007/s10457-012-9492-5

Smith, P., Bustamante, M., Ahammad, H., Clark, H., Dong, H., Elsiddig, E. A., . . . & Tubiello, F. N. (2013). How much land-based greenhouse gas mitigation can be achieved without compromising food security and environmental goals? *Global Change Biology*, *19*(8), 2285–2302. https://doi.org/10.1111/gcb.12160

Smith, P., et al. (2016). Global change and the impacts of climate change on agriculture. In *Climate Change and Food Security: A Framework Document* (pp. 38–45). Springer. https://doi.org/10.1007/978-3-319-26883-5_8

Smith, S. E., & Read, D. J. (2008). *Mycorrhizal symbiosis* (3rd ed.). Academic Press.

Sodhi, N. S., Brook, B. W., & Bradshaw, C. J. A. (2010). Causes and consequences of species extinctions. In N. S. Sodhi & P. R. Ehrlich (Eds.), Conservation Biology for All (pp. 514–536). Oxford University Press.

Sodhi, N. S., & Ehrlich, P. R. (Eds.). (2010). *Conservation biology for all*. Oxford University Press.

Sohi, S. P., So, J. R., & Amonette, J. E. (2010). *Biochar in agriculture – A review*. Soil Use and Management, 26(3), 194-201. https://doi.org/10.1111/j.1475-2743.2010.00280.x

Spierling, S., Azzone, G., & Bianchessi, S. (2018). Bioplastics in the circular economy: Production, recycling, and sustainability. *Springer*.

Spierling, S., Knüpffer, E., Behnsen, H., Mudersbach, M., Krieg, H., Springer, S., . . . & Endres, H. J. (2018). Bio-based plastics – A review of environmental, social and economic impact assessments. *Journal of Cleaner Production*, *185*, 476–491. https://doi.org/10.1016/j.jclepro.2018.03.014

Stede, J., Neuhoff, K., Hardadi, G., & Pauliuk, S. (2021). Carbon pricing of basic materials: Incentives and risks for the value chain and consumers. *Ecological Economics*, 189, 107168. https://doi.org/10.1016/j.ecolecon.2021.107168

Stichnothe, H., & Azapagic, A. (2013). Life cycle assessment of recycling PVC window frames. *Resources, Conservation and Recycling*, 71, 40–47. https://doi.org/10.1016/j.resconrec.2012.12.005

Staples, M., Hill, J., & Wurster, P. (2018). The use of hydrotreated vegetable oils (HVO) in aviation: A review. *Renewable and Sustainable Energy Reviews*, *90*, 558–573. https://doi.org/10.1016/j.rser.2018.03.040

Stiftung Wissenschaft und Politik. (2010). *Konkurrenz um knappe Resourcen*. https://www.swp-berlin.org/publications/products/projekt_papiere/projpap_resourcen_20100211_ks.pdf

Sun, Y., Liu, X., & Li, X. (2020). Review on biobatteries and their applications in energy storage: A pathway to sustainability. *Journal of Energy Chemistry, 35,* 85–100. https://doi.org/10.1016/j.jechem.2020.01.014

Steffen, W., Richardson, K., Rockström, J., Cornell, S. E., Fetzer, I., Bennett, E. M., . . . & Sörlin, S. (2015). Planetary boundaries: Guiding human development on a changing planet. *Science, 347*(6223), 736–748. https://doi.org/10.1126/science.1259855

Staehel, W. R. (2016). The circular economy. *Nature, 531*(7595), 435–438. https://doi.org/10.1038/531435a

Standard & Poor's. (2020). Sovereign ratings methodology. https://www.spglobal.com

Statista. (2021). Prognose der weltweiten fossilen Brennstoffreserven. https://www.statista.com

Stern, D. I. (2011). The role of energy in economic growth. *Annals of the New York Academy of Sciences, 1219*(1), 26–51. https://doi.org/10.1111/j.1749-6632.2010.05921.x

Stiglitz, J. E. (2012). *The price of inequality: How today's divided society endangers our future.* W. W. Norton & Company.

Sunderland, E. M., Kannan, K., & Brehm, S. (2019). PFAS chemicals in the environment and human health. *Environmental Health Perspectives, 127*(6), 960–976. https://doi.org/10.1289/EHP4638

SVR – Sachverständigenrat für Integration und Migration. (2023). *Jahresgutachten 2023: Klimawandel und Migration.* https://www.svr-migration.de/wp-content/uploads/2023/05/SVR_Jahresgutachten_2023-1.pdf

Sullivan, P. L., Blankinship, J. C., & Hart, S. C. (2015). Climate change and nutrient cycling in soils: A review of nitrogen mineralization responses to temperature. *Geoderma, 237–238,* 210–220. https://doi.org/10.1016/j.geoderma.2014.08.007

Sutton, M. A., et al. (2013). Our nutrient world: The challenge to produce more food with less pollution. The Global Partnership on Nutrient Management and the International Nitrogen Initiative.

TATuP. (2015). *Auswirkungen eines verstärkten Anbaus nachwachsender Rohstoffe auf die Landnutzung.* Zeitschrift für Technikfolgenabschätzung in Theorie und Praxis. https://www.tatup.de/index.php/tatup/article/view/1027/1920Tatup+1Tatup+1

Tanzi, M. C., Farè, S., Candiani, G., & Riva, F. (2019). *Foundations of Biomaterials Engineering.* Academic Press

Tainter, J. A. (1988). *The collapse of complex societies.* Cambridge University Press.

Tainter, J. A. (1988). *The collapse of complex societies.* Cambridge University Press.

Taiz, L., Zeiger, E., Møller, I. M., & Murphy, A. (2015). *Plant Physiology and Development* (6th ed.). Sinauer Associates.

Tautges, N. E., Chiartas, J. L., Gaudin, A. C., O'Geen, A. T., Herrera, I., & Scow, K. M. (2019). Deep soil inventories reveal that impacts of cover crops and compost on soil carbon sequestration extend beyond the surface. *Scientific Reports, 9,* 1–10. https://doi.org/10.1038/s41598-019-48047-6

Tilman, D. (1997). *Biodiversity and ecosystem functioning: Maintenance of ecosystem integrity and processes. Nature, 387*(6630), 46–49. https://doi.org/10.1038/387046a0

Tilman, D., Hill, J., & Lehman, C. (2009). Carbon-negative biofuels from low-input high-diversity grassland biomass. *Science, 314*(5805), 1598–1600. https://doi.org/10.1126/science.1133306

Tilman, D., Cassman, K. G., Matson, P. A., Naylor, R., & Polasky, S. (2006). Agricultural sustainability and intensive production practices. *Nature, 418*(6898), 671–677. https://doi.org/10.1038/nature01014

Tilman, D., et al. (2002). Agricultural sustainability and intensive production practices. *Nature, 418*(6898), 671–677. https://doi.org/10.1038/nature01014

Tilman, D., et al. (2011). Global food demand and the sustainable intensification of agriculture. *Proceedings of the National Academy of Sciences, 108*(50), 20260–20264. https://doi.org/10.1073/pnas.1116437108

Tilman, D., Cassman, K. G., Matson, P. A., Naylor, R., & Polasky, S. (2002). Agricultural sustainability and intensive production practices. *Nature,* 418(6898), 671–677. https://doi.org/10.1038/nature01014

Tilton, J. E. (2003). On borrowed time? Assessing the threat of mineral depletion. *Resources Policy, 29*(1), 37–50. https://doi.org/10.1016/S0301-4207(02)00060-9

Tilman, D., Balzer, C., Hill, J., & Befort, B. L. (2011). Global food demand and the sustainable intensification of agriculture. *Proceedings of the National Academy of Sciences, 108*(50), 20260–20264. https://doi.org/10.1073/pnas.1116437108

Timperley, J. (2021). Why fossil fuel subsidies are so hard to kill. *Nature*, 598, 403–405. https://doi.org/10.1038/d41586-021-02919-1

Thornton, P. K., Ericksen, P. J., Herrero, M., & Challinor, A. J. (2014). Agriculture and food systems in global change. *Global Food Security*, 3(2), 83–94. https://doi.org/10.1016/j.gfs.2014.02.001

Thornton, P. K., et al. (2014). Agriculture and food systems in sub-Saharan Africa in a 4°C+ world. *Environmental Science & Policy*, 40, 1–12. https://doi.org/10.1016/j.envsci.2014.02.003

Trostle, R. (2010). Global agricultural supply and demand: Factors contributing to the recent increase in food commodity prices. Economic Research Service, U.S. Department of Agriculture. https://www.ers.usda.gov/webdocs/publications/43808/8447_wrs0807.pdf

Tritscher, S., Ziegler, F., & Müller, J. (2018). Biobased and biodegradable plastics: The new wave of plastics technology. *Journal of Cleaner Production*, 185, 1510–1521. https://doi.org/10.1016/j.jclepro.2018.03.255

Trnka, M., Hlavinka, P., Semerádová, D., Dubrovský, M., & Žalud, Z. (2011). Climate change impacts on agriculture in Central Europe and adaptation options. *Agricultural and Forest Meteorology*, 150(3), 327–334. https://doi.org/10.1016/j.agrformet.2010.09.009

Trump, D. J. (2017). Inaugural address. The White House. https://www.whitehouse.gov/briefings-statements/inaugural-address/

Trnka, M., et al. (2011). Climate change and the impacts on agriculture in Europe: Results from the PESETA project. *Climatic Change*, 108(3), 379–406. https://doi.org/10.1007/s10584-011-0216-3

Umweltbundesamt UBA (2023). *Emissionsbilanz erneuerbarer Energieträger 2023*. https://www.umweltbundesamt.de/publikationen/emissionsbilanz-erneuerbarer-energietraeger-2023

UBA. (2021). *Emissionsbilanz erneuerbarer Energieträger*. Abgerufen von https://www.umweltbundesamt.de/sites/default/files/medien/1410/publikationen/2019-11-07_cc-37-2019_emissionsbilanz-erneuerbarer-energien_2018.pdf

UBA – Umweltbundesamt. (2020). *Bodenversiegelung: Folgen und Maßnahmen*. https://www.umweltbundesamt.de

UBA – Umweltbundesamt. (2020). *Kunststoffe in der Umwelt: Quellen, Senken, Lösungsansätze*. https://www.umweltbundesamt.de

UBA (Umweltbundesamt). (2022). *Flächennutzung und Flächenverbrauch in Deutschland*. https://www.umweltbundesamt.de

Umweltbundesamt. (2021). Resourcennutzung und ihre Folgen. https://www.umweltbundesamt.de/themen/abfall-resourcen/resourcennutzung-ihre-folgen:contentReference{index=33}

Umweltbundesamt. (2023). *Plastik und Recycling in Deutschland – Zahlen und Fakten*. https://www.umweltbundesamt.de

Umweltbundesamt. (2021). *Nachhaltige Nutzung biotischer Rohstoffe*. Abgerufen am 10. April 2025, von https://www.umweltbundesamt.de/nachhaltige-nutzung-biotischer-rohstoffe

Umweltbundesamt. (o. J.). *Ökologische Bewertung textiler Fasern*. Abgerufen von https://www.umweltbundesamt.de/sites/default/files/medien/11850/publikationen/117_2024_texte_oekologische_bewertung_textiler_fasern.pdf

Umweltbundesamt. (2021). *Bioenergie | Umweltbundesamt*. https://www.umweltbundesamt.de/themen/klima-energie/erneuerbare-energien/bioenergie

Umweltbundesamt. (2025). *Struktur der Flächennutzung*. https://www.umweltbundesamt.de/daten/flaeche-boden-land-oekosysteme/flaeche/struktur-der-flaechennutzungUmweltbundesamt

Umweltbundesamt. (2024). *Bioenergie*. https://www.umweltbundesamt.de/themen/klima-energie/erneuerbare-energien/bioenergie

Umweltbundesamt. (2019). *Übergreifende Aspekte der Biomassenutzung.* https://www.umweltbundesamt. de/sites/default/files/medien/1410/publikationen/2019-08-19_texte_87-2019_be_biomassenutzung_ue bergreifende-aspekte.pdfUmweltbundesamt

Unterwegs. (o. J.). *Hanf, Leinen - Unterwegs.* Abgerufen von https://www.unterwegs.biz/hanf-leinen.html

UNEP (United Nations Environment Programme). (2011). Decoupling natural resource use and environmental impacts from economic growth.

UNEP (United Nations Environment Programme). (2021). *Making peace with nature: A scientific blueprint to tackle the climate, biodiversity and pollution emergencies.* https://www.unep.org

UNEP. (2022). *Global environment outlook 2022: Sustainable production and consumption.* United Nations Environment Programme.

UNDP. (2021). *Human development report 2021: The next frontier.* United Nations Development Programme.

UNEP – United Nations Environment Programme. (2011). *Decoupling natural resource use and environmental impacts from economic growth.* https://www.resourcepanel.org/reports/decoupling-natural-resource-use-and-environmental-impacts-economic-growth

UNEP. (2019). *Global environment outlook – GEO-6: Healthy Planet, Healthy People.* Cambridge University Press.

Umweltbundesamt. (2019). *Bioenergy and sustainability: Land use and nitrogen balance.* Dessau-Roßlau: Umweltbundesamt. https://www.umweltbundesamt.de

U.S. Department of Energy (DOE). (1998). *Biodiesel from algae: Research highlights from the aquatic species program 1978–1996.* National Renewable Energy Laboratory. https://www.nrel.gov/docs/legosti/fy98/24190.pdf

U.S. Department of the Interior. (2024). *Mineral commodity summaries 2024.* U.S. Geological Survey. https://pubs.usgs.gov/periodicals/mcs2024/mcs2024.pdf

U.S. Energy Information Administration (EIA). (2020). *Annual energy outlook 2020.* https://www.eia.gov/outlooks/aeo/

U.S. Environmental Protection Agency (EPA). (2021). *State bioenergy primer: Information and resources for states on issues, opportunities, and options for advancing bioenergy – Chapter: Land resource considerations.* https://www.epa.gov/sites/default/files/2021-06/documents/bioenergy_state_primer_2021.pdf

U.S. Geological Survey. (2020a). *Mineral commodity summaries 2020.* https://pubs.usgs.gov/periodicals/mcs2020/mcs2020.pdf

U.S. Geological Survey. (2020b). Mineral resources and the environment: U.S. Geological Survey's work on scarce materials. https://www.usgs.gov/centers/nmic/critical-mineral-resources

U.S. Geological Survey. (2022). *U.S. Geological Survey releases 2022 list of critical minerals.* Abgerufen am 10. April 2025, von https://www.usgs.gov/news/national-news-release/us-geological-survey-releases-2022-list-critical-minerals

U.S. Geological Survey (USGS). (2023). *Mineral commodity summaries 2023.* U.S. Department of the Interior. https://doi.org/10.3133/mcs2023

Van der Zee, S., et al. (2012). Impacts of energy and nutrient markets on food security: The case of nitrogen fertilizers. *Food Policy, 37*(6), 586–595. https://doi.org/10.1016/j.foodpol.2012.09.001

van Dijk, K. C., Lesschen, J. P., & Oenema, O. (2016). Phosphorus flows and balances of the European Union Member States. *Science of The Total Environment, 542*, 1078–1093. https://doi.org/10.1016/j.scitotenv.2015.08.048

van der Putten, W. H., et al. (2010). The role of soil microbial communities in plant–soil interactions. *Ecology Letters, 13*(5), 545–559. https://doi.org/10.1111/j.1461-0248.2010.01456.x

Van Vuuren, D. P., et al. (2010). Long-term perspectives on sustainability and food security. *Global Environmental Change, 20*(1), 9–24. https://doi.org/10.1016/j.gloenvcha.2009.09.007

Vargas, J. R., Parra, R., & García, P. (2014). *Biomethanol as a biofuel: Current developments and future perspectives*. Renewable and Sustainable Energy Reviews, 39, 573–587. https://doi.org/10.1016/j.rser.2014.07.055

Vas, M. (2019). Hydrotreatment of biofuels for aviation and marine applications. *Fuel Processing Technology*, *192*, 1–10. https://doi.org/10.1016/j.fuproc.2019.05.027

Vargas-Tah, J. A., Ferrer, R., & García, J. A. (2021). Advances in synthetic biology for the production of bio-based chemicals: Challenges and future perspectives. *Bioengineering*, *8*(2), 33. https://doi.org/10.3390/bioengineering8020033

VDI. (2020, 26. November). Weniger Emissionen durch Pflanzenöl im Tank. *VDI Nachrichten*. Abgerufen von https://www.vdi.de/news/detail/weniger-co2-durch-pflanzenoel-im-tank

Vellend, M., Baeten, L., & Becker, A. (2013). Global biodiversity loss and the crisis of extinction. *Biological Conservation*, *163*, 1–15. https://doi.org/10.1016/j.biocon.2013.04.030

Vitousek, P. M., & Howarth, R. W. (1997). Human alteration of the global nitrogen cycle: Sources and consequences. *Ecological Applications*, *7*(3), 737–750. https://doi.org/10.1890/1051-0761(1997)007[0737:HAOTGN]2.0.CO;2

Vörösmarty, C. J., Green, P., Salisbury, J., & Lammers, R. B. (2000). Global water resources: Vulnerability from climate change and population growth. *Science*, 289(5477), 284–288. https://doi.org/10.1126/science.289.5477.284

Walker, T. R., & Rothman, R. (2020). Life cycle assessment of bioplastics: A critical review. *Sustainability*, *12*(15), 6215. https://doi.org/10.3390/su12156215

Walker, S., & Rothman, R. (2020). Life cycle assessment of bio-based and fossil-based plastic: A review. *Journal of Cleaner Production*, *261*, 121158. https://doi.org/10.1016/j.jclepro.2020.121158

Walker, P., Pavia, S., & Mitchell, R. (2014). Materials for sustainable construction. In K. J. D. Macdonald (Ed.), *Materials for sustainable sites* (pp. 1–25). Wiley.

Wang, L., Wei, L., & Zhang, W. (2022). Microbial fuel cells: A promising technology for renewable energy production and waste treatment. *Renewable and Sustainable Energy Reviews*, *137*, 110437. https://doi.org/10.1016/j.rser.2020.110437

Wang, Q., & Li, Y. (2020). Economic and environmental challenges of scaling up bio-battery production. *Journal of Power Sources*, *444*, 227251. https://doi.org/10.1016/j.jpowsour.2019.227251

Wang, X., Zhang, Y., & Li, Q. (2020). Sustainable fermentation strategies for microbial enzyme production: Minimizing environmental impact through substrate optimization and nutrient recycling. *Journal of Cleaner Production*, 250, 119545. https://doi.org/10.1016/j.jclepro.2019.119545

Wang, X., Zhang, L., & Song, Z. (2022). Recycling bio-based materials in circular economy: Impact on sustainable battery production. *Journal of Cleaner Production*, *314*, 128079. https://doi.org/10.1016/j.jclepro.2021.128079

Wang, Z., Guo, M., & Wei, W. (2021). The application of bio-derived materials in next-generation sodium-ion batteries. *Journal of Power Sources*, *482*, 228941. https://doi.org/10.1016/j.jpowsour.2020.228941

Walker, B., Holling, C. S., Carpenter, S. R., & Kinzig, A. (2004). Resilience, adaptability and transformability in social–ecological systems. *Ecology and Society*, *9*(2), Article 5. https://www.jstor.org/stable/26267673

Waschbär-Magazin. (2023). Hanf: Infos zur Textilfaser. Abgerufen von https://www.waschbaer.de/magazin/hanf-materialkunde/

Wardle, D. A., Bardgett, R. D., Klironomos, J. N., Setälä, H., Van der Putten, W. H., & Wall, D. H. (2004). Ecological linkages between aboveground and belowground biota. *Science*, 304(5677), 1629–1633. https://doi.org/10.1126/science.1094875

Weizsäcker, E. U. von. (2002). *Faktor fünf: Die Formel für nachhaltiges Wachstum*. Droemer Knaur.

Welthungerhilfe. (2020). *Politische Optionen für nachhaltige Bioökonomie*. Welthungerhilfe. Abgerufen von https://www.welthungerhilfe.de/welternaehrung/rubriken/klima-resourcen/politische-optionen-fuer-nachhaltige-biooekonomie

Werpy, T., & Petersen, G. (2004). Top value added chemicals from biomass: Volume I—Results of screening for potential candidates from sugars and synthesis gas (No. DOE/GO-102004-1992). National Renewable Energy Laboratory. https://doi.org/10.2172/15008859

White, P. J., & Brown, P. H. (2017). Plant nutrition for sustainable development and global food security. Plant and Soil, 395(1), 5–24. https://doi.org/10.1007/s11104-017-3122-0

Wien Energie. (2022). *Wien steigert erneuerbare Energieproduktion um 8%*. Retrieved from https://www.bio masseverband.at/wien-steigert-erneuerbare-energieproduktion-um-8/

Wikipedia. (2023). Bioethanol. *Wikipedia, Die freie Enzyklopädie*. Abgerufen von https://de.wikipedia.org/wiki/Bioethanol

Wikipedia. (2024 a). Polylactide. In *Wikipedia – Die freie Enzyklopädie*. Abgerufen am 11. April 2025, von https://de.wikipedia.org/wiki/Polylactide

Wikipedia. (2024 b). Polylactid-Polyhydroxyalkanoat-Blends. In *Wikipedia – Die freie Enzyklopädie*. Abgerufen am 11. April 2025, von https://de.wikipedia.org/wiki/Polylactid-Polyhydroxyalkanoat-Blends

Wikipedia. (2024). Hempcrete. In *Wikipedia – The Free Encyclopedia*. Abgerufen am 11. April 2025, von https://en.wikipedia.org/wiki/Hempcrete

Wikipedia. (2024). *Bioethanol*. Abgerufen von https://de.wikipedia.org/wiki/Bioethanol

Wikipedia. (n.d.). Schafwolle als Dämmstoff. In *Wikipedia*. Retrieved April 11, 2025, from https://de.wikipe dia.org/wiki/Schafwolle_als_D%C3%A4mmstoff

Wikipedia. (2023). Bio-basierter Kunststoff. Abgerufen von https://de.wikipedia.org/wiki/Bio-basierter _Kunststoff

Wikipedia. (2023). *Algenkraftstoff*. https://de.wikipedia.org/wiki/Algenkraftstoff

Wikipedia. (2023). *Biodiesel*. https://de.wikipedia.org/wiki/Biodiesel

Wikipedia. (2023). *Bioethanol*. https://de.wikipedia.org/wiki/Bioethanol

Wissenschaftsjournalismus. (2022, September 14). Biobasiertes Upcycling: Vom Rest- zum Wertstoff. *Bioökonomie.de*. https://biooekonomie.de/themen/dossiers/biobasiertes-upcycling-vom-rest-zum-wertstoff

Wilson, E. O. (2016). *Half-earth: Our planet's fight for life*. Liveright Publishing Corporation.

Williams, G., Zhao, T., & Zhang, Y. (2022). Land use conflicts in bioplastic production: Food versus fuel. *Renewable and Sustainable Energy Reviews, 21*(9), 187–195. https://doi.org/10.1016/j.rser.2021.11.004

Wirtschaftsdienst. (2022). Ukrainekrieg: Stopp der Gaslieferungen spaltet EU. https://www.wirtschafts dienst.eu/inhalt/jahr/2022/heft/6/beitrag/ukrainekrieg-stopp-der-gaslieferungen-spaltet-eu.html:con tentReference{index=39}

Wissenschaftlicher Beirat der Bundesregierung Globale Umweltveränderungen (WBGU). (2008). *Zukunftsfähige Bioenergie und nachhaltige Landnutzung*. https://www.wbgu.de/fileadmin/user_upload/wbgu/publikationen/hauptgutachten/hg2008/pdf/wbgu_jg2008.pdf

WMO (World Meteorological Organization). (2018). *Scientific assessment of ozone depletion: 2018*.

Worrell, E., & Reuter, M. A. (2020). *Handbook of Recycling: State-of-the-Art for Practical Applications*. Elsevier.

Woertz, E. (2013). *The political economy of the global food system: The Gulf States' agricultural investments in Africa and Pakistan*. Middle East Institute. https://www.mei.edu/publications/political-economy-global-food-system

World Bank. (2022). *Commodity markets outlook: The impact of the war in Ukraine on commodity markets*. https://doi.org/10.1596/978-1-4648-1803-4

World Food Programme (WFP). (2022). *Hunger hotspots: FAO-WFP early warnings on acute food insecurity*. https://doi.org/10.4060/cb9992en

World Bank. (2022). *Food security and nutrition: The World Bank's role in global food crises*. World Bank Group. https://www.worldbank.org/en/topic/foodsecurity

World Bank. (2022). *Food security update: World Bank response to rising food insecurity*. https://www.world bank.org/en/topic/agriculture/brief/food-security-update

World Bank. (2023). *Global food commodity price report 2023*. World Bank Group.

World Bank. (2023). *Food price inflation in developing countries: Trends and consequences*. The World Bank Group.

World Bank. (2020). *Doing Business 2020: Comparing Business Regulation in 190 Economies*. Abgerufen von https://www.worldbank.org/en/publication/doing-business

World Bank. (2024). *World Development Indicators Database*. Abgerufen von https://databank.worldbank.org/source/world-development-indicators

World Fertilizer Report. (2023). *Global fertilizer market outlook 2023*. World Fertilizer.

World Oil Resources. (2019). *World oil resources and production*. World Oil. https://www.worldoil.com

World Steel Association. (2021). *The steel industry's role in the circular economy*. https://www.worldsteel.org

World Steel Association. (2022). *Steel statistical yearbook 2022*. https://worldsteel.org

Wübbeke, J. (2013). Rare earth elements in China: Policies and narratives of reinventing an industry. *Resources Policy, 38*(3), 384–394. https://doi.org/10.1016/j.resourpol.2013.03.004

WWF (World Wide Fund for Nature). (2020). *Living planet report 2020: Bending the curve of biodiversity loss*. https://www.wwf.org

WWF (World Wide Fund for Nature). (2021). *Deutschland auf großem Fuß: Der ökologische Fußabdruck*. https://www.wwf.de

Xie, D., Yang, X., & Zhao, J. (2020). *Organic electrolytes and ionic liquids for biobatteries: A review of recent progress*. Journal of Power Sources, 457, 228102. https://doi.org/10.1016/j.jpowsour.2020.228102

Xu, S., et al. (2017). Algae and biocoke: Potential sustainable sources of fertilizers for crop production. *Bioresource Technology, 245*, 1309–1317. https://doi.org/10.1016/j.biortech.2017.07.144

Xu, Y., & Ramanathan, V. (2017). Well below 2°C: Mitigation pathways for avoiding dangerous to catastrophic climate changes. *Proceedings of the National Academy of Sciences, 114*(39), 10315–10323. https://doi.org/10.1073/pnas.1618481114

Yang, H., Zhang, X., & Luo, L. (2019). Recent advances in microbial fermentation for the production of biochemicals. *Biotechnology Advances, 37*(1), 1–19. https://doi.org/10.1016/j.biotechadv.2018.11.006

Yang, S., & Liu, S. (2021). *Advances in biocatalysts for energy production: Land use and environmental impact assessment*. Renewable and Sustainable Energy Reviews, 132, 110043. https://doi.org/10.1016/j.rser.2020.110043

Yang, Y., Zhang, H., & Li, W. (2021). *Sustainable bioelectrochemical systems for energy production: Applications and challenges*. Energy & Environmental Science, 14(3), 1234–1249. https://doi.org/10.1039/d0ee03765k

Yang, B., Wyman, C. E., & Dale, B. E. (2009). *Cellulose to biofuels: Advances in enzymatic hydrolysis*. Springer.

Yuan, J., Liu, Y., & Wang, M. (2018). Production and applications of bioplastics: A review of PLA and PHA materials. *Renewable Materials and Resources, 24*(8), 90–102. https://doi.org/10.1016/j.rmr.2018.08.003

Zhang, R., Li, Q., & Chen, Z. (2021). Bio-polymer-based energy storage devices: Recent advances and challenges. *Journal of Materials Science & Technology, 65*, 32–40. https://doi.org/10.1016/j.jmst.2020.09.047

Zhang, X., Chen, H., & Liu, Y. (2020). Thermochemical conversion of biomass: The latest developments and applications. *Fuel Processing Technology, 197*, 106235. https://doi.org/10.1016/j.fuproc.2019.106235

Zhang, Y., Li, Y. (2020). Sustainability assessment of bioplastics production from agricultural resources. *Renewable Resources Review, 13*(5), 134–143. https://doi.org/10.1016/j.rrr.2020.05.004

Zhang, Y., Li, Q., & Zhao, M. (2020). Advances in microbial fermentation technology for bioethanol production. *Biotechnology for Biofuels, 13*, 56. https://doi.org/10.1186/s13068-020-01831-3

Zhao, M., Li, Z., & Wang, L. (2021). The future of bio-batteries in renewable energy systems: Current state and prospects. *Energy & Environmental Science, 14*(9), 4989–5001. https://doi.org/10.1039/d1ee01404c

Zhao, M., Zhang, Q., & Song, S. (2020). Biodegradable plastics: Challenges and prospects for their production and application. *Environmental Science and Pollution Research, 27*(28), 35157–35176. https://doi.org/10.1007/s11356-020-09876-x

Zhao, R., & Li, Y. (2019). Land use and sustainability issues in bioplastic production from renewable sources. *Agricultural Sustainability, 13*(7), 305-314. https://doi.org/10.1016/j.agres.2019.04.001

Zhao, X., & Liu, P. (2023). Environmental and economic impacts of substituting cobalt in lithium-ion batteries with bio-materials. *Sustainable Energy & Fuels, 7*(2), 254–263. https://doi.org/10.1039/d2se01854a

Zhou, J. (2025). China's rare earth dominance could crumble in 10 years, CAS study forecasts. *South China Morning Post.* https://www.scmp.com/news/china/science/article/3303167/chinas-rare-earth-dominance-could-crumble-10-years-cas-study-forecasts:contentReference{index=35}

Zhu, J., Wang, X., & Zhang, Y. (2014). *Biomethanol production from biomass: Technology, economics, and environmental impact.* Journal of Cleaner Production, 71, 1–9. https://doi.org/10.1016/j.jclepro.2014.01.004

Zantow, A. (2013). Zwischen Resourcengerechtigkeit und Biopiraterie. Deutschlandfunk. https://www.deutschlandfunk.de/zwischen-resourcengerechtigkeit-und-biopiraterie-100.html:contentReference{index=35}

Zeng, X., Li, J., & Xie, L. (2018). Circular economy and e-waste recycling: Challenges and opportunities. *Current Opinion in Green and Sustainable Chemistry, 13*, 61–67. https://doi.org/10.1016/j.cogsc.2018.06.003

Zhang, Y., et al. (2020). Impact of reduced fertilizer use on crop yields and nutrient losses: A global analysis. *Environmental Science & Technology, 54*(3), 1534–1542. https://doi.org/10.1021/acs.est.9b04299

Zhao, C., He, X., Wang, Y., & Mu, Y. (2020). Recent advances in biodegradable materials for biomedical applications. *Journal of Materials Chemistry B, 8*(46), 10426–10443. https://doi.org/10.1039/d0tb01667a

Zhao, C., et al. (2015). Soil acidification and its impacts on soil ecosystem. *Soil Biology & Biochemistry, 90*, 37–45. https://doi.org/10.1016/j.soilbio.2015.08.012

Zhao, C., et al. (2017). Climate change impacts on global agricultural yields. *Nature, 534*(7609), 98–102. https://doi.org/10.1038/nature18309

Zhao, F. J., et al. (2018). Phosphorus-efficient crop varieties: A new approach to sustainable phosphorus management in agriculture. *Field Crops Research, 225*, 60–69. https://doi.org/10.1016/j.fcr.2018.05.017

Zhoary, M. (2017). *The vegetation of the Sahara: The desert, its plants and its people.* Cambridge University Press.

Zhou, L., & Zhao, X. (2020). Circular economy strategies for sustainable materials: Review and prospects. *Waste Management & Research, 38*(12), 1300–1311. https://doi.org/10.1177/0734242X20942830

Zhu, J., et al. (2014). Biological potassium mobilization in soil. *Soil Biology and Biochemistry, 75*, 1–10. https://doi.org/10.1016/j.soilbio.2014.03.022

Zilberman, D., Hochman, G., Rajagopal, D., Sexton, S., & Timilsina, G. (2013). *The impact of biofuels on commodity food prices: Assessment of findings.* American Journal of Agricultural Economics, 95(2), 275–281. https://doi.org/10.1093/ajae/aas093

General Literature on the Subject

1. Park, J. B., & Bronzino, J. D. (Eds.). (2002). *Biomaterials: Principles and applications* (2nd ed.). CRC Press. ISBN: 9780849314917

2. Hench, L. L., & Polak, J. M. (2002). Third-generation biomedical materials. *Science, 295*(5557), 1014–1017. https://doi.org/10.1126/science.1067404

3. Williams, D. F. (2009). On the nature of biomaterials. *Biomaterials, 30*(30), 5897–5909. https://doi.org/10.1016/j.biomaterials.2009.07.027

4. Williams, D. F. (2008). *Essentials of biomaterials science*. Cambridge University Press. ISBN: 9780521869584
5. Black, J., & Hastings, G. W. (2013). *Handbook of biomaterial properties* (2nd ed.). Springer. https://doi.org/10.1007/978-1-4614-4202-8
6. Guelcher, S. A., & Hollinger, J. O. (Eds.). (2006). *An introduction to biomaterials*. CRC Press. ISBN: 9780849390263

List of Figures and Tables

List of Figures

https://doi.org/10.1515/9783112218747-012

List of Tables

About the Author

After training as a skilled chemical worker, Michael Has studied physics at the University of Regensburg and marketing at INSEAD in Fontainebleau. He received his doctorate in 1991 from the University of Regensburg in the field of biophysics with a thesis on pressure and temperature effects on hydrophobic interaction. At the FOGRA Institute in Munich, Dr Has headed the Innovation Research and Prepress departments.

After joining industry, he held senior and executive positions in research and development as well as in partner management, marketing and product line management, and product strategy. He led several new product/portfolio developments and successful market entries in software and hardware.
In parallel to his activities in industry, "Habilitation à Diriger des Recherches" Dr Has habilitated at the Institut National Polytechnique of the University of Grenoble in 1998.

Since 1998 he has been teaching in Grenoble as distinguished professor. He served as visiting professor at Alpe Adria Universität Klagenfurt. Teaching topics included new technologies and business and portfolio strategy. Starting with the analysis of the circular economy of paper, he has been working on the focus of sustainability since 2011. In this context, he deals with non-financial reports and the assessment of data (KPIs, footprints, and risks) to be included in reports, including measures to reduce footprints and ecodesign. He consults companies and organizations on sustainability-related topics. In doing so, he worked with companies and products from very different sectors such as the automotive, chemical, or printing industries to establish the internal structures required and reduce footprints.

His scientific work on industry development led to publications in numerous fields and led to several awards.

Dr Has is a frequent speaker at international events and conferences.

In recognition of his contribution and his way of guiding students, he was appointed as a trusted lecturer of the Hans Böckler Foundation – the foundation of the German Trade Union Confederation. Dr Has heads the Foundation Diversity of Cultures of the World (Stiftung Vielfalt der Kulturen der Welt) – a foundation dealing with and supporting human rights activities and projects for ethnic and religious minorities. At de Gruyter Michael Has published the books Sustainable Products (2022 and 2024); Klima, Rohstoffe und Prognosen (2025); Grenzen von Nachhaltigkeit und Ecodesign (2024).

https://doi.org/10.1515/9783112218747-013

Persons- and Key Words

https://doi.org/10.1515/9783112218747-014

www.ingramcontent.com/pod-product-compliance
Lightning Source LLC
Chambersburg PA
CBHW061347210326
41598CB00035B/5903